Power Systems

Electrical power has been the technological foundation of industrial societies for many years. Although the systems designed to provide and apply electrical energy have reached a high degree of maturity, unforeseen problems are constantly encountered, necessitating the design of more efficient and reliable systems based on novel technologies. The book series Power Systems is aimed at providing detailed, accurate and sound technical information about these new developments in electrical power engineering. It includes topics on power generation, storage and transmission as well as electrical machines. The monographs and advanced textbooks in this series address researchers, lecturers, industrial engineers and senior students in electrical engineering.

Power Systems is indexed in Scopus

Vasily Ya. Ushakov · Alexey V. Mytnikov · Ikromjon U. Rakhmonov

High-Voltage Equipment of Power Systems

Design, Principles of Operation, Testing, Monitoring and Diagnostics

Springer

Fig. 1.4 Photograph of a 1150 kV power transformer

- **From the end of the third quarter of the twentieth century up to now—** unification of IPS into UESR by means of OTL 500, 750 kV.
- **From the beginning of the twenty-first century up to now—**development of Smart Grid and Distributed Generation on the base of digitalization.

The key provisions characterizing the situation in the global electrical power engineering can be formulated as follows:

- most of the electricity is produced at large thermal power plants (TPPs) and nuclear power plants (NPPs). Their share in the world and Russia exceeds 80%. Only in a small number of countries (Denmark, Germany, Spain) this share is significantly lower;
- this means that the transmission of electricity from powerful PP to consumers located at a great distance is a critical task and will remain so for many decades; as before, OTL and cable lines (to a lesser extent) will be the main tool for solving this task;
- an increase in the capacity of individual electric generators, the length, capacity and responsibility of power lines stimulates the use of the most modern devices and systems for their control and management based on digitalization;
- similar requirements are imposed on distributed generation in connection with the growing trend towards an increase in its role in energy production. This refers to PPs using both mineral fuel and renewable energy sources. Modern digital technology is also required to monitor and control such a complex system.

Vasily Ya. Ushakov · Alexey V. Mytnikov ·
Ikromjon U. Rakhmonov

High-Voltage Equipment of Power Systems

Design, Principles of Operation, Testing, Monitoring and Diagnostics

Springer

Vasily Ya. Ushakov
Power Engineering School
Tomsk Polytechnic University
Tomsk, Russia

Alexey V. Mytnikov
Power Engineering School
Tomsk Polytechnic University
Tomsk, Russia

Ikromjon U. Rakhmonov
Department of Power Supply
Tashkent State Technical University
Tashkent, Uzbekistan

ISSN 1612-1287 ISSN 1860-4676 (electronic)
Power Systems
ISBN 978-3-031-38251-2 ISBN 978-3-031-38252-9 (eBook)
https://doi.org/10.1007/978-3-031-38252-9

This Springer imprint is published by the registered company Springer Nature Switzerland AG
The registered company address is: Gewerbestrasse 11, 6330 Cham, Switzerland

Preface

The continuing growth of the world's consumption of electrical energy due to the growth of the population and its per capita consumption sets the task of increasing the capacity of generation, transmission and distribution systems. At the same time, high requirements are placed on the reliability (uninterrupted) operation of these systems and on the quality of electricity supplied to consumers. Among the most serious challenges is the aging of the entire power infrastructure and, primarily, the aging of electric power grids (EPG) and substations high-voltage equipment.

According to analysts, a distinctive feature of the state of fixed assets of the electric power complex in most countries is a high degree of wear and tear of power electrical equipment and a large number of power plants with an exhausted standard operating life. As follows from the Russian General Scheme for the Location of Electricity Facilities until 2035, about 46% of the power generating facilities as part of the installed capacity of the Unified Energy System Russia (UESR) were put into operation before 1980 and have already celebrated their 40th anniversary. The ninety GW of steam turbine equipment have a large number of power plants with an exhausted standard resource of operation. By 2025, this figure will increase to 120 GW.

Despite the measures taken, including programs for power supply agreements and the introduction of distributed energy technologies, in recent years the share of new power generation equipment has remained relatively low. The economic indicators of power companies indicate the need to take measures aimed at increasing the service life of existing power equipment. This is due primarily to the high cost of high-voltage equipment.

An increase in the share of distributed (small-scale) generation in the total production and consumption of electrical energy, as a current trend in the development of the energy sector, allows achieving a number of positive effects: (1) reducing the construction time and payback period of an energy facility (power plants, transmission and distribution systems of electrical energy), (2) modularity and scalability, (3) the ability to control and manage the quality of electricity, (4) simplification of solving the problem of daily unevenness of the load schedule, (5) greater resistance to uncertainty in the energy market, (6) an expanded range of consumed primary energy resources (both mineral non-renewable and renewable) and (7) greater stability of energy supply in cases of hostilities or terrorist acts. Nevertheless, the problem of ensuring the reliability of high-voltage equipment remains relevant.

As one of the effective ways to ensure the reliable and safe operation of such installations, experts call the use of non-destructive testing and diagnostic methods that allow you to monitor and analyze the state of electrical equipment, quickly identify emerging faults and prevent possible failures.

In this textbook, three ways to ensure the reliable operation of the EPS are considered in sufficient detail: (1) testing of new equipment for compliance with passport characteristics (parameters), (2) its constant monitoring and (3) periodic diagnostics of the state of the equipment during operation

In order to help students master this basic material, it is preceded by information related to: (1) the history of the creation of electrical power systems, (2) the purpose, design and principles of operation of the main high-voltage equipment, (3) dielectric materials and insulating structures used in this equipment and (4) factors affecting the equipment in normal operating modes and in force majeure circumstances, causing their aging and failure.

Scientific bases and technologies for carrying out testing, monitoring and diagnostics operations, as well as their practical application for assessing the technical condition of the main high-voltage equipment, are presented in chapters 5–12. All material is arranged in the form of twelve chapters and two Appendixes.

Each chapter contains a brief annotation, tasks and questions to test and consolidate knowledge, a list of recommended literature for those wishing to expand and deepen their knowledge in this area.

Appendix 1 is a list of abbreviations and their decoding, which will help the student when reading the textbook.

The authors hope that the photographs of the main equipment, placed in the relevant sections of the book, and high-voltage test equipment (high-voltage laboratories), placed in the Appendix 2, will contribute to the understanding of the problems considered in the textbook. There are very few high-voltage laboratories all over the world and the students may not have an opportunity to visit such a laboratory.

We consider it our pleasant duty to note the financial support in the preparation of the manuscript of this textbook from the Agency for Innovative Development under the Ministry of Higher Education, Science and Innovation of the Republic of Uzbekistan.

With gratitude will be accepted comments and suggestions that readers can send to the e-mail addresses: vyush@tpu and mytnikov66@ mail.ru.

Tomsk, Russia — Vasily Ya. Ushakov
Tomsk, Russia — Alexey V. Mytnikov
Tashkent, Uzbekistan — Ikromjon U. Rakhmonov

Contents

Chapter 1
Past, Present and Future of Electric Power Systems (Brief Excursion)

Abstract This chapter introduces the history of transmission lines and their integration into power systems. The existing energy systems and plans to create new ones are briefly described. Given the vast scale of the UESR and the pioneering efforts of the Union of Soviet Socialist Republics (USSR) to create the world's first 1150 kV AC and 1500 kV DC transmission lines, the development of the Russian/USSR energy industry has received somewhat more attention.

1.1 Brief History of the Creation and Development of Power Systems

Electrification in the history of science and technology dates back to 1891 when a three-phase electric transmission system was demonstrated and tested in the International Electrical Engineering Exhibition in Frankfurt am Main. In August 1891, 1000 incandescent lamps supplied by a current from the Laufen hydroelectric power plant (HPP) were first lit in the exhibition; on September 12, 1891 the M. O. Dolivo-Dobrovolskii engine put into operation a decorative waterfall.

In the Laufen HPP, the electric power generated by a turbine was transmitted through a conical gear to the shaft of a three-phase synchronous generator (with an output power of 230 kV·A, rotation speed of 150 rpm, and voltage of 95 V in which windings were connected in star). In Laufen and Frankfurt there were three three-phase transformers that were immersed in tanks filled with oil.

The three-wire transmission line supports were made from wooden with an average span of about 60 m. Pin-type porcelain-oil insulators carried copper wires 4 mm in diameter. An interesting feature of this transmission line was the installation of fuses on the high-voltage side: a segment 2.5 m long consisting of two copper wires each having a diameter of 0.15 mm was connected in the gap of each wire at the beginning of the line. To switch off the line in Frankfurt, a simple device was used to short the three-phase circuit. The fuses blew, the turbine started to increase its speed, and when the machine operator noticed this, he stopped the machine. In the exhibition site in Frankfurt, a step-down transformer was placed from which 1000 incandescent lamps arranged on a huge board were supplied at a voltage of 65 V.

V. Y. Ushakov et al., *High-Voltage Equipment of Power Systems*, Power Systems,
https://doi.org/10.1007/978-3-031-38252-9_1

The Dolivo-Dobrovolskii three-phase asynchronous motor that drove a hydraulic pump with an output power of about 100 hp that put into operation a small artificial waterfall was also demonstrated in this exhibition. Simultaneously with this high-power motor, M. O. Dolivo-Dobrovolskii exhibited a 100-W three-phase asynchronous motor with a fan fitted to its shaft and a 1.5-kW motor with a dc generator fitted to its shaft.

Tests of transmission system conducted by the International Commission gave the following results: the minimum electricity transmission efficiency (the ratio of the power on the secondary terminals of the transformer in Frankfurt to the power on the turbine shaft in Laufen) was 68.5%, the maximum efficiency was 75.2%, the linear voltage during the tests was about 15 kV and increased to 25.1 kV at a higher voltage, and the maximum efficiency was 78.9%.

These tests not only demonstrated the capabilities of electrical energy transmission but also resolved the old debates: the alternating current conquered the direct current. The creation of the three-phase system was the most important stage in the development of electric power engineering and electrification [1–4].

In 1892, the Bulakh–Erlikon line was put into operation in Switzerland. A HPP with three three-phase generators having a power of 150 kW each was built at a waterfall in Bulakh. The electric power was transmitted at a distance of 23 km to supply a plant. After these first stations, a number of electric power stations were built rather fast. Most of them were built in Germany.

In USA, the first three-phase station was built only in late 1893 inCalifornia. At the beginning, the rate of use of three-phase systems in USAwas much less than in Europe. This was due to competition between leading energy companies of USA, each of them offered its own solution to the problem of energy transmission—along direct current lines, two-phase alternating current lines of different frequencies, etc. Two three-phase PPs were first combined in 1892 inSwitzerland. In Russia, the first enterprise with three-phase power supply was the Novorossiisk elevator. This was a huge set of buildings, and the problem of power distribution over floors and buildings could be solved in the best way only with the help of electricity. Drawings of three-phase machines were ordered to the Swiss Brown-Bowery Plant in summer of 1892. In 1893 the elevator was electrified. It is interesting to note that all machines designed abroad were manufactured in workshops of the elevator. Four synchronous generators having power of 300 kV·A each were mounted in the PP built near the elevator. Thus, the total power of the PP was 1200 kW·A, that is, it was the most powerful three-phase electric power station in the world at that time. Three-phase engines having a power of 3.5–15 kW worked in elevator rooms. They actuated various machines and mechanisms. A portion of energy was used for illumination.

Over the past century, the electric power industry continues to shape and contribute to the welfare, progress, and technological advances of the human race. The growth of electric energy consumption in the world has been nothing but phenomenal. In the USA, for example, electric energy sales have grown to well over 400 times in the period between the turn of the century and the early 1970s. This growth rate was 50 times as much as the growth rate in all other energy forms used during the same period. It is estimated that the installed kW capacity per capita in the USA is close to 3 kW.

The growth in size of PPs and in the higher voltage equipment was accompanied by interconnections of the generating facilities. These interconnections decreased the probability of service interruptions, made the utilization of the most economical units possible, and decreased the total reserve capacity required to meet equipment-forced outages. This was accompanied by use of sophisticated analysis tools such as the network analyzer. Central control of the interconnected systems was introduced for reasons of economy and safety. The advent of the load dispatcher heralded the dawn of power systems engineering, an exciting area that strives to provide the best system to meet the load requirements reliably, safely, and economically, and utilizing state-of-the-art computer facilities. At present, switchgears are being transformed unmanned stations, and with the advent of digital relays, the system becomes completely computerized and all actions are taken out of the outdoor switchgear due to remote operation control (for example, SCADA system, etc.).

The main vectors for the development of electric power systems are formed and discussed by scientists and power engineers around the world at the International Council on Large High Voltage Electric Systems (CIGRE), which meets every 2 years in Paris.

The need to increase the power and length of the transmission line forces power engineers to increase the level of operating voltage, as shown in Fig. 1.1.

In the USSR in 1988, a world-record power transmission line-1150 kV was put into operation, which was supposed to transmit a power of 5–6 GW from Kuzbass (Itat village, Kemerovo region) to the Urals (Chelyabinsk city). The route of this line is shown on the map of Russia and Kazakhstan, Fig. 1.2, and the appearance of the

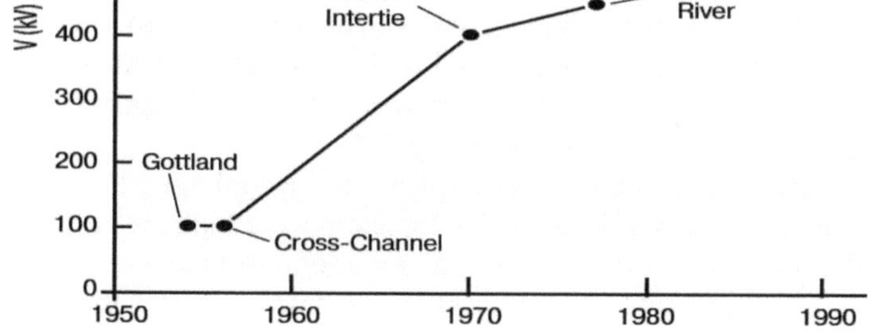

Fig. 1.1 Chronology of commissioning of AC transmission lines of record voltage classes

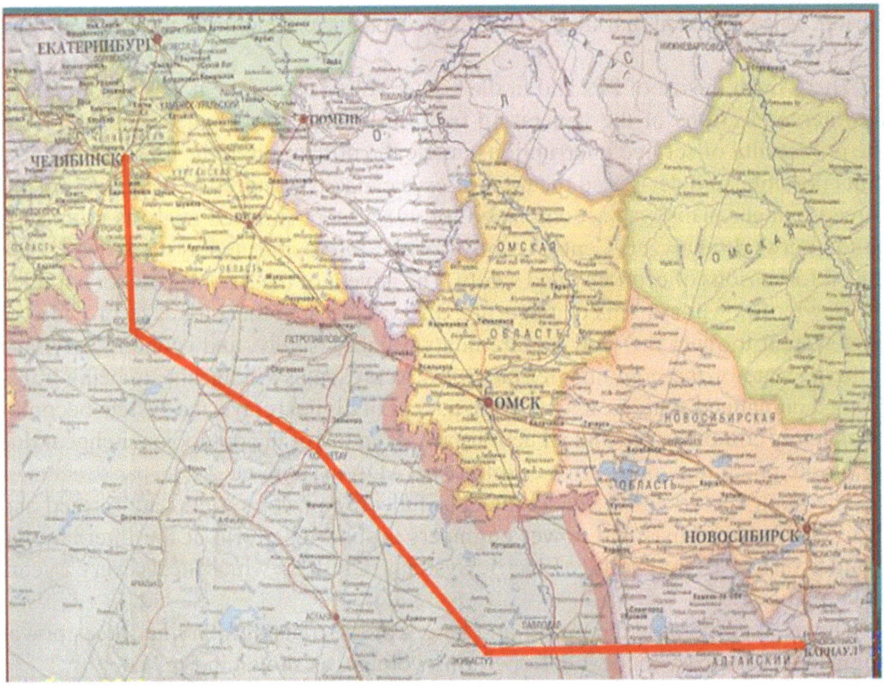

Fig. 1.2 Route of 1150 kV transmission line on the map of Russia and Kazakhstan

section of the power transmission line and the power transformer for a voltage of 1150 kV in Figs. 1.3 and 1.4 respectively.

After the collapse of the USSR, most of the line ended up outside of Russia—in Kazakhstan. On the territory of Kazakhstan, the section of the power transmission line-1150 kV "Ekibastuz-Kokchetav-Kustanai" operated at a nominal voltage of 1150 kV from 1988 to 1991. In the rest of the section, the line operated at a voltage of 500 kV. Now the entire line operates at a voltage of 500 kV.

In 1978, the construction of a direct current transmission line with a rated voltage of 1500 kV between Ekibastuz city in Kazakhstan and Tambov city in Russia began in the USSR. It was planned to build a line with a length of 2,414 km, which would have made it at that time the longest and highest voltage transmission line in the world. The maximum transmission power was to be 6 GW. Approximately 4,000 supports 41 m high were to be installed, Fig. 1.5.

Due to economic problems in the country and, accordingly, in the Ministry of Energy of the USSR, several hundred kilometers were built from the entire line, including the line crossing the river Volga in the form of pylons 124 m high. Since 2012, pylons have been used for a new 500 kV power transmission line, which makes it possible to organize the output of power from the Balakovo NPP to the integrated power systems of the Volga and the Center [2].

Currently, in Russia there are about 30 thousand km power lines 500–750 kV.

Fig. 1.3 Photograph of AC OTL—1150 kV

Today, in USA and in some other countries, the following voltage drops are accepted as nominal: 230, 287, 345, 500, 735, and 765 kV. In the countries of the European Union, in Russia and a number of other countries, the following voltage levels of high-voltage power lines are accepted classes: 110–220 kV—high—voltage (HV), 330–750 kV—ultra-high voltage (UHV), above 750 kV—super-high voltage (SHV). The trend is motivated by economy of scale due to the higher transmission capacities possible, more efficient use of right-of-way, lower transmission losses, and reduced environmental impact [5–7].

The stages of development of the electric power engineering from autonomous low-power PPs to unified energy systems (in a number of countries, unified/nationwide systems) can be traced on the example of the Russian energy sector.

From the end of the nineteenth century to first quarter of the twentieth century—unification of PP for parallel working (at the beginning—on cities scale, further—on regions scale).

- **From the first to the third quarter of the twentieth century—**formation Interconnected Power Systems (IPS) on the basis of overhead transmission line (OTL) 110 and 220 kV.
- **From second to the third quarter of the twentieth century**—creation Unified IPS on the basis of OTL 330 and 500 kV.

Fig. 1.4 Photograph of a 1150 kV power transformer

- **From the end of the third quarter of the twentieth century up to now**—unification of IPS into UESR by means of OTL 500, 750 kV.
- **From the beginning of the twenty-first century up to now**—development of Smart Grid and Distributed Generation on the base of digitalization.

The key provisions characterizing the situation in the global electrical power engineering can be formulated as follows:

- most of the electricity is produced at large thermal power plants (TPPs) and nuclear power plants (NPPs). Their share in the world and Russia exceeds 80%. Only in a small number of countries (Denmark, Germany, Spain) this share is significantly lower;
- this means that the transmission of electricity from powerful PP to consumers located at a great distance is a critical task and will remain so for many decades; as before, OTL and cable lines (to a lesser extent) will be the main tool for solving this task;
- an increase in the capacity of individual electric generators, the length, capacity and responsibility of power lines stimulates the use of the most modern devices and systems for their control and management based on digitalization;
- similar requirements are imposed on distributed generation in connection with the growing trend towards an increase in its role in energy production. This refers to PPs using both mineral fuel and renewable energy sources. Modern digital technology is also required to monitor and control such a complex system.

Fig. 1.5 Photograph of DC power lines for voltage 1500 kV

1.2 Main Tendencies in the Development of Electric Power Systems

Among specific features of the future global electric power engineering will be centralization of energy distribution and a great diversity of generating sources. Its characteristic features are combination of large-sized powerful centralized and relatively small enterprises operating in a common grid, application of hybrid schemes of electric power and heat generation (cogeneration), compatibility of energy and production technologies with complete utilization of wastes and secondary resources, and formation of interconnected power systems. Known advantages of parallel operation of PPs led to intensive development of electric networks (grids) of higher voltage classes, expansion and consolidation of the electric power systems, and formation of large, geographically extensive, including interstate, power systems [1]. Positive experience on the creation and operation of large interconnected power systems in Western Europe, North America, the former USSR, and Eastern European countries has been accumulated.

There are all prerequisites for the development of the electric power engineering of the world economy in this direction.

The tendency observed in the past few decades, namely, formation of large national and international interconnected power systems is of particular importance. An important role in the formation of UPS in the Eurasian continent is played by Russia having large fuel and energy resources and the largest in the world centralized interconnected power system UESR (Joint-Stock Company).

Important milestones in the development of interstate power transmission systems are the creation of the three largest power interconnections on the Eurasian continent: (1) the Union for the Coordination of Transmission of Electricity **(UCTE)**, (2) the **Mir** power interconnection, and (3) the **Nordel** power interconnection.

In 1951 it was created one of the largest energy networks in the world—the UCTE including 12 countries of Western Europe (Belgium, Germany, Spain, France, Greece, Italy, Yugoslavia, Luxembourg, Netherlands, Austria, Switzerland, and Portugal). Over time, the number of national energy systems included in the UCTE increased, primarily due to the energy systems of Eastern Europe. By 2009 UCTE united 29 system operators from 24 countries of continental Europe. In the same year, it was transformed into the European Network of Transmission System Operators for Electricity. (ENTSO-E). This union is Asynchronously with the Great Britain Power System operated through a dc cable.

In 1959, the Permanent Commission of the Council for Mutual Economic Assistance (CMEA) on the electric power industry prepared recommendations for the construction of intersystem power lines, and already in 1960, such 220 kV lines connected the power systems of the GDR, Poland, Czechoslovakia and Hungary. In 1962, a number of 220 kV transmission lines were put into operation, uniting the power systems of the USSR, Hungary and Poland, through which electricity was exported from the USSR. In the same year, to ensure reliable parallel operation of the energy systems of Bulgaria, Hungary, the GDR, Poland, Romania, the USSR and Czechoslovakia, the Central Dispatching Office for the Unification of Energy Systems was created. Thus, the foundation was laid for the creation of one of the largest interstate energy systems—**Mir**.

In 1963, the power systems of Hungary, Romania, Czechoslovakia were connected to this system, and in 1965. Bulgaria was included in the unified energy system.

In 1978, the construction of a 750 kV power transmission line Vinnitsa (USSR)—Albertirsha (Hungary) was completed. This line connected the Mir energy system and the Unified Energy System of the USSR, and already in 1979 they began to work in parallel.

After the collapse of the CMEA, the Mir energy system ceased to function, but its infrastructure was preserved: 11 high-voltage power lines (through Ukraine and Belarus connecting Russia with the countries of Eastern and Southern Europe) and 3 direct current voltage synchronization inserts worth $150 million each (in Austria and Germany).

In 1963, the Nordel energy association was created, which included Finland, Sweden, Norway and Eastern Denmark. For the transmission of electricity between the countries of the region, underwater high-voltage direct current cables were laid.

Figure 1.6 shows the existing and projected submarine cable power lines connecting European countries with each other and with neighboring countries [1].

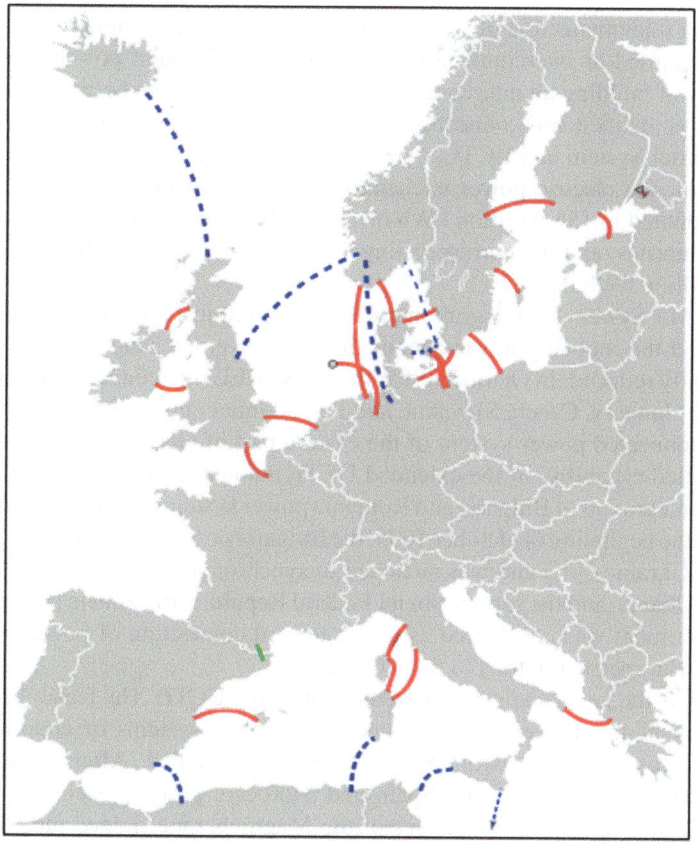

Fig. 1.6 DC cable lines in Europe (Red marked existing lines, green—under construction, blue—the proposed/discussed)

Successful attempts were made to interact with these three energy associations.

The installed capability of PPs included in the UCTE amounted to more than 390 million kW; it amounted to 85 million kW for the Nordel System and to more than 400 million kW for the Mir Power System. The Mir Power System was connected to the UCPE through three dc inserts having net capability of 1750 MW and with the Nordel System by a dc insert having a capability of 1100 MW. Power systems of Eastern European countries and the Unified Power Grid (UPG) of the USSR were connected by three 750-kV overhead transmission lines (OTL), four 400-kV OTL, and four 220-kV OTL to deliver electric power from the USSR to Eastern European countries. The energy supply amounted to ~40 billion kWh in separate years.

The integration processes in the UCTE and Nordel System are intensified. In 1994, a dc cable line between Switzerland and Germany having a length of about 250 km and capability of 600 MW was put into service. Two projects of connecting the Norway Power Grid to the continental Europe are being considered. The first project is aimed at interconnection of Norway and Germany power systems, and the second project is aimed at interconnection of Norway and Holland power systems. The feasibility of building an interconnecting dc transmission line between Sweden and Poland was justified and connecting Latvia, Lithuania, and Estonia power systems to the Nordel System and UCTE.

In 1994, the electric power exchange in the UCTE, including the third countries, amounted to 155.9 billion kWh or 10% of the net electric power generated by UCTE countries, and the corresponding exchange in the Nordel System amounted to 39.3 billion kWh or 11.2%.

In the Mir System, disintegration processes started just after disintegration of the USSR, and the mutually advantageous electric power exchange inside the system significantly reduced. In October 1995, the CENTREL interconnected power system including Hungary, Czech, Slovakia, and Poland interconnected power systems and the interconnected power system of the eastern part of Germany joined the UCTE. The installed capability of the extended UCTE amounted to more than 470 million kW. The integration of Bulgaria and Romania power systems to the UCTE is also was made. At the beginning of October 1995, the Bulgaria power system was disconnected from the Ukraine UPG and was switched to synchronous operation with Romania, Greece, Albania, and the former Soviet Federal Republic Yugoslavia power systems. This experiment was considered as a step toward connection of power systems of Southern European countries to the UCTE.

Turkey is the next candidate for connection to the UCTE. The feasibility of interconnection of the Turkey power system with power systems of countries of the Mashreq economic zone (from Syria to Egypt) is studied. After commissioning a deep-sea ac cable connecting Spain and Morocco, power systems of Morocco, Algeria, Tunis, and Libya (countries of the Maghreb zone) joined the UCTE. The feasibility of interconnection of power systems of Mashreq and Maghreb countries is studied. Thus, a large interconnected power system of countries of the Mediterranean Sea basin is being built. This interconnected power system will operate in parallel with the UCTE. It is also planned to study the feasibility of joint operation of the Turkey power system with power systems of Trans-Caucasian republics including Armenia, Georgia, and Azerbaijan.

At the same time, the UCTE continues to operate synchronously with power systems of Baltic countries, Byelorussia, Ukraine, Moldova, and Kazakhstan. Power systems of Azerbaijan, Armenia and Georgia have retained the feasibility of synchronous operation with the UCTE.

The best way of cooperation on the Eurasian continent is the creation of a common electric power market as a basis for an interconnected power system. A number of international projects are aimed at solving this problem. The most interesting of them are discussed below.

Baltic Electric Power Generating Ring. This project is aimed at creating a high-power grid connecting power systems of 11 Baltic countries including Denmark, Sweden, Norway, Finland, Russia, Estonia, Latvia, Lithuania, Byelorussia, Poland, and Germany, Fig. 1.7. Another project of the East–West Power Bridge envisages the

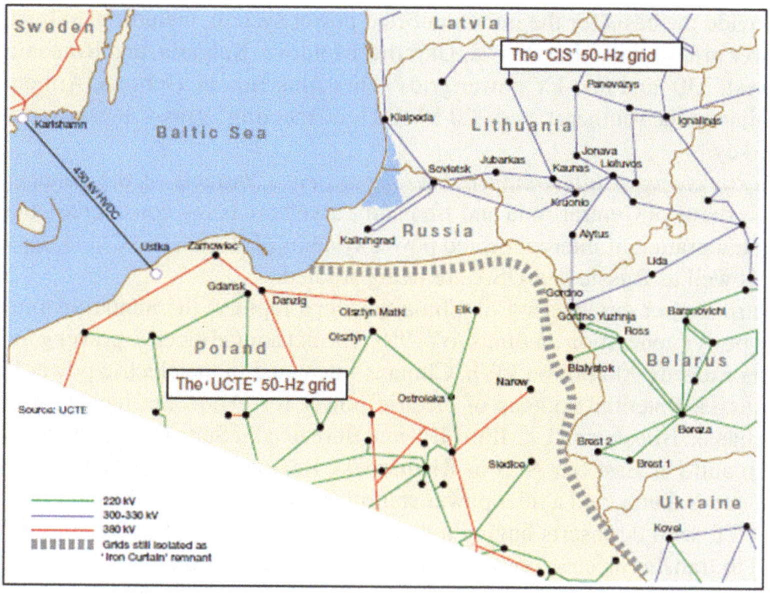

Fig. 1.7 Baltic electric power generating ring

building of a 4000-MW dc transmission line connecting power systems of Russia, Byelorussia, Poland, and Germany and is conceptually a part of the first project [2].

It is assumed that the Baltic Ring will allow the operation of power systems of participating countries to be improved and as a whole, will foster the economic development of Baltic countries.

A meeting of 17 electric power-generating companies from 11 countries of this region devoted to the creation of the Baltic Electric Power Generating Ring was held.

It should be noted that positive experience on cooperation of the UCTE with the Nordel System has already been accumulated. Work is underway on an increase in the capability of the dc insert with Finland up to 1400 and then to 2000 MW. The feasibility of connection of the Karelian and Kola power systems with power systems of countries included in the Nordel System is being studied.

Black Sea Interconnected Power System. The majority of countries involved in the Black Sea Economic Community (BSEC) including Ukraine, Romania, and Bulgaria supported the proposal of the UCTE to create the BSEC United Power Grid. Its formation is aimed at interconnection of regional power systems into the high-power grids; some of them already exist. Such an interconnected power system will allow BSEC countries to develop electric power engineering in the entire region in the best way, to use rationally the available power resources, to increase reliability of electric power supply to customers, to exchange electric power to mutual advantage, and to affect positively the economic development of all countries in the region. High-voltage grids built by countries-members of the Council for Mutual Economic Assistance will provide the basis for the interconnected power system, including 400- and 750-kV power grids connecting Russia, Ukraine, Moldova, Bulgaria, and Romania in the southwest, 330- and 500-kV power grids connecting Russia, Georgia, Armenia, and Azerbaijan in the southeast, and 220-kV OTL connecting Trans-Caucasian countries and Turkey.

Other projects of interconnected power systems. Variants of the connection of power systems of Central Asia and Iran and Turkey are being considered, and problems of integration of interconnected power systems of Russia and China, Japan, and Korea as well as Russia and USA are being studied.

Electric power engineering of China develops rapidly; the annual increment of electric power generation amounts to 7–9%. The net annual electric power generation in China exceeds 900 billion kWh. China is interested in the electric power transfer from Russia. Potential sources of electric power for export are located in Siberia (Boguchansk, Bratsk, and Ust-Ilim HPS and Berezovskii State District Power Station [SDPS]) and Far East (an APS in Khabarovsk Region, HPS and TPS in the Amur Region and Yakutia, and a tidal power station in the south of the Okhotsk Sea). 500-kV ac OTL with dc inserts having a dc carrying capacity of 1.5–2 million kW can be used to transfer electric power. In the East interconnected power systems, Amur, Khabarovsk, and Far East power systems are considered as transmitting one. OTL at voltages up to 500 kV inclusively can be used to export electric power.

The main premises for electric power import by Japan consist in the absence of its own fuel and energy resources and extremely high population density. Sakhalin TPS burning a shelf gas or South-Sakhalin coal, HPS and APS of the Far Eastern Interconnected Power System, and a tidal power station in the south of the Okhotsk Sea can be considered as potential sources to export electric power from Russia to Japan. For this aim transmission lines can be built through Sakhalin Island and two shallow and narrow channels (Tatarskii and Laperuza) or through the territories of China and Korea and the Korean Channel 200 km wide.

With allowance for long transmission lines, the electric power transfer to the USA is planned in small amount provided that the main expenses on building of the OTL through the Bering channel and mastering the hard-to-reach coastal zone of this channel will be covered by the headquarters of building of transcontinental railway through the Bering channel.

The implementation of the above-considered international projects as well as of the suggested variants of interconnected power systems will allow the Japan–China–Siberia–Kazakhstan–European part of Russia–Western Europe powerful extended electric power system to be formed and will be an important stage in the creation of the Interconnected Power System on the Eurasian continent with the net electric power of the order of 60% of the electric power of all power stations in the world in which this United Power Grid (UPG), by virtue of its geopolitical position, can become a central link [2].

The required transfer capability of intersystem connections can be estimated based on the practical recommendations of the UPG of the former USSR according to which the net transfer capability of intersystem connection in cross sections dividing the powerful interconnected power system into two parts should be of the order of 2–3% of the maximum load of the smaller part of the examined interconnected power system. With allowance for this condition, the required transfer capability of intersystem connections for the Euroasian Interconnected Power System in the territory of Russia and Kazakhstan should exceed 10 GW. Such transfer capability can be obtained only with the use of the super-high voltage (1150 kV of ac and 1500 kV of dc) transmission lines.

The North American transmission system is interconnected into a large power grid known as the North American Power Systems Interconnection [7]. The grid is divided into several pools. The pools consist of several neighboring utilities which operate jointly to schedule generation in a cost-effective manner. A privately regulated organization called the North American Electric Reliability Council (NERC) is responsible for maintaining system standards and reliability. NERC works cooperatively with every provider and distributor of power to ensure reliability. NERC coordinates its efforts with FERC as well as other organizations such as the Edison Electric Institute. Figure 1.8 shows a diagram of high-voltage transmission lines of the United States.

NERC currently has four distinct electrically separated areas. These areas are the Electric Reliability Council of Texas (ERCOT), the Western States Coordination Council (WSCC), the Eastern Interconnect, which includes all the states and

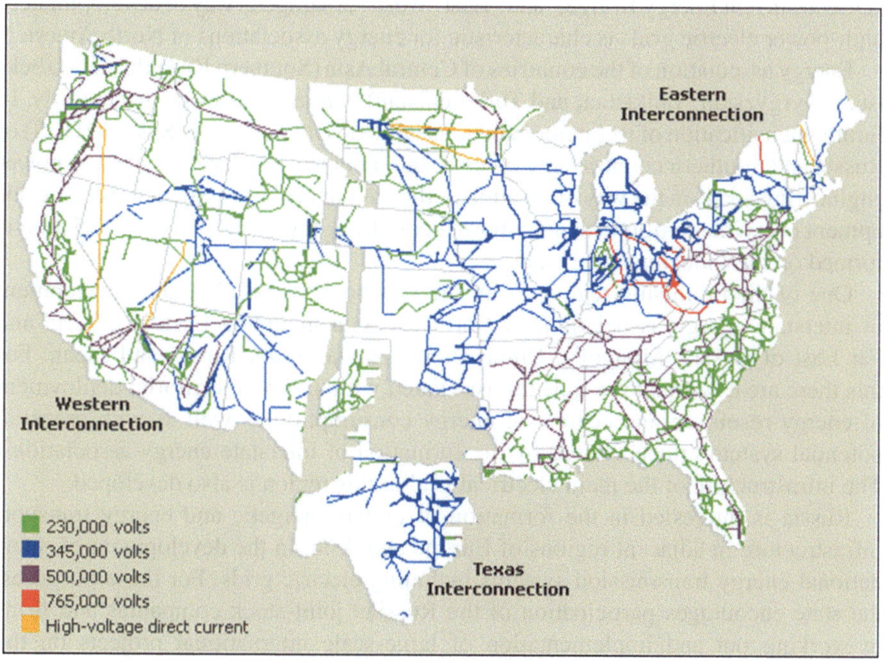

Fig. 1.8 Diagram of high-voltage transmission lines of the United States

provinces of Canada east of the Rocky Mountains (excluding Texas), and Hydro-Quebec. These electrically separate areas exchange with each other but are not synchronized electrically [3].

The electric power industry in the US is undergoing fundamental changes since the deregulation of the telecommunication, gas, and other industries. The generation business is rapidly becoming market-driven. The power industry was, until the last decade, characterized by larger, vertically integrated entities. The advent of open transmission access has resulted in wholesale and retail markets.

Trends in the development of electric power systems of the USA united at the state and interstate levels formed in the second half of the XXth century acquired clear-cut temporal and territorial guidelines to which the world electric power engineering adheres in the new century.

The EPS integration and formation of large electric energy associations are underway in Asia, Africa, and Southern and Central America.

The Trance-European Synchronously Interconnected System (TESIS) is formed by Trans-European Western, Central, and South-Eastern European countries. Further development of this electric energy association can proceed in collaboration with CIS electric energy association (established in 1992) based on intensification of interconnections on alternating current and construction of electricity transmission lines on direct current within the limits of widely discussed complex projects "Baltic Ring," "Direct Current Energy Bridge," and "East–West." Analogous way of development of high-power electric grids is characteristic for energy associations of North America.

Energy association of the countries of Central Asia (Southern Kazakhstan, Uzbekistan, Kyrgyzstan, Tajikistan, and Turkmenistan) continues to work successfully. In future intensification of its interconnections with the United Energy System (UES) of Russia and Southern countries is expected. India intensively develops electric power engineering and is in the way of creation of the national energy association and development of its integration with neighboring countries. Interstate energy association is formed on the Indochina peninsula.

One of the promising regions from the viewpoint of creation and development of interstate energy associations is Northeast Asia including Eastern Siberia and Far East of Russia, Mongolia, China, North Korea, South Korea, and Japan. For this there are essential prerequisites associated with different territorial deployment of energy resources and centers of energy consumption as well as with essential potential system effects caused by the formation of interstate energy associations. The infrastructure of the main electrical grid in this region is also developed.

Russia is interested in the formation of united energetic and energy transport infrastructure in adjacent regions of Europe and Asia, in the development of international energy transmission systems including electric grids. For these purposes, the state encourages participation of the Russian joint-stock companies and firms in working out and implementation of large-scale international projects on the creation of electric energy associations "Baltic Ring," "United Energy System of Caspian and Black Sea Countries" and energy bridges "East–West," "Siberia–China," "Sakhalin – Hokkaido," etc., Figs. 1.9 and 1.10.

Fig. 1.9 Main OTL of the USSR and Russia various voltage classes: OTL-220, 330, 500, 750, and 1150

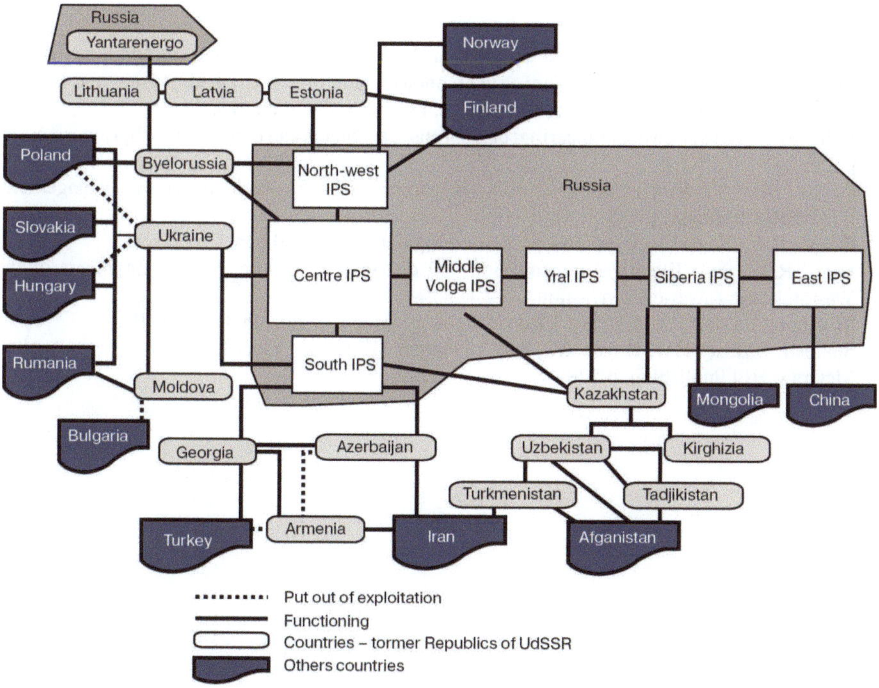

Fig. 1.10 One of the options for creating the Asian Energy Super Ring

It should be emphasized that the fate of existing and planned energy associations of an interstate and global scale, unfortunately, is controlled not only by economic interests and technological capabilities, but also by political interests. The influence of political processes in the world on the state of the global fuel and energy complex, which was especially pronounced in 2022, convincingly confirms this conclusion.

Question and Tasks

1. What are the motives for increasing the operating voltage level of electric power systems?
2. What goals can be achieved by an association local energy systems into a unified one?
3. In which city was the prototype of the modern electric power system first demonstrated?
4. What are the main trends in the development of the electric power industry?
5. What is Asian Energy Super Ring?

References

1. European Commission Directorate-General for Research Information and Communication Unit European Communities (2006) European technology platform smart grids, vision and strategy for Europe's electricity networks of the future. European Communities
2. Ushakov V (2018) Power engineering: current state, problems and perspectives. Springer Verlag, p 258
3. Garg A, Bhoi AK, Sanjeevikumar P (2006) Advances in power systems and energy management, ETAEERE. Springer, p 735
4. Grigsby LL (2006) Electrical power engineering handbook, 2nd edn. CRC Press
5. Evans RD (1964) Electrical transmission and distribution. Central Station Engineers of the Westinghouse Electric Corporation.
6. Mohamed E (2018). El Hawary. Electrical energy systems, 2nd edn. CRC, p 403
7. Santoso S, Beaty HW, Dugan RC, McGranaghan MF (2002) Electric power systems quality. McGraw-Hill Publishing, p 528

Chapter 2
Characteristics of the Main Elements of Electric Power Systems

Abstract This chapter provides a description of main function, design and principle of operation of main high-voltage equipment, such as: power and instrument transformers, cables, whole spectrum of commutation facilities including circuit breakers, disconnectors, circuit switchers and switchgear, main types of power capacitors. Also, rotating machine function, bushing design, whole range of cable constructions and their insulation structure development are considered.

2.1 Power and Instrument Transformers

2.1.1 Purpose of Power Transformers

In accordance with the purpose, two types of transformers are distinguished: **power transformers** and **instrument/measuring transformers**.

The main purpose of the **power transformers** is to increase or decrease the AC voltage.

Appointment of **instrument transformers:** Supply accurately scaled current and voltage quantities for measurement while insulating the relay from the high voltage and current of the power system [1–3].

In turn, **instrument transformers** are subdivided in two types:

- **Current Transformer**—a transformer intended to have its primary winding connected in series with the conductor carrying the current to be measured or controlled.
- **Potential/Voltage Transformer**—a transformer intended to have its primary winding connected in shunt (parallel) with the voltage to be measured or controlled.

A brief description and scope of high-voltage transformers can be formulated as follows:

1. A transformer is a static electrical component with no moving parts that is used for stepping voltage up or down or isolating one circuit from another.

V. Y. Ushakov et al., *High-Voltage Equipment of Power Systems*, Power Systems,
https://doi.org/10.1007/978-3-031-38252-9_2

2. Transformers have the ability to convert low-voltage, high-current AC to high voltage, low-current AC, or vice versa, with minimal energy losses.
3. Transformers work only with AC in accordance with the physical laws of magnetic induction, and they are inherently low-loss components.
4. The simplest low-voltage transformers can be made by winding separate coils of insulated wire around a ferromagnetic core, typically a stack of steel laminations.
5. When one coil or winding, called the *primary* or *input coil,* is energized, the core is magnetized so that the resulting magnetic flux induces a voltage in the second winding, called the *secondary* or *output coil.*
6. The change in voltage (voltage ratio) between the primary and secondary coils depends on the number of turns in each winding. Transformers are widely used in electrical power circuits. The large transformers in power generation stations step up the output voltage of AC generators to higher values for more efficient transmission over transmission lines.
7. Somewhat smaller transformers at electrical substations step the transmitted voltage down to the values more useful for regional and local distribution to customers.
8. Unlike transformer, which has two electrically isolated windings called: the primary and the secondary, an **autotransformer** has only one single voltage winding which is common to both sides (the primary and secondary circuits).
9. This single winding is "tapped" at various points along its length to provide a percentage of the primary voltage supply across its secondary load.
10. Therefore, in an autotransformer the primary and secondary windings are linked together both electrically and magnetically.
11. The main advantage of this type of transformer design is that it can be made a lot cheaper for the same VA rating.
12. The biggest disadvantage of the autotransformer is that the primary and secondary windings are not isolated from each other, as is done in a conventional transformer (double winding).

Figure 2.1 shows the main elements of a modern transformer, and below are the innovations applied in it.

1. **Core loss (no-load loss)** is minimized by using laser-deposited crystalline steel. (Super crystalline steel is a soft magnetic material that is used as a core material in electrical transformers. It is characterized by the preferred crystal orientation 110 <001>).
2. **Customized laminating widths** chosen to achieve a near-perfect round core cross-section allows for the most efficient use of materials and the ability to create a lighter and more compact transformer.
3. **Coil assembly rigidly braced** in a high-strength frame that distributes clamping forces) around the full circumference of the windings.
4. **Submerged-arc welding** process produces deep penetration welds, virtually eliminating leakage from welded tank joints.
5. **Inside tank surfaces are painted white** to facilitate internal inspections.

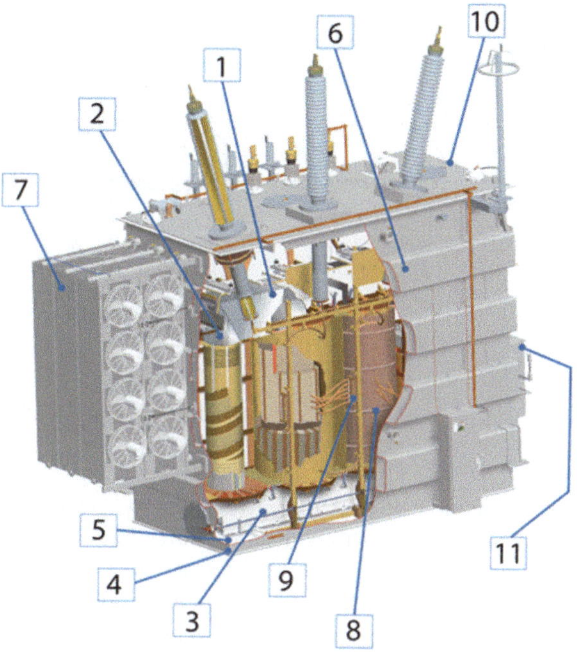

Fig. 2.1 The main elements of a modern power transformer

6. **Exterior tank coated** to a minimum thickness of 3 mils (76.2 μm); this coating has superior endurance characteristics ANSI C57.12.28 standards.
7. **Galvanized radiators** provide excellent corrosion resistance and require minimal maintenance. Fan guards and blades also galvanized.
8. **Material-stabilized coils** are pressure-fit within the core frame.
9. **Patented De-Energized Tap Changer (DETC)** features simple and compact in-line contact arrangement.
10. **Load Tap Changer** is designed to withstand up to a half-million operations without need for contact replacement.
11. **Control Box** is an enclosure for customer interface with transformer monitoring equipment, LTC and transformer cooling controls. Standard equipment includes anti-condensation heaters, an enclosed filing cabinet, as well as external stainless steel hardware and a large, removable bottom panel for customer conduits and cables.

The design of commercial transformers requires the selection of a simple yet suitable form of construction so that the coils are easy to wind and the core is easy to build. At the same time, the mean length of the windings and magnetic circuit must be as short as possible for a given cross-sectional area, so that the amount of material required and resulting losses are minimized. The core must provide a continuous path for magnetic flux while its lamination pattern must be easy to cut and stack. The windings should be insulated in a simple and economical manner, should permit

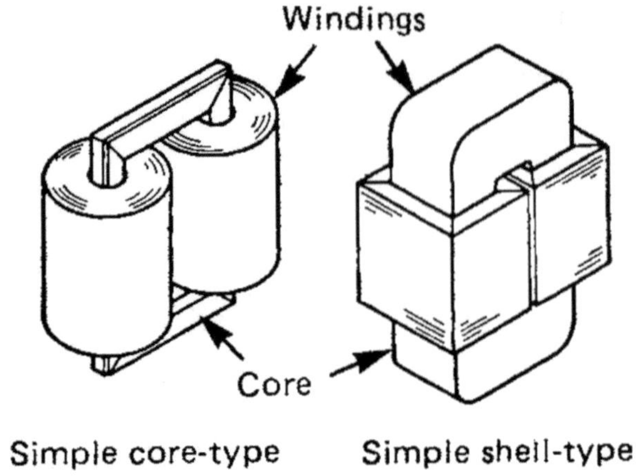

Fig. 2.2 Types of transformer magnetic circuits

the dissipation of heat (due to losses) by means of cooling ducts, and should be mechanically strong to withstand short-circuit forces.

There are two types of transformer in common use. When the magnetic circuit takes the form of a single ring encircled by two or more groups of primary and secondary windings distributed around the periphery of the ring, the transformer is termed a core-type transformer. When the primary and secondary windings take the form of a common ring which is encircled by two or more rings of magnetic material distributed around its periphery, the transformer is termed a shell-type transformer (Fig. 2.2). Actually, core-type (or "core-form") in U.S. power-transformer engineering usage means that the coils are cylindrical and concentric (the outer winding over the inner) whereas shell-type (or "form") denotes large pancake coils that are stacked or interleaved to make primary-secondary (P-S) groups. Except for certain extremes of current rating, the choice between the core- and shell-type constructions is largely a matter of manufacturing facilities and of individual preference.

Core-form transformer characteristic features are a long mean length of magnetic circuit and a short mean length of windings. Commonly used core constructions for 3-phase units are shown in Fig. 2.2. The simple concentric primary (inside) and secondary (outside) winding arrangement is common for all small- and medium-power transformers. However, large MVA transformers frequently have some degree of interleaving of windings, such as secondary-primary-secondary. The core-form construction can be used throughout the full size range of power transformers.

A photograph of a modern three-phase transformer is shown in Fig. 2.3. Power transformers are one of the most complex in design, expensive and widespread elements of EPG [3–9].

For example, the total number of power transformers in all voltage classes (except instrument transformers) as a whole across the UESR and DGC of Russia amounted as approximately 700 thousand units, their installed capacity is approximately 900 thousand MVA. The main part of transformers (in terms of quantity) consists of transformers with voltage up to 20 kV (94.7%). The UESR facilities account for

Fig. 2.3 Power three-phase transformer

about 0.8% of transformers in terms of quantity, and for about 37.8% in terms of capacity, while the DGC facilities account for 99.2% of transformers in terms of quantity and 62.2% in terms of capacity. Of the total number of power transformers installed in UESR networks with voltage of 110 kV and higher, about 55.3% of the units have exhausted their resource, with the accepted standard resource of service life for basic electrotechnical equipment being 25 years.

Approximately the same situation is typical for all countries with a developed electric power industry. As in Russia, despite the fact that many transformers have completed their standard service life, they continue to be operated due to the reason of saving the finances needed for replacement.

The problems of choosing and applying insulation in transformers will be discussed below. This problem must be given special attention, since the performance of transformers largely depends on the solution of this problem [10].

2.1.2 Transformers with Different Insulation

Insulation systems in power transformers consist of a fluid—either liquid or gas—together with solid materials. Petroleum-based oils have been used to insulate power transformers since 1886 and are still used in virtually all medium and large transformers. Askeral was used from 1932 through the mid-1970s when the flammability of mineral oil was a concern, but it has since been completely phased out of transformer production because of environmental concerns (Askeral is a generic term for a group of synthetic, fire-resistant, chlorinated aromatic hydrocarbons used as electrical insulated fluids). It has been replaced by any of a wide variety of high-flash-point fluids (silicones, high-flash-point hydro-carbons, chlorinated benzenes, or chlorofluorocarbons). Gas systems include nitrogen, air, and fluorogases, predominantly SF_6. (Additional positive properties of SF_6 will be named in Sect. 2.2). The fluorogases are used to avoid combustability and limit secondary effects of internal failure. Some transformers have been constructed using low boiling-point liquids such as Freon which allows improved heat transfer using a 2-phase cooling system.

Within the core and coil assembly, insulation can be divided into two fundamental groups: major insulation and minor insulation. Major insulation separates the high- and low-voltage windings, and the windings to core. Minor insulation may be used between the parts of individual coils or windings depending on construction. Finally, turn insulation is applied to each strand of conductor and/or groups of strands forming a single turn.

For small rating, the coils are made of super-enameled copper wire. For layer to layer, coil to coil and coil to ground (iron core) craft paper is used. However, for large size transformers paper or glass tape is rapped on the rectangular conductors whereas for coil to coil or coil to ground, insulation is provided using thick radial spacers made of press board or fiberglass.

In oil-filled transformers, the transformer oil is the main insulation. However between various layers of low voltage and high voltage winding oil-impregnated press boards are placed.

SF_6 gas insulated power transformers make use of sheet aluminum conductors for windings and turn to turn insulation is provided by a polymer film. The transformer has annular cooling ducts through which SF_6 gas circulates for cooling the winding. SF_6 gas provides insulations to all major gaps in the transformer. This transformer is used where oil filled transform is not suitable e.g., in cinema halls, high rise buildings and some especial circumstances: The end turns of a large power transformer are provided with extra insulation to avoid damage to coil when lighting or switching surges of high frequency are incident on the transformer winding.

The terminal bushings of large size power transformer are made of condenser type bushing. The terminal itself consists of a brass rod or tube which is wound with alternate layers of treated paper and tin foil, so proportioned, as to length, that the series of condensers formed by the tin foil cylinders and the intervening insulation have equal capacitances, thereby the dielectric stress is distributed uniformly.

2.1.2.1 Oil-Insulated Transformers

Low cost, high dielectric strength, excellent heat transfer characteristics, and ability to recover after dielectric overstress make mineral oil the most widely used transformer insulating material. The oil is reinforced with solid insulation in various ways. The major insulation usually includes barriers of wood-based paperboard (pressboard), the barriers usually alternating with oil spaces. Because the dielectric constant of the oil is 2.2 and that of the solid is approximately 4.0, the dielectric stress in the oil ends up being higher than that of the pressboard, and the design of the structure is usually limited by the stress in the oil.

The insulation on the conductors of the winding may be enamel or wrapped paper which is either wood- or nylon-based. The use of insulation directly on the conductor actually inhibits the formation of potentially harmful streamers in the oil, thereby increasing the strength of the structure. Again, the limit of dielectric strength is usually that of the oil.

Heavy paper wrapping is also usually used on the leads coming from the winding. In this case, the insulation serves to reduce the stress in the oil by moving the interface from the surface of the conductor (where the stress is high) to a distance away from the conductor (where the stress is considerably lower). Again, the stress in the oil determines the amount of paper required, and the thermal considerations establish the minimum size of the conductor for the necessary insulation.

2.1.2.2 Askeral-Insulated Transformers

These transformers have constructions similar to the oil-insulated transformers. The relatively high dielectric constant of the askeral aids in transferring the dielectric stress to the solid elements. Askeral has limited ability to recover after dielectric overstress, and thus the strength is limited in nonuniform dielectric fields. Askerals are seldom used over 34.5 kV operating voltage. They are powerful solvents; their products of decomposition are so harmful that they have been completely abandoned in transformers manufactured after the mid-1970s.

2.1.2.3 Gas-Filled Transformers

(a) *Nitrogen and air-insulated transformers*

These are generally limited to 34.5 kV and lower operating voltages. Air-insulated transformers in clean locations are frequently ventilated to the atmosphere. In contaminated atmospheres a sealed construction is required, and nitrogen is generally used at approximately 1 atm and some elevated operating temperatures.

(b) *Fluorogas-Insulated Transformers*

For applications where low flammability is paramount, designs have been developed in which the transformer is insulated and cooled with SF6 gas. This provides an alternative to dry-type construction where the risk of fire must be eliminated and the possible contamination of the environment by oil spillage must be avoided. High-voltage SF_6 transformers are available at ratings up to 300 MVA at 275 kV and prototype designs have been tested at up to 500 kV. Gas-filled transformers and reactors are more expensive than oil-filled units but the costs may be justified to eliminate a risk of fire, particularly at a site where the cost of land is high and where the overall 'footprint' of the unit can be reduced by the elimination of fire-fighting equipment.

Fluorogases have better dielectric strength than nitrogen or air. Although their heat transfer characteristics are poorer than oil, they are better than nitrogen or air because of their higher density. Both dielectric strength and heat transfer capability increase with pressure; in fact, the dielectric strength at 3 atm gage pressure, where some fluorocarbon-insulated transformers operate, can approach that of oil. (The advantages of sulfur hexafluoride as insulation are especially pronounced when used in high-voltage circuit breakers, see Sect. 2.2).The gas insulation is reinforced with solid insulation used in the form of barriers, layer or disk insulation, turn insulation, and lead insulation similar to oil-immersed transformers.

It is usually economical to operate fluorogas-insulated transformers at higher temperatures than oil-insulated transformers. Suitable solid insulating materials include glass, asbestos, mica, high-temperature resins, and ceramics.

Dielectric stress on the gas is several times higher than in the adjacent solid insulation; care must be taken to avoid overstressing the gas.

2.1.2.4 Dry-Type Transformers

A dry-type construction is possible where a higher-temperature class of insulation is required than is offered by cellulose and a class 'O' or class 'K' fluid. Dry-type transformers use non-cellulosic solid insulation and the windings may be varnish dipped to provide a class 'C' capability, or vacuum encapsulated in epoxy resin to form a class 'F' or class 'H' system. Ratings are generally up to 30 MVA at voltages up to 36 kV, but cast resin transformers have recently been successfully manufactured at 110 kV using a novel winding design. Overload performance is limited but it can be augmented by the use of cooling fans.

This type is more expensive than a fluid-filled equivalent, and because of the reduced fire risk they are used in special applications where the public are involved, such as underground tunnels, residential blocks of flats or oil-rigs.

2.1.2.5 Design of Insulation Structures

Three factors must be considered in the evaluation of the dielectric capability of an insulation structure—the voltage distribution must be calculated between different parts of the winding, the dielectric stresses are then calculated knowing the voltages and the geometry, and finally the actual stresses can be compared with breakdown or design stresses to determine the design margin.

Voltage distributions are linear when the flux in the core is established. This occurs during all power frequency test and operating conditions and to a great extent under switching impulse conditions. (Switching impulse waves have front times in the order of tens to hundreds of microseconds and tails in excess of 1000 μs.) These conditions tend to stress the major insulation and not inside of the winding. For shorter-duration impulses, such as full-wave, chopped-wave, or front-wave, the voltage does not divide linearly within the winding and must be determined by calculation or low voltage measurement. The initial distribution is determined by the capacitative network of the winding. For disk and helical windings, the capacitance to ground is usually much greater than the series capacitance through the winding. Under impulse conditions, most of the capacitive current flows through the capacitance to ground near the end of the winding, creating a large voltage drop across the line end portion of the coil.

The capacitance network for shell form and layer-wound core form results in a more uniform initial distribution because they use electrostatic shields on both terminals of the coil to increase the ratio between the series and to ground capacitances. Static shields are commonly used in disk windings to prevent excessive concentrations of voltages on the line-end turns by increasing the effective series capacitance within the coil, especially in the line end sections. Interleaving turns and introducing floating metal shields are two other techniques that are commonly used to increase the series capacitance of the coil.

Following the initial period, electrical oscillations occur within the windings. These oscillations impose greater stresses from the middle parts of the windings to ground for long-duration waves than for short-duration waves. Very fast impulses, such as steep chopped waves, impose the greatest stresses between turns and coil portions. Note that switching impulse transient voltages are two types—aperiodic and oscillatory. Unlike the aperiodic waves discussed earlier, the oscillatory waves can excite winding natural frequencies and produce stresses of concern in the internal winding insulation. Transformer windings that have low natural frequencies are the most vulnerable because internal damping is more effective at high frequencies.

Dielectric stresses existing within the insulation structure are determined using direct calculation (for basic geometries), analog modeling, or most recently, sophisticated finite-element computer programs.

Allowable stresses are determined from experience, model tests, or published data. For liquid-insulated transformers, insulation strength is greatly affected by contamination and moisture. The relatively porous and hygroscopic paper-based insulation must be carefully dried and vacuum impregnated with oil to remove moisture and gas to obtain the required high dielectric strength and to resist deterioration at operating temperatures. Gas pockets or bubbles in the insulation are particularly destructive to the insulation because the gas (usually air) not only has a low dielectric constant (about 1.0), which means that it will be stressed more highly than the other insulation, but also air has a low dielectric strength.

High-voltage dc stresses may be imposed on certain transformers used in terminal equipment for dc transmission lines. Direct-current voltage applied to a composite insulation structure divides between individual components in proportion to the resistivities of the material. In general the resistivity of an insulating material is not a constant but varies over a range of 100:1 or more, depending on temperature, dryness, contamination, and stress. Insulation design of high-voltage dc transformers in particular requires extreme care.

2.1.3 Purpose and Features of Instrument Transformers (ITs)

We know that the voltages and currents within a power system are very large. Thus, direct measurement of voltage and current with high magnitude is not possible. So we need measuring instruments which have a high range of measurements or there is another technique like using the property of conversion within AC currents as well as voltages. A transformer is used to transform the current or voltage down when turns ration is known after that determining the stepped down magnitude using traditional ammeter and voltmeter. The magnitude of current or voltage of interest to us is determined by simply multiplying the result by the transformation ratio. So such kind of transformer with a precise turn ratio is known as ***Instrument transformer***. General view of the Instrument Transformer is shown on Fig. 2.4.

Its have a number of advantages over other instruments used in measuring high currents and voltages. These include the following:

- ITs use ordinary ammeter and voltmeter (low-voltage and low-current measuring devices) to measure high currents and voltages.
- By using ITs, several protecting devices can be operated like relays.

Fig. 2.4 General view of the Instrument Transformer

- ITs are cheaper than power transformers.
- Damaged parts can be easily replaced.
- ITs offer electrical isolation among measuring instruments and high voltage power circuits. So that electrical insulation requirements can be reduced in protective circuits and measuring instruments. By using IT, various measuring instruments can be connected to a power system.
- Low power consumption will be there in protective and measuring circuits because of the low level of voltage and current. These transformers are mainly used with relays to protect the power system. In general, voltmeters and ampermeters are mainly designed for 5 A and 110 V.
- ITs play an important role in monitoring and protecting power supply systems. They Supply accurately scaled current and voltage quantities for measurement while insulating the relay from the high voltage and current of the power system.
- The only disadvantage of instrument transformer is, these can be used simply for AC circuits but not for DC circuits.

As noted above, Instrument transformers are classified into two types such as

- Current transformer
- Potential transformer

2.1.3.1 Current Transformer

This type of transformer can be used in power systems to step down the voltage from a high level to a low level with the help of a 5A ammeter. CTs are mainly used to use secondary (reduced) current in relays, meters, control devices and other devices. This transformer includes two windings like primary and secondary. CTs intended to have its primary winding connected in series with the conductor carrying the current to be measured or controlled. The current in the secondary winding is proportional to the current in the primary winding as it generates current in the secondary winding. The circuit diagram of a typical current transformer is demonstrated in Fig. 2.5.

In this transformer, the primary winding consists of few turns and it is connected with the power circuit in series. Therefore, it is called a series transformer. Likewise, the secondary winding includes a number of turns and it is connected to an ammeter directly because the ammeter includes small resistance. Thus, the secondary winding of this transformer works almost in the condition of a short circuit. This winding includes two terminals where one of its terminals is connected to ground to evade the huge current. So insulation breakdown chances will be reduced to guard the operator from huge voltage.

The secondary winding of this transformer in the above circuit is short-circuited before disconnecting the ammeter with the help of a switch to avoid the high voltage across the winding [1].

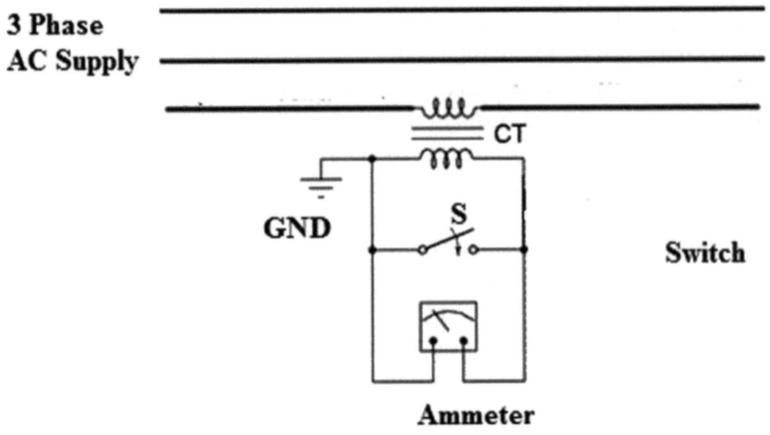

Fig. 2.5 The circuit diagram of a current transformer

Types of insulation and types of designs of current transformers (CT)
According to the type of dielectric, CT insulation can be divided into solid and paper-oil. For voltages up to 35 kV, insulation is most often made from solid materials: porcelain, glass, epoxy.

Epoxy resin is mainly used in indoor CTs.

By design, oil-paper-insulated transformers can be divided into the following main types, which are shown in Fig. 2.6a–c.

(1) U-shaped. The insulation is carried out by continuous winding of a paper tape around the curved primary winding. High-voltage insulation is not applied to the magnetic circuit with the secondary winding.
(2) Eye-shaped I (rimoid form I). Insulation is carried out by winding paper tape on the primary winding. High-voltage insulation is not applied to the magnetic circuit with the secondary winding.
(3) Eye-shaped II (rimoid form II). In this design, insulation is applied to both the primary and secondary windings. High-voltage insulation is not applied to the magnetic core of the primary winding.

2.1.3.2 Potential Transformer

This type of transformer can be used in power systems to step down the voltage from a high level to a lower level with the help of a small rating voltmeter which ranges from 110 to 120 V. A potential transformer typical circuit diagram is illustrated below.

This transformer includes two windings like a normal transformer like primary and secondary. The primary winding of the transformer includes a number of turns and it is connected in parallel with the circuit. So it is called a parallel transformer, Fig. 2.7.

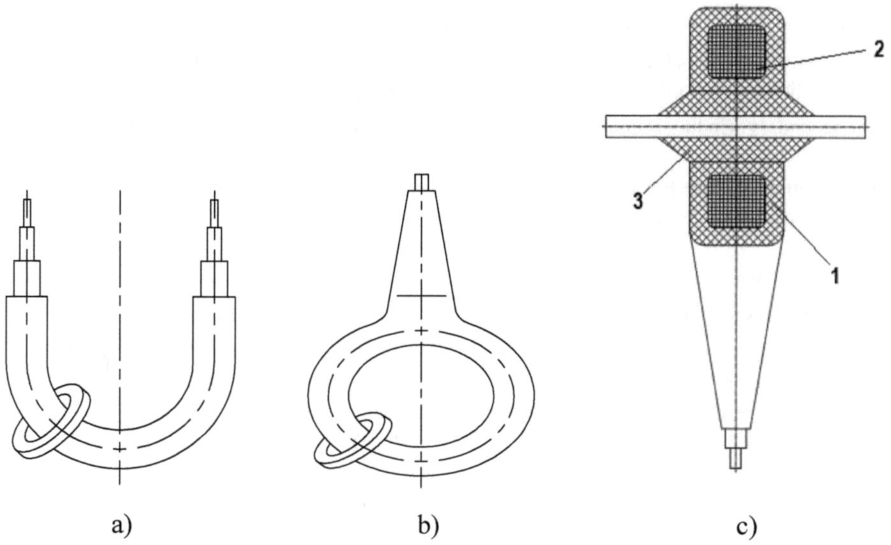

Fig. 2.6 **a** U-shaped current transformer; **b** eye-shaped current transformer with insulation on the primary only; **c** eye-shaped current transformer with insulation on both windings: 1—primary winding, 2—secondary winding, 3—insulation

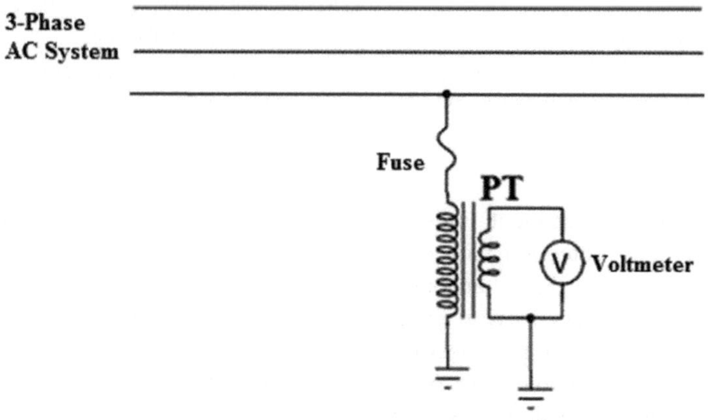

Fig. 2.7 The circuit diagram of a potential transformer

Similar to the primary winding, the secondary winding includes fewer turns and that is connected to a voltmeter directly because it includes huge resistance. Therefore the secondary winding works approximately in open circuit condition. One terminal of this winding is connected to the earth to maintain the voltage with respect to the earth to protect the operator from a huge voltage.

The difference between the current transformer and potential transformer is discussed below. Comparative analysis of instrument transformers is in Table 2.1.

Table 2.1 Comparative analysis of instrument transformer

The connection of this CT can be done in series with the power circuit	The connection of PT can be done in parallel with the power circuit
The secondary winding is connected to an ammeter	The secondary winding is connected to a voltmeter
The design of this can be done by using the lamination of silicon steel	The designing of PT is can be done by using high-quality steel which operates at low-flux densities
The primary winding of CT carries the current	The primary winding of PT carries the voltage
It includes less number of turns	It includes a number of turns
The secondary winding of CT works in the condition of a short circuit	The secondary winding of PT works in the condition of an open circuit
The primary current mainly depends on the flow of current within the power circuit	The primary current mainly depends on the secondary load
The insulation breakdown can be avoided by connecting the secondary winding of CT to the earth	The secondary winding can be connected to the earth to protect the operator from a huge voltage
The range of CT is 1A or 5A	The range of PT is 110v
CT ratio is high	PT ratio is low
The input of CT is the constant current	The input of **PT** is a constant voltage
CT is classified into two types like wound type and closed core	PT is classified into two types like electromagnetic and capacitor voltage
The impedance of CT is low	The impedance of PT is high
CTs are used to measure current, power, monitoring the operation of power grid and protective relay	PTs are used to measure, operating protective relay and power source

Testing of an instrument transformer is essential when metering, mixing up connections and protection fault occurs. Otherwise high degree of exactness can be reduced drastically. Simultaneously, changes within a CT will occur. Due to these reasons, it is necessary to verify and adjust current transformers along with their connected devices at normal intervals. There are some electrical tests are employed for these transformers to ensure exactness and optimal service reliability like ratio, polarity, excitation, insulation, winding and burden test [1, 3].

2.1.3.3 Non Conventional Instrument Transformers (NCIT)

NCIT technologies may be based on optical techniques, or Rogowski coils, and overcome the limitation of iron-cored transformers. Rogowski coil installation in GIS (gas insulated switchgear) is shown on Fig. 2.8.

Some pilot applications using NCIT have been implemented in France and UK on actual 245 and 420 kV GIS. These field trial installations have confirmed the performances of these modern sensors, as well as the robustness of a comprehensive

Fig. 2.8 Rogowski coil installation in GIS (gas insulated switchgear)

Protection and Metering system governed by the former applications of the IEC 61850. Making use of protection relays from different vendors, these pilots also proved the perspectives of interoperability, absolutely mandatory for the end users.

Despite the maturity of these technologies, the limitation of their utilization is a reality and some reasons can be reminded to explain that.

Three ways of reasons can be mentioned and many works are undertaken to overcome the issues in order to start a large deployment in the coming years:

1. Technology acceptance,
2. Standardization of the interface,
3. Testing methods.

View of a very compact three-phased GIS with NCIT is shown on Fig. 2.9.

The current trends address a large number of different technologies and applications of the sensors like, for instance, Rogowsky or optical-type current transformers, electronic or magnetic-type core for metering, capacitive effect voltage transformers of different technologies, together with demanding high specification in term of Reliability, Availability and Maintenance performance criteria. We limit ourselves here to a reminder of the more common NCIT technologies in high voltage substations, which are suited to GIS substations.

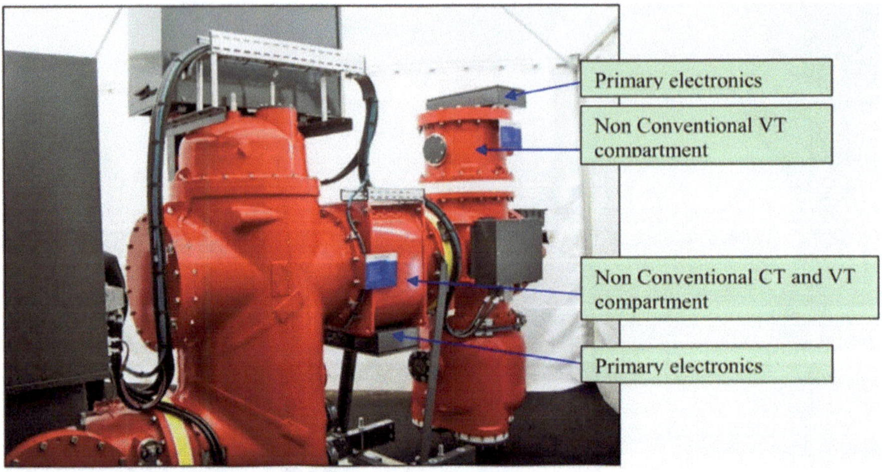

Fig. 2.9 View of a very compact three-phased GIS with NCIT

Merging units present signals such as power system voltages and currents to IEDs within the substation, in the form of numerical values adhering to standardized definitions.

The use of NCIT sensors has made it possible for raw measurement information to be fed into so-called Merging Units for further distribution. It is these merging units that are one of the main contributors to the digital substation. Common view of Siemens GIS up to 145 kV Switchgear bay with Non-conventional Instrument Transformer is shown on Fig. 2.10.

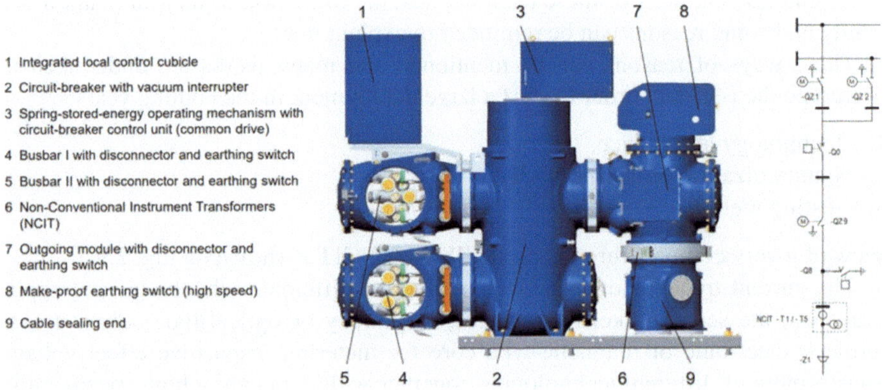

Fig. 2.10 Siemens GIS up to 145 kV Switchgear bay with Non-conventional Instrument Transformer

2.1.4 The Faults, Which Occurs Inside a Power Transformer

There are five main reasons for the failure of the transformer:

1. Insulation breakdown between winding and earth
2. Insulation breakdown between different phases
3. Insulation breakdown between adjacent turns i.e. inter—turn fault
4. Transformer core fault
5. Tank rupture.

Regarding the first three reasons, the following should be noted. Insulation breakdowns and the resulting short circuits inside a transformer place extreme physical stresses on the transformer windings and are a major cause of transformer failures [4]. (The mechanical effects of alternating current flowing through the windings of the transformer, even in normal mode, produce a buzzing noise that is amplified by current overload).

1. Overloads rarely result in transformer failures, but do cause thermal aging of winding insulation. When a transformer becomes hot, the insulation on the windings slowly breaks down and becomes brittle over time. The rate of thermal breakdown approximately doubles for every 10 °C. 10 °C is referred to as the *Montsinger Factor* and is a rule describing the Arrhenius theory of electrolytic dissociation. Because of this exponential relationship, transformer overloads can result in rapid transformer aging. When thermal aging has caused insulation to become sufficiently brittle, the next fault current that passes through the transformer will mechanically shake the windings, a crack will form in the insulation, and an internal transformer fault will result, as shown on Fig. 2.11.

2. Extreme hot-spot temperatures in liquid-filled transformers can also result in failure. This is because the hot spot can cause free bubbles that reduce the dielectric strength of the liquid. Even if free bubbles are not formed, high temperatures will increase internal tank pressure and may result in overflow or tank rupture.
3. Many transformers are fitted with load tap changers (LTCs) for voltage regulation. These mechanically moving devices have historically been prone to failure and can substantially reduce the reliability of a transformer. Manufacturers have addressed this problem and new LTC models using vacuum technology have succeeded in reducing failure rates.

2.1.5 Disconnectors

A high voltage disconnector is a device that ensures the safety of an electrical consuming installation or electrical circuit. It has the task of opening a circuit or a line, so physically and visibly perceptible, separating two points electrically connected to each other so there is no more metallic continuity between them. A disconnector can be associated to a switch, but a disconnector has a limit in terms of maximum current

Fig. 2.11 Typical view of the destroyed coil of a three-phase transformer

opening. The disconnector task is to disconnect safely the voltage from the electrical consuming installation, in order to enable people to access the system itself for performing work, involving contact with live parts or activity close to the electrical consuming installation.

2.1.5.1 Horizontal Double Break Disconnector

Double Break Disconnector can be installed in locations, which have minimal overhead clearance, as blades swing open to the side rather than lifting upwards. Successfully tested for SC level of 40, 50 and 63 kA. Common view of horizontal disconnector is shown on Fig. 2.12.

2.1.5.2 Horizontal Centre Break Disconnector

Centre Break Disconnector as well as Horizontal Double Break Disconnector can be installed in locations, which have minimal overhead clearance, as blades swing open to the side rather than lifting upwards. Successfully tested for SC level of 40 kA, 50 kA. Common view of the centre break disconnector is shown on Fig. 2.13.

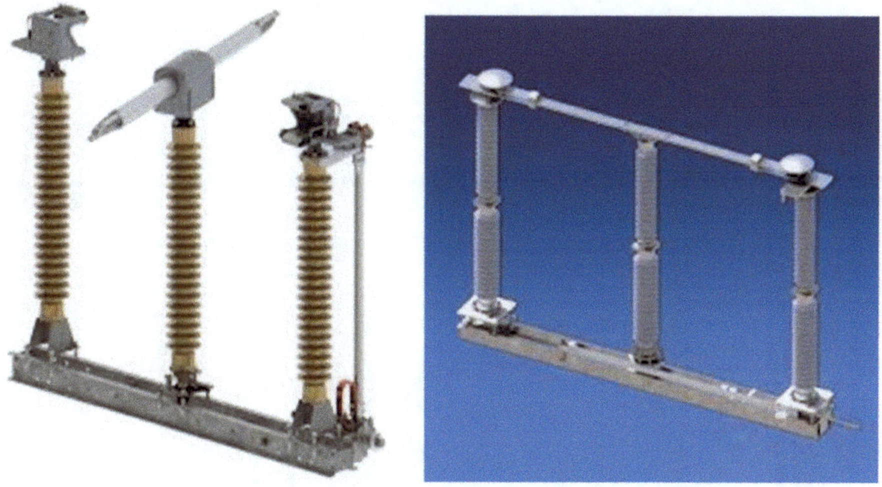

Fig. 2.12 Common view of the horizontal disconnector

Fig. 2.13 Common view of the centre break disconnector

Fig. 2.14 Common view of the pantograph disconnector in working position

2.1.5.3 Pantograph Disconnector

Pantograph disconnectors can be installed in locations where coupling of buses for current transfer are required at two different heights. Current transfer will take place from the top-level bus to the bottom level bus through this type of disconnector. Successfully tested for SC level of 40 kA. View of pantograph disconnector in working position is shown on Fig. 2.14.

2.1.5.4 Horizontal Knee Type Disconnector

A particular advantage of using horizontal knee type disconnector is reduced vertical dimension and these disconnectors can be used for any conventional application. These disconnectors can withstand SC current rating up to 50 kA/1 s. Common view of the horizontal knee type disconnector is shown on Fig. 2.15.

Fig. 2.15 Common view of the horizontal knee type disconnector

2.1.6 Circuit Switchers and Switchgear Assemblies

Circuit switchers are mechanical devices that combine the features of a disconnect switch and circuit breaker. They typically combine sulfur hexafluoride (SF_6) as an arc-interrupting medium and a trip device connected to a relay to open the circuit switcher automatically, with an air-break disconnect switch, under certain abnormal conditions. Blade-type circuit switchers with fault-interrupting ratings up to 15.000 A are being manufactured, and they are capable of interrupting voltages of 38–230 kV.

The disconnector provides visual isolation, meaning that when the circuit has been interrupted it can be seen that the disconnect blade, part of the switching mechanism, has sprung open, leaving a large air gap between the blade end and its closed position.

Circuit switchers are used primarily for transformer protection. They can also provide load-switching capability, line and loop switching, capacitor or reactor switching, and load management, usually with protection features. They combine the functions of a circuit breaker (without its high-speed reclosing capability) and a disconnecting switch, to make, carry, and break normal load currents to make, carry, and break normal load currents. Circuit switchers must be able to prevent damage to key system components within a defined temperature range not seen as endangering

Fig. 2.16 Circuit-switcher model 2010—horizontal interrupters and vertical break disconnect

the integrity of the system such, as contacts, linkage, terminals, and isolators. They must also be able to make and carry load currents for predetermined lengths of under certain abnormal conditions is normally open and will be closed only upon the activation of a control device, and they must be able to break currents under overcurrent or fault conditions. View of circuit switchers and switchgears of different types are shown on Figs. 2.16, 2.17 and 2.18.

2.1.7 Switchgear Assemblies

Switchgear assemblies cover a wide range of low-voltage and high-voltage structures that are generally factory assembled and are divided into the following main groups: (1) metal-enclosed low-voltage power circuit breaker switchgear, (2) medium-voltage metal-clad switchgear, (3) metal-enclosed interrupter switchgear, (4) metal-enclosed bus, and (5) switchboards. How compact the complete switchgear becomes when using SF6 insulation (instead of air) can be seen in Fig. 2.19a.

Fig. 2.17 Circuit switcher for outdoor transmission, 69 kV through 138 kV

2.1.7.1 Metal-Enclosed Low-Voltage Power Circuit Breaker Switchgear

Metal-enclosed low-voltage power circuit breaker switchgear indicates a design, which contains low-voltage ac or dc power circuit breakers in individual grounded metal compartments. The circuit breakers can be either stationary or drawout; manually or electrically operated; fused or unfused; and either 3-pole, 2-pole or single-pole, construction. The switchgear may also contain associated control, instruments, metering, protective and regulating equipment as necessary. Definitions, ratings, design and production tests, construction requirements, and guidelines for application, handling, storage, and installation are covered in IEEE C37.20.11.

Low-voltage metal-enclosed switchgear is typically installed in industrial plants, utility and cogeneration facilities, and commercial buildings for the protection and distribution of power for loads such as lighting, machinery, motor control centers, elevators, air conditioning, blowers, compressors, fans, pumps, and motors. Low-voltage switchgear is available in ac ratings up to 635 V and 5000 A continuous and in dc ratings up to 3200 V and 12,000 A continuous. Short-circuit current ratings are available up to 200 kA.

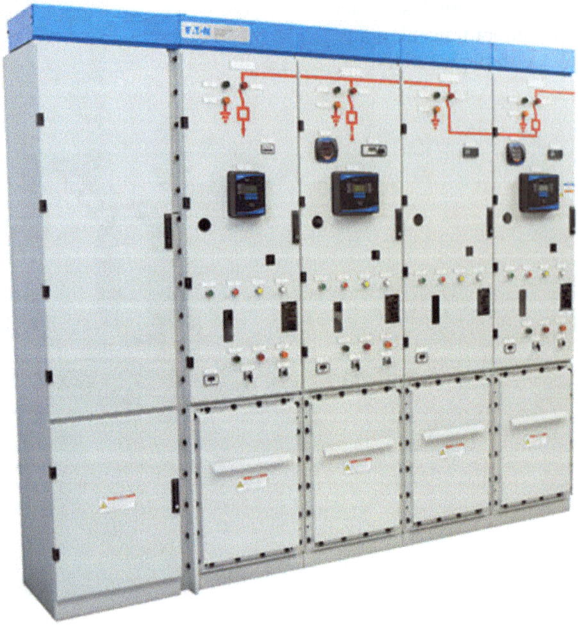

Fig. 2.18 Power Xpert XGIS gas insulated medium voltage switchgear

2.1.7.2 Metal-Clad Switchgear

The term "metal-clad switchgear" indicates a design providing metal barriers between primary sections of adjacent vertical sections and between major primary sections of each circuit. Primary sections comprise the bus compartment, the primary entrance compartment, the removable element compartment, the voltage transformer(s) compartment, and the control power transformer(s) compartment. Low-voltage control equipment such as metering, relays, instruments, and controls are located in compartments separate from the primary voltage components. To minimize the possibility of communicating faults between primary sections, the barriers between primary sections shall have no intentional openings. Barriers may be provided to segregate the voltage transformers for each poly-phase circuit but not to segregate them individually. Where buses penetrate barriers, suitable bushings or other insulation is required. Definitions, ratings, design and production tests, construction requirements, and guidelines for application, handling, storage and installation are covered in IEEE C37.20.22.

Circuit breakers are generally the vacuum type, although air-magnetic circuit breakers were used for many years. Circuit breaker dis-connection is accomplished by horizontal draw-out design, illustrated in Fig. 2.19a,b, respectively. Interlocks are provided in metal-clad assemblies to prevent disconnecting or connecting the circuit breaker while in the closed position and to prevent breaker operation while moving between disconnected and connected position or vice versa. The metalclad assembly

(a)

(b)

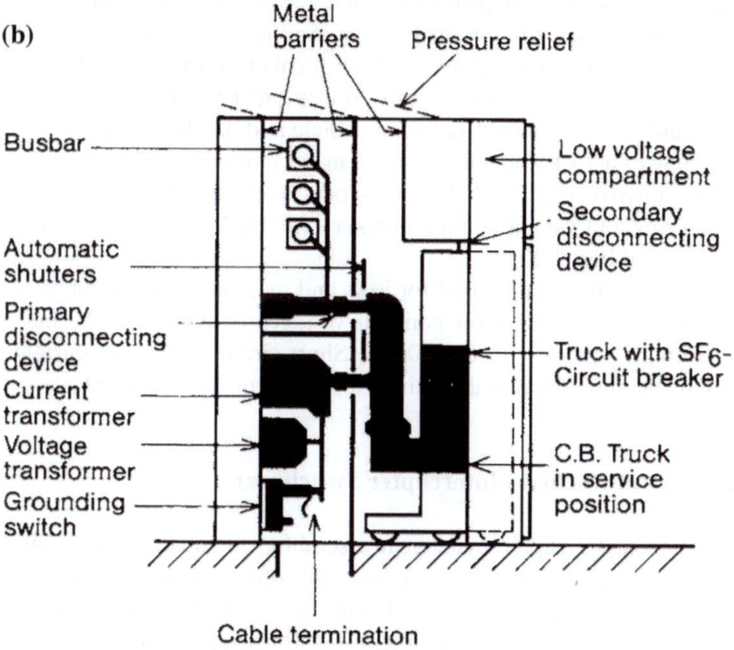

Fig. 2.19 **a** Complete switchgear with SF_6 Insulation (GIS). **b** Side view of a 15-kV metal-clad switchgear unit with horizontal-drawout circuit breaker

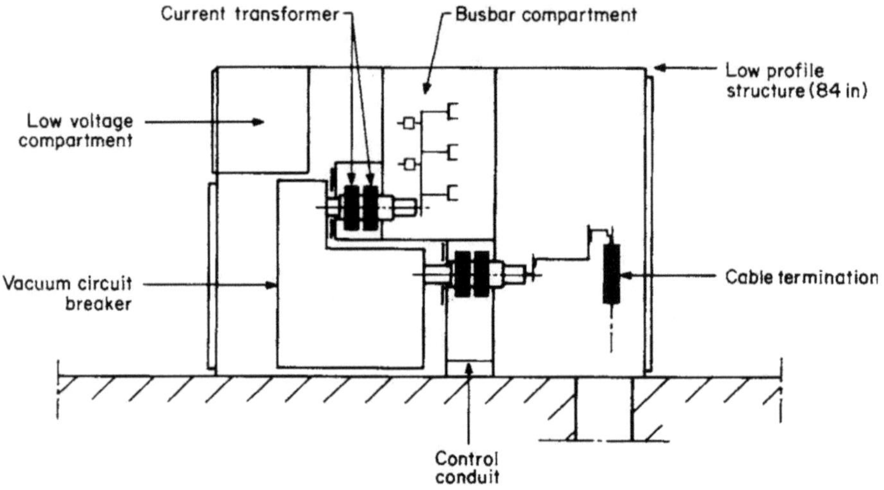

Fig. 2.20 Side view of a 38-kV metal-clad switchgear unit with horizontal-drawout vacuum circuit breaker, type HKV, up to 3000 A, fan cooled, 22 kA

is equipped with shutters to protect personnel from coming in contact with the high-voltage circuits when the circuit breaker is removed from the cubicle. A circuit breaker test position is standard to allow breaker control with the main contacts (primary disconnecting devices) removed from the primary circuit, but maintaining auxiliary and ground contacts between cubicle and breaker truck.

Side view of a 15-kV metal-clad switchgear unit with horizontal-drawout circuit breaker is shown on Fig. 2.19b. Side view of a 38-kV metal-clad switchgear unit with horizontal-drawout vacuum circuit breaker, type HKV, up to 3000 A, fan cooled, 22 kA is shown on Fig. 2.20.

Metal-clad switchgear is used for low- and medium-capacity circuits, for indoor and outdoor installations with nominal voltages of 2.4–34.5 kV and continuous current ratings typically up to 3000 A. Short-circuit withstand current ratings of the switchgear should equal the ratings of the circuit breaker used.

2.1.7.3 Metal-Enclosed Interrupter Switchgear

Metal-enclosed interrupter switchgear assemblies include the following equipment as required: interrupter switches, bare bus and connections, selector switches, power fuses (current-limiting or noncurrent-limiting), control and protective equipment, instrumentation, meters, and instrument transformers. The interrupter switches and power fuses may be stationary or removable (drawout). When switches and fuses are removable, mechanical interlocks are provided for proper operating sequence. Also, automatic shutters are provided which cover primary circuit elements when the removable device is in the disconnected, test or removed position. Definitions, ratings, design and production tests, construction requirements, and guidelines for application, handling, storage, and installation for metal-enclosed interrupter switchgear are covered in IEEE C37.20.33.

Metal-enclosed interrupter switchgear is typically used in industrial or institutional environments where continuous load currents are low and frequent switching is not required. Interrupter switches will interrupt load currents up to their rated continuous current capability. Fuses can be installed to provide short-circuit protection. For example, if the interrupter switchgear is connected to other switching equipment fuses can be installed in the connection between the two to prevent an interruption of one assembly for a fault in the other assembly.

Typical applications for interrupter switchgear include main service disconnect, transformer primary and secondary switching, medium voltage switchgear primary and feeder circuit switching. The switching device may be manually operated or motor operated. Motor operated designs are often applied in an automatic transfer scheme.

Metal-enclosed interrupter switchgear is typically available in ac ratings above 1 kV to up to 38 kV and 2000 A continuous current. Short-circuit withstand ratings have to be equivalent to the ratings of the switching and protective equipment used or to the rating of the current transformers used.

2.1.7.4 Metal-Enclosed Bus

Metal-enclosed bus is an assembly of rigid conductors with associated connections, joints and insulating supports with a grounded metal enclosure. Metal enclosed buses have three basic types of construction: (1) nonsegregated-phase, (2) segregated-phase, and (3) isolated phase.

Rated voltages of ac metal-enclosed bus assemblies range from 635 V through 38 kV, and dc metal-enclosed bus assemblies range from 300 through 3200 V. Definitions, service conditions, ratings, testing, construction requirements, and application guidelines for metal-enclosed bus are covered in IEEE C37.23. An informative guide for calculating losses in isolated-phase bus is also included.

Nonsegregated-Phase Metal-Enclosed Bus. Nonsegregated-phase metal-enclosed bus is a type of design in which all phase conductors, with their associated connections, joints, and insulating supports, are enclosed in a common metal housing without barriers between phases, see Fig. 2.21. When associated with metal-clad switchgear, the phase conductors of a non-insulated bus assembly entering the switchgear assembly and connecting to the switchgear bus shall be covered with insulating material equivalent to the switchgear insulation system. Enclosures that are totally enclosed are preferred, but ventilated enclosures can be provided in indoor applications. Typical nonsegregated-phase metal enclosed bus is shown on Fig. 2.21.

Nonsegregated-phase metal-enclosed bus is utilized on circuits which require higher reliability than can be obtained with the application of power cables. Typical applications are the connections between transformers and switchgear assemblies, connections from switchgear assemblies to rotating apparatus, tie connections

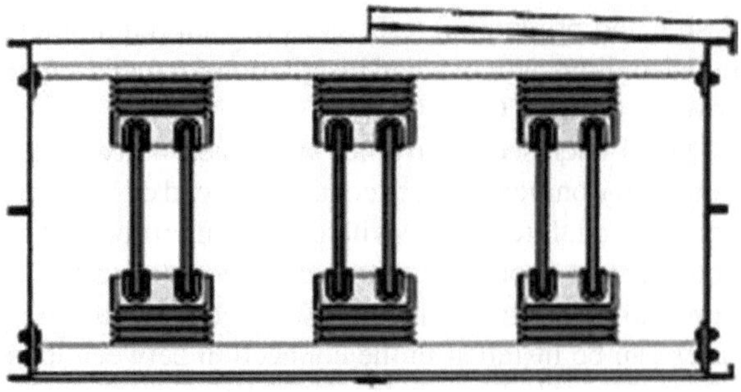

Fig. 2.21 Typical nonsegregated-phase metal enclosed bus

between switchgear assemblies, connections between motor control centers and large motors, and as main generator leads for small generators. Preferred continuous self-cooled current ratings for nonsegregated-phase are available up to 12,000 A for 635 V ac and all dc voltage ratings, 6000 A for 4.75 through 15.5 kV, and 3000 A above 15.5 through 38 kV. Short-time withstand current ratings up to 85 kA rms symmetrical are available for ac ratings, and up to 120 kA for dc ratings.

Segregated-Phase Metal-Enclosed Bus. Barriers may be installed between the phase conductors to segregate the conductors and the assembly is then referred to as "segregated-phase metal-enclosed bus". This design is also used on circuits which require a higher degree of reliability. Segregated-phase bus is primarily used as generator leads in power plants, but it is also applied in heavy industrial environments and as tie connections in metal-enclosed substations. Preferred continuous self-cooled current ratings for segregated-phase are available up to 12,000 A for 635 V ac and all dc voltage ratings, 6000 A for 4.75 through 15.5 kV, and 3000 A above 15.5 through 38 kV. Short-time withstand current ratings up to 85 kA rms symmetrical are available for ac ratings, and up to 120 kA for dc ratings.

Isolated-Phase Metal-Enclosed Bus is a type of design in which each phase is enclosed in an individual metal housing, and an air space is provided between the housings. It is considered to be the safest, most practical, and most economical way of preventing phase-to-phase short circuits by means of construction methods. The bus may be self-cooled or force-cooled by circulating air or liquid. Definitions, ratings, design and production tests, construction requirements, and application guidelines for metal-enclosed bus are covered in IEEE C37.23.

Briefly, the isolated-phase bus duct has the following features:

- Proof against contact; locked electrical premises not necessary
- Faults only in the form of ground faults; protection against fault spreading to more than one phase
- Field forces, static and dynamic, only between enclosures and conductor, not between phases
- Protection against contamination and moisture
- No losses in surrounding conductive material (grilles, railings, concrete reinforcements, lines, etc.).

The range of bus ducts are available up through 38 kV and includes continuous current ratings from about 5 up to 25 kA self-cooled, or 40 kA with forced cooling. The momentary current ratings have to match the rating of attached equipment. With high current ratings, more attention must be paid to the following:

- Progressive rise of conductor temperature due to skin effects.
- Heating of surrounding conducting material by the magnetic field of conductors.
- High forces on main or component conductors in the event of a short circuit.

In an enclosure with sections of tube insulation (sectional enclosure), eddy currents exist with values as large as the conductor current. These give rise to heat losses, and so the magnetic field of the main conductor is not always compensated for sufficiently. An important technical feature of the bus duct, therefore, is the electrically continuous enclosure. The tubes enclosing each phase have electric conducting joints throughout their length and are short-circuited across the three phases at both ends. The enclosure thus constitutes a secondary circuit to the conductors (Fig. 2.4). The currents in the enclosures reach almost the corresponding conductor currents, depending on the resistance of the duct, but are of the opposite direction. The magnetic field outside the enclosure is almost completely eliminated, and thus there are no external losses or field forces between the phases. Connections to machines and switchgear must be adaptable and removable. Current transformers for measurement and protection are of the bushing type or are integrated into the bus duct at a suitable place. Voltage transformers can be contained in the bus duct or mounted in separate instrument boards. The same applies to protective capacitors. Care must be taken that branch lines are adequately dimensioned with regard to thermal short-circuit strength. Three-phase arrangement of an isolated-phase bus duct and principle of enclosure connection; according to Kirchhoff's law sum of conductor currents (+) and sum of duct currents (−) is zero is shown on Fig. 2.22.

The reliability of generator bus ducts can be enhanced by employing means to maintain the air pressure in the duct. Although, generally bus ducts are leakproof, the large number of dismantleable joints may cause a slight leakage and might lead to moisture condensation during a plant shutdown. Supplying the bus duct with filtered, precompressed air at slight pressure ensures that the air flow is only outward; contamination of the conductors is not possible. Drying the air by precompressing prevents condensation.

Short-circuiting and grounding facilities are required in the bus duct to protect the generator and also for maintenance grounding purposes. Manually positioned links and straps are sufficient for small unit ratings; motor-operated grounding switches are recommended for higher capacities. A typical isolated-phase bus arrangement of a power station including generator circuit breaker is shown in Fig. 2.23.

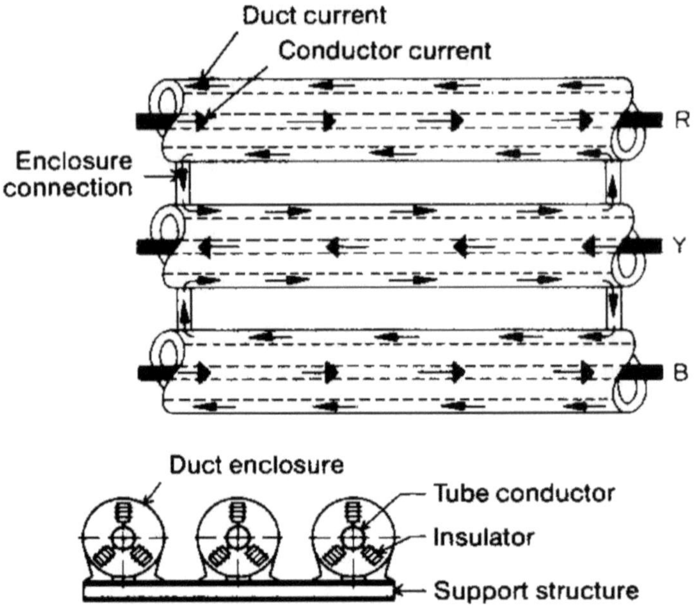

Fig. 2.22 Three-phase arrangement of an isolated-phase bus duct and principle of enclosure connection; according to Kirchhoff's law sum of conductor currents (+) and sum of duct currents (−) is zero

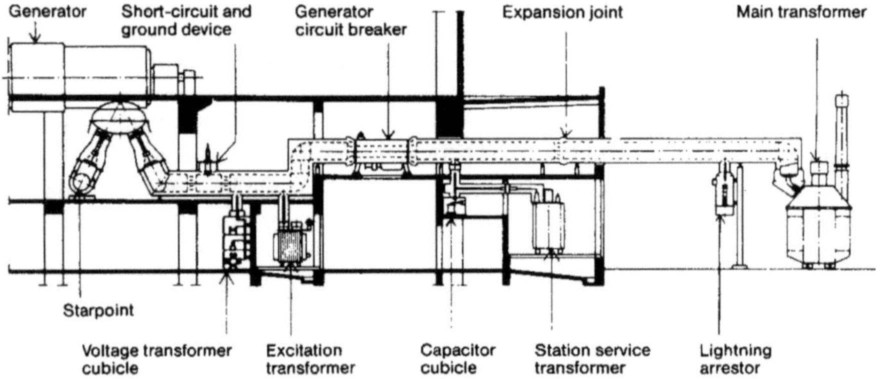

Fig. 2.23 Generating-plant isolated-phase bus-duct arrangement with generator circuit breaker type DR

2.1.7.5 Switchboards

Floor-mounted deadfront "*switchboards*" typically consist of an enclosure, molded case or low-voltage power circuit breakers, fusible or nonfusible switches, instruments, metering equipment, monitoring equipment and/or control equipment, and are fitted with associated interconnections and supporting structures. Switchboards can consist of one or more sections which are electrically and mechanically interconnected. Main disconnect devices can be mounted individually or be an integral part of a panel assembly. Definitions, ratings, design and production tests, construction requirements, and guidelines for application, handling, storage and installation are covered in NEMA PB-2.

Switchboards are typically installed in industrial plants, utility and cogeneration facilities, and commercial and residential buildings for the distribution of electricity for light, heat, and power. They are typically available in voltage ratings of 600 V or less, continuous current ratings of 6000 A or less, and short-circuit current ratings up to 200 kA.

2.1.7.6 Arc-Resistant Metal-Enclosed Switchgear

The term "arc-resistant switchgear" indicates a design in which the equipment has met the requirements of ANSI/IEEE C37.20.7. The switchgear assembly is subjected to an internal arcing fault in key locations throughout the assembly for a specified current level and duration and the equipment performance is evaluated against five basic criteria. The arcing fault is initiated by a small wire placed across the primary conductors which vaporizes when current flows, providing an ionized air path for the arc. The preferred current level for this test is the short-circuit rating of the equipment and the preferred duration for current flow is 0.5 s. The equipment is evaluated for its ability to mitigate conditions which could be hazardous to personnel working nearby. Definitions, ratings, test requirements, and guidelines for application and installation are covered in IEEE C37.20.7.

Station-Type Switchgear

Another type of switchgear assembly that was previously used is "*station-type switchgear*." Station-type switchgear is no longer manufactured, but is briefly discussed here for historical purposes.

The term "station-type switchgear" indicates a design in which the major component parts of a circuit, such as buses, circuit breakers, disconnecting switches, and current and voltage transformers, are in separate metal housings, and the circuit breakers are of the stationary type. Phase segregation in metal-enclosed switchgear is a type of design in which a 3-phase metal housing is divided into three single-phase compartments by means of single metal barriers.

Metal-enclosed station-type switchgear was used in industrial, commercial, and utility installations, generally for voltages of 14.4–69 kV, and continuous current ratings up to 5000 A. Scheme of metal-enclosed station-type switchgear cubicle for outdoor installation; equipped with a heavy-duty, air-blast circuit breaker, 14.4 kV, 3000 A, 50 kA is shown on Fig. 2.24.

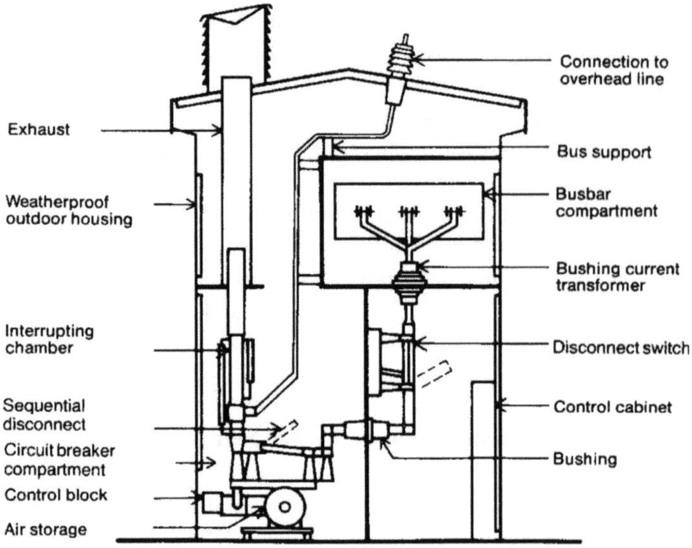

Fig. 2.24 Metal-enclosed station-type switchgear cubicle for outdoor installation; equipped with a heavy-duty, air-blast circuit breaker, 14.4 kV, 3000 A, 50 kA

2.1.8 Design of an Air-Insulated Switchgear Substation Based on New Technology

As cities grow, so does their hunger for electrical power. This calls for extra-high-voltage (EHV) substations to act as load hubs in the city. However, such developments lead to a conflict between substation footprint and power requirement. Today's technology is making these objectives attainable.

Firstly, new primary equipment permits a drastic reduction in the footprint area. A revolution in circuit-breaker design is allowing switchgear configuration or even integration to be redefined, decreasing costs both in terms of land acquisition and equipment cost.

Secondly, innovative optical sensors are replacing the traditional expensive and large current transformers (CT). These non-traditional CTs are so small that they can be easily integrated with the breaker in the same circuit.

Thirdly, the use of such sensors is an integral aspect of a digital substation. This not only enhances substation performance, but also saves on investment, e.g., cables for secondary instrument transformer circuits and for the control of circuit breakers, disconnectors, and earthing switches.

In an urban substation, the key factor is land-related cost.

DCBs can save not only equipment cost, but also reduce the footprint and related construction costs while increasing availability. A layout comparison between conventional and Combined 145 kV based switchyards is shown in Fig. 2.25.

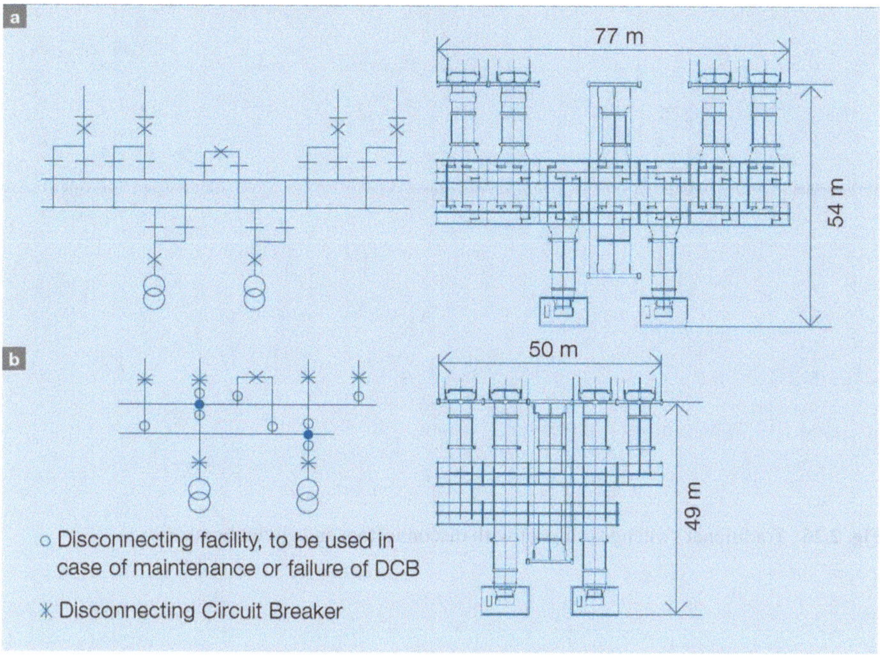

Fig. 2.25 Space requirement of 145 kV substation **a** traditional, **b** with DCB

2.1.9 Layout Reconfiguration and Optimization

Based on the new technologies above, the substation footprint can be reduced by more than 50%, while at the same time increasing its availability and reliability. Moreover, the use of DCB not only permits the disconnector to be eliminated, but also allows the busbar topology to be improved. Since the maintenance interval for the combined DCB is 15 years, replacing a double busbar system with a sectioned single busbar system is justified without compromising availability. In addition, because Fiber Optic Current Sensor (FOCS) and an intelligent interface are used, the cable and related auxiliaries can be done away with, resulting in a much simplified control setup and simplifying connections to the primary equipment. Traditional switchgear layout with disconnectors and circuit breakers is shown on Fig. 2.26. Switchgear with disconnecting circuit breakers is shown on Fig. 2.27.

Short summary for Sect. 2.2

- Plain air circuit breakers (ACB) are used in low voltage and medium voltage up to 15 kV Plain ACB is the simplest form of air circuit breakers. Their point of contact is in the shape of 2 horns, in breakers of this type, the arc is interrupted inside the switch chamber.
- For low and medium voltages, fuses can be also used, but the main disadvantage is that they must be replaced after fault clearing.

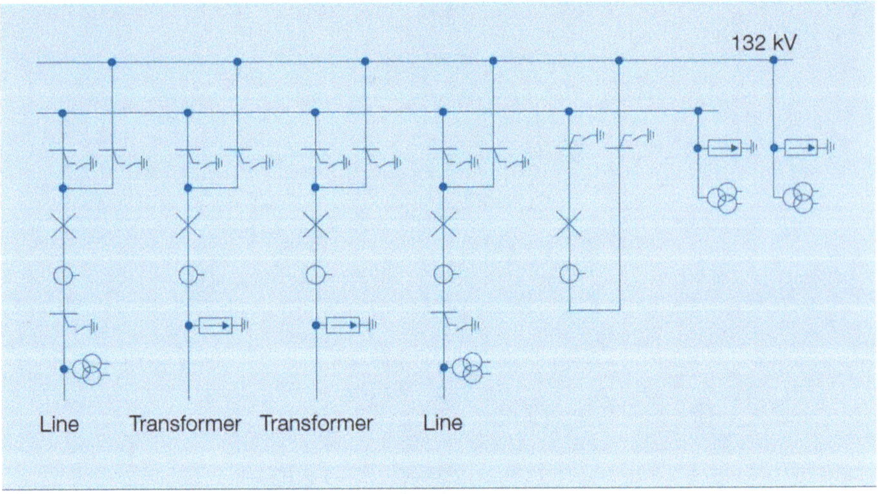

Fig. 2.26 Traditional switchgear layout with disconnectors and circuit breakers

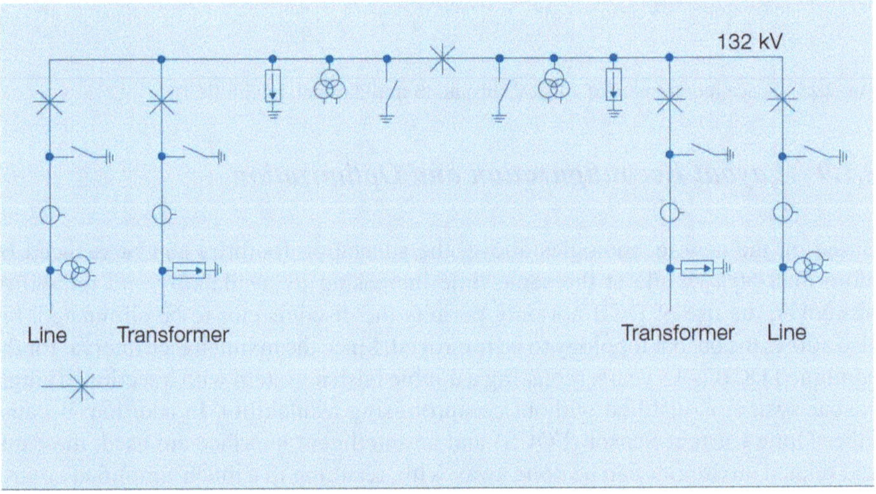

Fig. 2.27 Switchgear with disconnecting circuit breakers

- In medium voltage systems minimum oil, SF_6 and vacuum breakers are also being used.
- For high voltages minimum oil, SF_6 and blast-air breakers are used, but always with multiple interrupters in series.
- The maximum voltage per interrupter is 100 kV for air-blast and SF_6 breakers, 170 kV for minimum oil breakers.

- At present, switchgears are being transformed unmanned stations, and with the advent of digital relays, the system becomes completely computerized and all actions are taken out of the outdoor switchgear due to remote operation control (for example, SCADA system, etc.).

2.2 Circuit Breakers and Other Switching Devices

2.2.1 Circuit Breakers

2.2.1.1 Purpose and Classification

The most dangerous malfunction in the electrical network is a SC. It is an electrical circuit that allows a current to travel along an unintended path, often where essentially no (or a very low) electrical impedance is takes place. In circuit analysis, a SC is a connection between two nodes that forces them to be at the same voltage. The short circuit current may damage the equipment and networks permanently [5–7].

For saving equipment, the fault current should be cleared from the system as quickly as possible. Again, after the fault is removed, the system must come to its normal working condition as soon as possible for supplying reliable quality power to the receiving ends. In addition to that, for proper controlling of the power system, different switching operations are required to be performed. View of vacuum circuit breaker is shown on Fig. 2.28a and outline of oil circuit breaker on Fig. 2.28b.

Therefore, for timely disconnecting and reconnecting different parts of power system network for protection and control, there must be some special type of switching devices, which can be operated safely under huge current carrying condition. During the interruption of large current, there would be large arcing in between switching contacts, so care should be taken to quench these arcs in circuit breaker in a safe manner. An electric arc is a final stage of electrical breakdown resulting from a current through normally nonconductive media such as air. An electric arc or arc discharge has the highest current density. Only the external circuit limits the maximum current through an arc, not by the arc itself. Arcing also erodes the surfaces of the contacts, wearing them down and creating high contact resistance when closed.

CB is a switching device, which can be operated both manually and automatically for controlling and protection of any electrical power system. As the modern power system deals with huge currents, the special attention should be given during designing of CB to safe interruption of arc produced during the opening/closing operation of CB.

Standard short circuit currents range from 12.5 to 63 kA, with some applications requiring 80 kA. Commercially available continuous current ratings range from 600 to 5000 A. Ratings differentiate between indoor and outdoor service. Preferred ratings are further established for capacitive current switching, dielectric withstand and external insulation, transient recovery volt age capabilities, switching surge

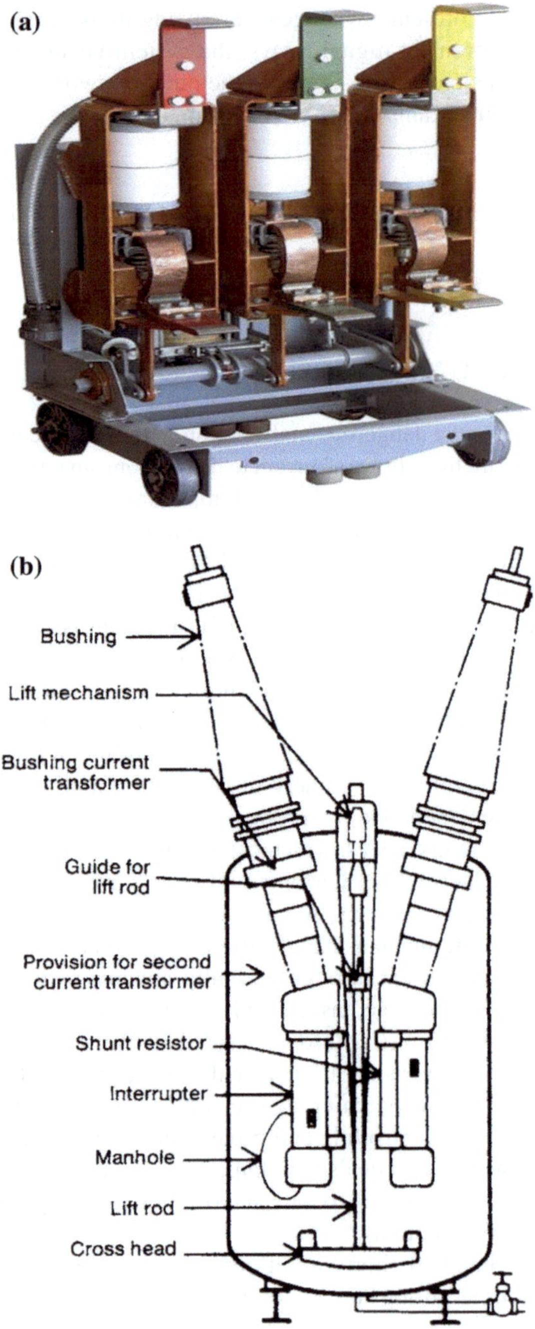

Fig. 2.28 **a** View of vacuum circuit breaker. **b** Outline of a dead-tank 161-kV out-door oil circuit breaker

factors for line closing, control voltages, reclosing times, and operation endurance capabilities.

The most important function of CBs is to interrupt SC currents. This is to protect generators, transmission lines, transformers, and other components of the transmission system. Typical SC requirements of high voltage systems are 25–63 kA, though there is an increasing need for 80 kA. During a SC (fault), the CB is subjected to both high currents and voltages at the same time. Designing for this capability involves engineering simulations and computational fluid dynamic analysis.

To date, a large number of high-voltage circuit breakers have been developed. Some of them (the latest developments) have just begun to be used in electric power systems, others are recognized as obsolete, but continue to be used in order to save money needed to replace them.

1. According to their arc ***quenching (rapid cooling) media*** the CB can be divided as:

   (1) Air CB
   (2) Oil CB
   (3) Vacuum CB
   (4) SF_6 (sulfur hexafluoride) circuit breaker

2. According to ***the voltage level*** of installation types of CB are referred as:

   (1) High voltage CB (>72 kV)
   (2) Medium voltage CB (1–72 kV)
   (3) Low voltage CB (<1 kV)

3. According to their ***services***, the CB can be divided as:

   (1) Outdoor CB
   (2) Indoor CB

4. According to ***the operating mechanism*** they can be divided as:

   (1) Spring operated CB
   (2) Pneumatic CB
   (3) Hydraulic CB

High-voltage CBs have been applied from 1 kV to up to 110 kV. They are available in either live tank or dead tank design, and may incorporate various insulating and interrupting media.

In today's market, SF_6 circuit breakers dominate above 38 kV, with vacuum circuit breakers being dominant in the "medium" voltage range between 15 and 38 kV. Air magnetic circuit breakers are also common up to 15 kV. Historically, oil circuit breakers were prevalent in these voltage ranges. Even though oil circuit breakers are out of production today, many remain in service.

Today vacuum CDs dominate for indoor application, but there are a few SF_6 types.

Designing for these voltages requires specialized engineering software and knowledge. Dimensions are set for insulating gaps between contacts, and between live parts and the grounded surrounding structures (tanks, shields, etc.). Electromagnetic finite element analysis software is the main tool for this task. It extinguishes the arc drawn between the contacts when the CB, opens.

Insulation of Circuit Breakers. The basic construction of any circuit breaker requires the separation of contacts in insulating mediums, which serve two functions here:

(i) It extinguishes the arc drawn between the contacts when the CB, opens.
(ii) It provides adequate insulation between the contacts and from each contact to earth.

Many insulating mediums are used for arc extinction and the medium chosen depends upon the rating and type of CB. The insulating mediums commonly used for circuit breakers are

1. Air at atmospheric pressure: Air break circuit breaker up to 11 kV.
2. Compressed air (Air blast circuit breaker between 220 and 400 kV)
3. Mineral oil which produces hydrogen for arc restriction (transformer oil):
   3.1. Plain break oil, CB 11–66 kV
   3.2. Controlled break oil CB or bulk oil CB between 66 and 220 kV
   3.3. Minimum oil CB between 66 and 132 kV.
4. Ultra high vacuum CB up to 33 kV.
5. SF_6 circuit breakers above 220 kV.

The controlled break and minimum oil circuit breakers enclose the breaker contacts in an arcing chamber made of insulating materials such as fiberglass, reinforced synthetic resins etc.

2.2.1.2 Bulk Oil Circuit Breaker (BOCB)

(a) ***The BOCB*** design is very simple. This type of circuit breaker uses a steel tank containing oil and the contacts are immersed in the oil. The steel tank is earthed (dead tank type). In this type construction the oil requirement is more as the oil is required to provide insulation to the contacts from the steel tank and insulation between the contacts (in open state). The oil also serves as the medium for extinguishing the arc formed when the moving contact separates from fixed contact. When the contacts separate, arc is formed between the contacts. The arc gives rise to formation of gas in the oil, which initiates oil circulation. This phenomenon helps in extinguishing the arc so breaking the circuit. In oil circuit breakers, the heat released during triggering decomposes the oil that boils at 658 K. The resulting gas mixture consists of: (1) hydrogen—70%, (2) acetylene—20%, (3) methane—5% and (4) ethane—five%.

The arc temperature is so high (more than 6000 K), that hydrogen goes into an atomic state, and hydrocarbons decompose into hydrogen and carbon. This mixture with the addition of copper of electrodes is the medium in which the arc is burning. High pressure in the gas "shell" and the presence of atomic hydrogen change the characteristics of the arc—there is a sharp reduction in the diameter of the arc channel, increasing the plasma temperature near the channel axis, thermal ionization of gas atoms. In some cases, thermal ionization can be so intense that due to the high conductivity of the channel, the field strength decreases to values insufficient to maintain the impact ionization (field ionization). The cessation of shock ionization leads to the extinguish of the arc. The greatest role in this process belongs to hydrogen and, therefore, the most extent it is manifested by oil circuit breaker.

The separation of the CB contacts which are carrying current gives rise to an arc without changing much the current wave form. Initially when the contacts just begin to separate the magnitude of current is very large but the contact resistance being very small; a small voltage appears across them. But the distance of separation being very small, a large voltage gradient is set up which is good enough to cause ionization of the particles between the contacts. Also it is known that with the copper contacts which are generally used for the circuit breakers very little thermal ionization can occur at temperature below the melting point. For effective field emission, the voltage gradient required is 10^6 V/cm. From this it is clear that the arc is initiated by the field emission rather than the thermal ionization. This high voltage gradient exists only for a fraction of a micro-second. But in this short period, a large number of electrons would have been liberated from the cathode and these electrons while reaching anode, on their way would have collided with the atoms and molecules of the gases. Thus, each emitted electron tends to create others and these in turn derive energy from the field and multiply. In short, the work has done by the initially-emitted electrons enables the discharge to be maintained. Finally, if the current is high, the discharge attains the form of an arc having a temperature high enough for thermal ionization, which results in lower voltage gradient. Thus, an arc is initiated due to field effect and then maintained due to thermal ionization.

For higher voltage, this very simple principle cannot be much effective. So an arc control device is usually used to facilitate arc extinction process.

(b) *The BOCB with improved (lightened) arc extinction*

Such a device has a special hydraulic mechanism for creating pressure and supplying oil to the place where the chain breaks.

Magnetic quenching in oil is carried out using special electromagnet elements that create a field that moves the arc into narrow channels to break the created circuit.

Oil switch with auto blow. The scheme of this type of oil switch provides for the presence of a special element in the system, which releases energy from the formed arc to move oil or gas in the tank.

2.2.1.3 Minimum Oil Circuit Breaker (MOCB)

MOCB require less oil as the purpose here is only to extinguish the arc and not for providing insulation to the contact. Arc interruption takes place inside the Interrupter.

The whole system is placed inside the porcelain housing. Because of this insulating porcelain, the insulation requirement of contacts is reduced very much. This is the reason of its smaller size.

As such the MOCBs are of oil tank outdoor type design and mainly used for voltage levels above 38 kV. The main disadvantage is that the MOCB can explode causing harm to the personnel and other equipment of the system. The tank type design is very bulky so making it difficult for transportation and handling and requires more space.

The MOCB requires more maintenance in comparison to vacuum and SF_6 breakers. Irrespective of these few disadvantages, the MOCBs are not going to vanish within few years.

2.2.1.4 SF_6 Circuit Breaker

In the past for higher transmission voltage Oil Circuit Breaker (OCB) and Air Blast Circuit Breaker (ABCB) were used.

These days for higher transmission voltage levels SF_6 circuit breakers are largely used. OCB and ABCB have almost become obsolete (old, out of use). In fact in many installations SF_6 CB is used for lower voltages like 11 kV, 6 kV etc.

SF_6 is a gas which satisfy the requirements of an ideal arc interrupting medium. So SF_6 is extensively used these days as an arc-interrupting medium in circuit breakers ranging from 3 kV up to 765 kV class.

In addition to this SF_6 is used in many electrical equipments for insulation. SF_6 CBs are mostly employed for high voltage applications. For protection and reliability of an electrical substation of an electricity generation, transmission and distribution system where voltage is transformed from high to low or the reverse using transformers.

The properties of SF_6, which is the reason of its extensive use in circuit breakers, are:

- SF_6 gas has high dielectric strength which is the most important quality of a material for use in electrical equipment and in particular for breaker it is one of the most desired properties. Moreover it has high rate of rise of dielectric strength after arc extinction. This property is very important for the circuit breaker in order to avoid reconnection.
- SF_6 is colorless, odorless and non toxic gas.
- SF_6 is an inert gas. So in normal operating condition the metallic parts in contact with the gas are not corroded. This ensures the life of the breaker and reduces the need for maintenance.
- SF_6 has high thermal conductivity which means the heat dissipation capacity is more. This implies greater current carrying capacity when surrounded by SF_6.
- SF_6 being non-flammable so there is no risk of fire hazard and explosion.

Fig. 2.29 Single SF_6 circuit breaker for voltage level 245 kV

The construction and working principles of SF_6 circuit breaker varies from manufacturer to manufacturer. The Puffer type SF_6 breakers of single pressure type are the most favored types prevalent in power industry.

Disadvantages of SF_6 circuit breakers are**:**

1. SF_6 breakers are costly due to high cost of SF_6
2. SF_6 gas is sensitive to low temperature (−35 °C and lower), heating system is required.

Views of SF6 circuit breakers are shown on Figs. 2.29 and 2.30.

2.2.1.5 Disconnecting Circuit Breaker (DCB)—Unconventional (State-of-the Art) Circuit Breaker

This apparatus combines the switching and disconnecting functions into one device. The breaking function is identical to that of a conventional CB. It uses the same breaking chamber and operating mechanism. The disconnection function for a DCB is obtained between the circuit breaker contact set. DCBs are available from 72.5 to 550 kV voltage level.

Fig. 2.30 Row of SF_6 circuit breakers for voltage level 800 kV

Substations equipped with DCB have higher availability in supplying power, they require less maintenance and space, and CO_2 emissions are greatly reduced in comparison to conventional substations. DCB improves the eco-efficiency of the equipment throughout the life cycle by using fewer primary apparatus and, consequently, fewer raw materials. DCBs also create lower electrical losses during the use phase due to lower electrothermal losses. Common Disconnecting Circuit Breaker is shown on Fig. 2.31.

2.3 Rotating Machines

Rotating machines are used in the production of electricity (turbo generators, hydro generators), in the generation of reactive power (synchronous compensators), as well as in the conversion of electrical energy into mechanical energy (electric motors). The most responsible of them are generators, and therefore the most stringent requirements are imposed on them. Currently, generators with a capacity of more than 1 MW with a rated voltage of 20–30 kV are in operation.

Fig. 2.31 Disconnecting circuit breaker

Despite the relatively low nominal voltages, the design, creation and monitoring of the state of generators are associated with significant difficulties. First of all, this is due to severe restrictions on the thickness of the stator winding insulation to reduce the dimensions and weight of the generator, the view of which is shown in Fig. 2.32.

In addition, the insulation of electrical machines experiences constant vibrations and large shock mechanical loads during short circuits. Large generated or converted electric power and, as a result, large currents in the active parts of electrical machines lead to significant thermal loads in the insulation. Therefore, in high-power machines, internal water, oil, or hydrogen cooling of the windings using copper hollow conductors is used.

The insulation of machines can be divided into body (main) and interturn (longitudinal). Modern high power turbogenerators, as a rule, have a winding with single-turn rods and, therefore, there is no turn insulation in the rod [1, 11, 12]. On Fig. 2.33 shows a diagram of a high-voltage insulation device for an electrical machine.

Fig. 2.32 Turbine generator stator

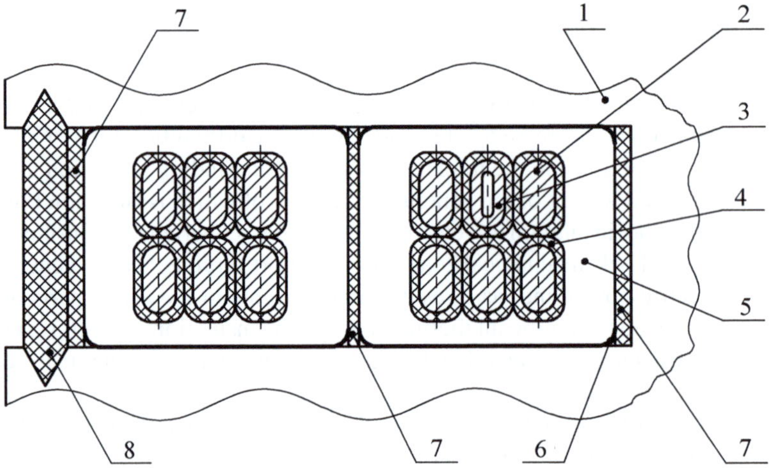

Fig. 2.33 Scheme of the device for high-voltage electrical insulation machines: 1—stator; 2—solid conductor; 3—hollow conductor; 4—coil (longitudinal) insulation; 5—main body insulation; 6—semi-conductive coating; 7—gaskets; 8—wedge

2.3.1 Main Insulation

The main (high-voltage) insulation is the insulation between the rod and the body and between the rods. Modern machines have continuous mica compound insulation covered with asbestos tape. Myconite tape consists of two layers of a special paper substrate, between which are mica plates. The mica in myconite is held between the substrates by an adhesive oil-bitumen varnish (compound). The rods consist of copper conductors separated by low voltage asbestos or fiberglass insulation.

Modern powerful machines use thermosetting high voltage insulation which does not soften during heating cycles (unlike thermoplastic insulation). This allows the use of such insulation at higher temperatures while maintaining the original dielectric strength and high reliability.

According to the design, the insulation of the rods is divided into *sleeve* (In old designs, mainly low-power machines) and *continuous*. The latter is made of the same material along the entire length of the rod and has almost the same electrical strength in the slot and frontal parts. After applying several layers of mica tape, the rods are compounded, i.e. placed in special boilers, where the insulation is dried, evacuated and impregnated under pressure with a molten compound.

Modern powerful turbogenerators have slots and rectangular bars [11]. For field alignment, semi-conductive varnished cloth is applied to the edges of the rod to increase the radius of curvature. It is electrically connected to the iron of the stator in order to avoid discharges between the rod and the wall of the groove.

To prevent corona at the point where the rod leaves the groove, semi-conductive coatings are applied on the insulation surface, which prevents its premature destruction. A layer of insulating material is applied to the semi-conductive coating, which increases the reliability of the coating. Sometimes capacitive field equalization is also used with the help of conductive or semi-conductive plates (screens) built into the thickness of the insulation.

2.3.2 Turn-to-Turn Insulation

In machines of low and medium power (up to 30 MW); inter-turn insulation has rods with several turns, the potential difference between which does not exceed several hundred volts. Materials of heat resistance class B with a permissible operating temperature of 130 °C are used as coil insulation. In case of connecting generators to OTLs through cables or transformers, it is not necessary to take into account lightning overvoltages when calculating their coil insulation. Generators for very high voltages (at the level of 35 kV) can be connected directly to OTLs, and therefore lightning surges must be taken into account in insulation calculations.

A promising direction in the development and operation of rotating machines is the use in them of the phenomenon of superconductivity in a number of ceramic materials at temperatures close to the temperature of liquid nitrogen (−196 °C). A technical and economic analysis has shown that the cost of deep cooling of large superconducting objects is less than the cost related to the prevention of the heating of conventional conductors [11].

For low voltage ac and dc machines, the winding wires are super enameled wire and the other insulation used are vulcanized rubber and varnished cambric and paper. For high voltage and large power capacity machines, the space limitations demand the use of insulating materials having substantially greater dielectric strength. Mica is considered to be a good choice not only due to space requirements but because of its ability to withstand higher temperatures. However, the brittleness of mica makes it necessary to build up the required thickness by using thin flakes cemented together by varnish or bakelite generally with a backing of thin paper or cloth and then baking it under pressure. Epoxy resin bounded mica paper is widely used for both low and high voltage machines. Multilayer slot insulation is made of press board and polyester film. However, for machines with high operating temperatures kapton polymide is used for slot insulation. Mica has always been used for stator insulations. In addition to mica, conducting non-woven polyesters are used for corona protection both inside and at the edges of the slots. Glass fibre reinforced epoxy wedge profiles are used to provide support between the winding bars, slots and the core laminations.

2.4 Power Capacitors

The electricity is a unique product produced by power plants—at any time it can and should be produced as much as consumers need. This significantly complicates the work of all links of the technological energy chain: electricity production—transmission—distribution—consumption. Electrical engineers dream of the time when energy storage facilities will be created that can store electricity in volumes that meet the requirements of the electric power industry. The prospect of creating such energy storage devices today seems to be just a dream, although there are serious successes on this path.

Despite the fact that high-voltage capacitors do not currently meet the needs of power systems in terms of the amount of stored energy, they are widely used in the power industry to ensure the reliability and efficiency of production and transmission of electricity. By appointment, there are 4 main types of high-voltage capacitors:

(1) cosine capacitors,
(2) coupling capacitors,
(3) capacitors used to protect rotating machines from surges,
(4) capacitors for power take-off, for the manufacture of capacitive voltage dividers.

Cosine capacitors are designed to increase the power factor of 50 Hz AC electrical installations for both group and individual compensation. They are installed at consumer substations on the side of high (6–10 kV) and low (0.22–0.66 kV) voltage, as well as at distribution points in workshops or individual receivers of electrical energy. Capacitors are connected in parallel with the inductive resistance of

Fig. 2.34 Cosine capacitors in a substation

the network (transverse compensation), in most cases the connection of capacitors in power compensation installations is a triangle. Cosine capacitors in a substation are shown on Fig. 2.34.

Coupling capacitors are used in high-frequency communication installations to separate high and low voltage circuits. Capacitors are made single-phase or three-phase for various voltages and capacities. Currently, capacitor units are made from individual capacitors of small capacity (about 70 microfarads), from which the total capacitance of the steps is collected. Moreover, each of the sections consists of separate elements that are controlled by several switching devices. Switching devices of one section are switched on simultaneously.

Capacitors used to protect rotating machines and other electrical installations from atmospheric overvoltages are installed on the high voltage side of indoor switchgear 10 kV and connected in parallel to valve arresters. In this case, three capacitors are assembled into a star and the common point of the circuit is connected to the ground.

Some types of capacitors are used to make capacitive voltage dividers and power take-offs.

Capacitor design economics suggests the use of individual unit assembled in appropriate series and parallel connected groups to obtain the desired bank voltage and reactive power ratings both in shunt and series capacitor equipments. Series capacitor duty usually requires that a unit designated for a series application be more conservatively rated than a shunt unit. However, there is no basic difference in the construction of the two capacitors.

The most commonly used capacitor for the purpose is the impregnated paper capacitor. This consists of a pair of aluminum foil electrodes separated by a number of Kraft paper tissues which are impregnated with chlorinated biphenyl and has a higher permittivity and results in reduction in the quantity of materials required for a given capacitance and the cost.

The working stress of an impregnated paper is 15–25 V/mm and papers of thickness 6–12 mcm are available and hence depending upon the operating voltage of the capacitor, a suitable thickness of the paper can be selected. Because of imperfection involved in the manufacturing process of the dielectric paper it is desirable to use at least two layers of tissues between metal foils so that the possibility of coincidence of weak spots is avoided.

The effective relative permittivity depends upon the paper and the impregnate. For an impregnate based on chlorinated biphenyl, the relative permittivity is from 5 to 6. Normally through past experience, the area of the plate for a particular material of paper and impregnate per microfarad of capacitance is known and hence it is possible to obtain the number of turns of paper to be wound on a given diameter of mandrel for a specified foil width and for the particular lay-up of foil and paper.

The method of laying up the paper and metallic foil and the connection of lugs is shown in Fig. 2.35a.

Two layers of dielectric are used as without it rolling would short circuit the plates. As a result of this, two capacitors in parallel are formed by the roll. The foil and the paper interleaved in this fashion are wound on to a mandrel which is split to allow easy removal of the finished roll. If the section of the container is same as that of the roll, minimum overall value for the capacitor is obtained. As a result of this, quantity of free impregnate is a minimum thereby the risk of leakage of impregnant with variation in temperature is reduced. Sometimes a high resistance (for discharge) is connected across the terminals of the capacitor for safety reasons.

The elements are placed in a metal enclosure with bushings for external connections and impregnated with a dielectric fluid, usually under vacuum. The unit is then hermetically sealed (see Fig. 2.35b).

The replacement of linen by the Kraft paper and oil by askarel made it possible to have individual unit ratings up to 15 kVAr by 1930. After making some costly refinements in basic paper/askarel dielectric 100 kVAr rating capacitor were manufactured by 1960.

General Electric Company designed a 150 kVAr unit using a paper/poly propylene film/askarel dielectric. Further advances in the manufacture of dielectric materials led to single unit of 600 kVAr even though the rating of a single unit based on economy ranges between 200 and 300 kVAr. Replacement of askarel with non-PCB fluids did not have much effect on unit sizes or ratings. The newer all polypropylene film dielectric units offer distinct advantages in reduced losses and probability of case rapture as well as improvement in unit ratings. The large size units have made it possible to reduce the physical equipment size and the site area requirements.

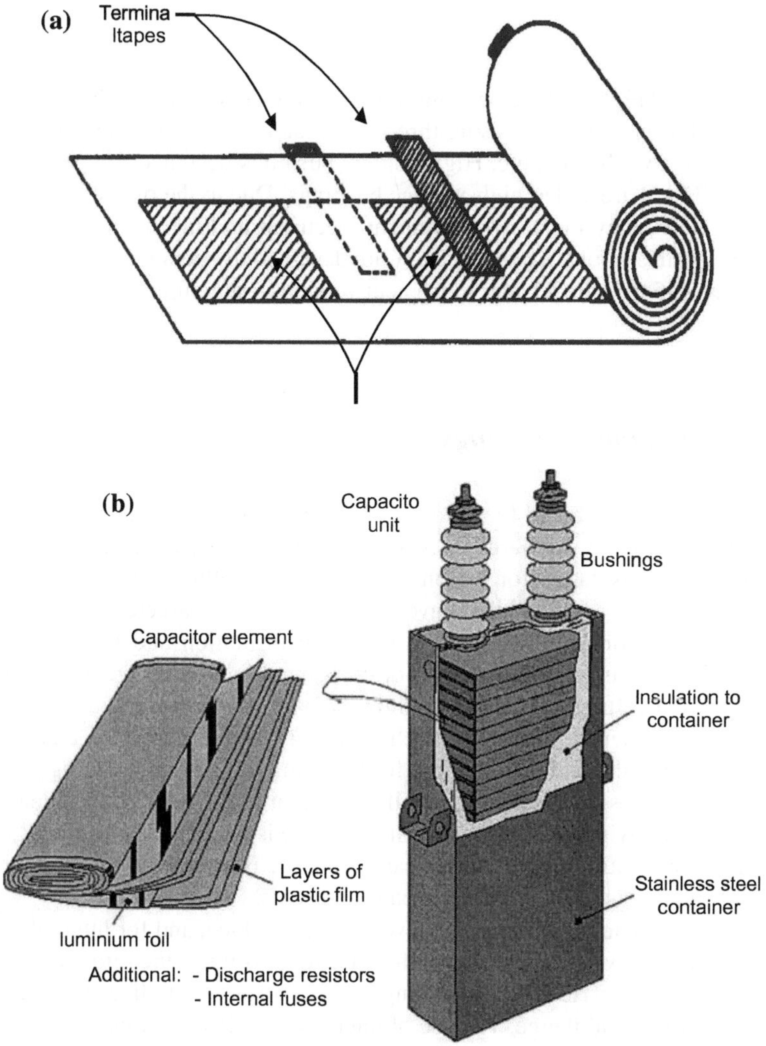

Fig. 2.35 **a** Impregnated paper capacitor-terminal tape type. **b** Typical high-voltage capacitor unit

With further development, it has now been possible to have series and shout capacitor rating upto 550 kV and bank rating of up to 800 MVAr. The average price of smaller units in terms of 100 kVAr is Rs. 100 per kVAr or $2 per kVAr. It is to be noted that aluminium foil are used in these capacitors as it has high thermal and electrical conductivity, has high tensile strength, high melting point, is light in weight, low cost and is easily available.

2.5 Bushings

Low-voltage and high-voltage bushings are used to insulate and mechanically fasten conductors (busbars, wires) passing through grounded covers of apparatuses, partitions, walls of switchgears, etc. High-voltage bushings designed to operate at voltages above 35 kV are called high-voltage bushings. Due to the operating conditions of these devices, they must have sufficient electrical and mechanical strength and provide a certain amount of current. They are designed to operate at ambient temperatures from −45 to +40 °C, and in tropical versions—at temperatures from −10 to +55 °C.

2.5.1 *Low Voltage Bushings*

Since bushings are designed for indoor and outdoor installations, they are rated for the voltage at which a discharge occurs on a wet surface (wet discharge voltage). When designing insulators for regions with a highly polluted atmosphere, special types of insulators are created that have higher discharge characteristics.

Bushing insulators with monolithic porcelain insulation and insulators with a cavity filled with gas (air), mineral oil or compound have been created, which are used mainly at a voltage of 6–10 kV. Solid porcelain insulator—Fig. 2.36—consists of a porcelain element 1, inside of which a current-carrying rod (bus) 2 passes, two flanges (caps) 3 at the ends of the input for fastening the current-carrying part and a support flange 4, designed for fastening the insulator to the body of the apparatus or to the wall of the switchgear, through which voltage is applied. The support flange is connected to porcelain using a cement-sand binder 5. Between the cap and porcelain are placed shock-absorbing cardboard pads and seals.

Insulators for outdoor installation have a ribbed surface, and for internal installation—smooth or wavy. To increase the dry-discharge voltage, the reference (middle) flange is shielded with ribs facing the flange, with a surface metallized between them, connected to the metal flange. The use of intermediate ribs also makes it possible to increase the discharge voltage. The lower part of the hardware insulators is placed in oil. To exclude ionization processes in the air gap between the current-carrying rod and the inner wall of the insulator, the latter is metallized and connected to the current-carrying rod.

The thickness of the porcelain wall at the middle flange of the bushing insulators for voltages of 6, 10, 20, 35 kV is 20.25, 30, and 40 mm, respectively.

Bushing insulators with a cavity are used in installations for voltages up to 35 kV. From the point of view of electrical strength, the weakest point of such an insulator is a cavity filled with gas, oil or compound. The cross section of the bushing with a cavity is shown in Fig. 2.37.

Breakdown of the cavity leads to a complete breakdown of the insulator. Therefore, the maximum field strength in the cavity of the insulator Emax should not exceed

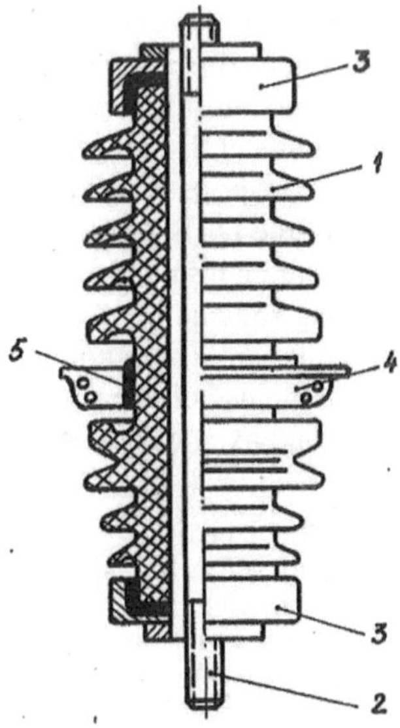

Fig. 2.36 Monolithic porcelain bushing

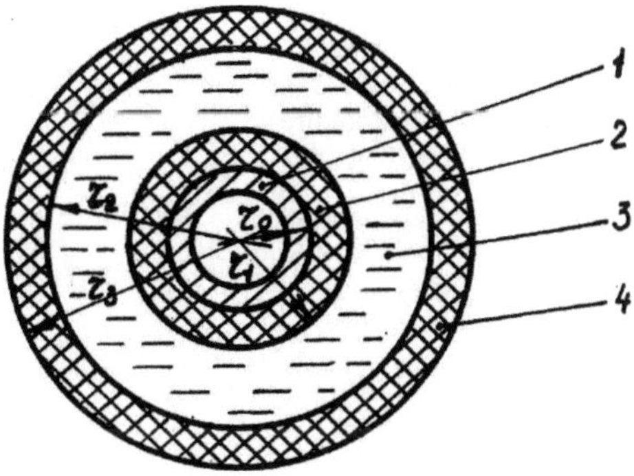

Fig. 2.37 Cross-section of a bushing insulator with a cavity: 1—current-carrying rod (pipe); 2—insulating coating; 3—cavity; 4—porcelain

the allowable values for the cavity-filling medium Eadm, which, in turn, should be less than the breakdown strength of the cavity Ebr. The length of the lower part of the bushings, immersed in oil, is approximately 2 times shorter than the outer part of the insulator. The number of ribs on the insulator is selected in accordance with the formula $n = 1.5 + 0.065\ Unom$.

2.5.2 *High Voltage Bushings*

At voltages above 35 kV, bushings with oil-barrier, paper-bakelite and paper-oil capacitor internal insulation are used. The outer insulating structure of such bushings are porcelain covers—top and bottom. Discharge gradients over the dry surface of a porcelain tire are 5.0–7.0 kVmax/cm for the case of a full impulse wave, 9.5–11 kVmax/cm for a cut wave, 4.0–5.0 kV/cm for power frequency voltage. When assessing wet discharge voltages, a gradient of 2.5–2.7 kV/cm is chosen.

Oil barrier bushings (OBB) are used in installations for voltages up to 400 kV. A view of the OBB g for a voltage of 220 kV is shown in Fig. 2.38.

The main insulation of such bushings is mineral oil, in which between the current-carrying rod and the flange there are paper-bakelite cylinders with capacitor plates made of copper or aluminum foil 0.01–0.015 mm thick with a paper coating 3–5 mm thick. The use of capacitor plates makes it possible to equalize the field both in the radial and axial directions.

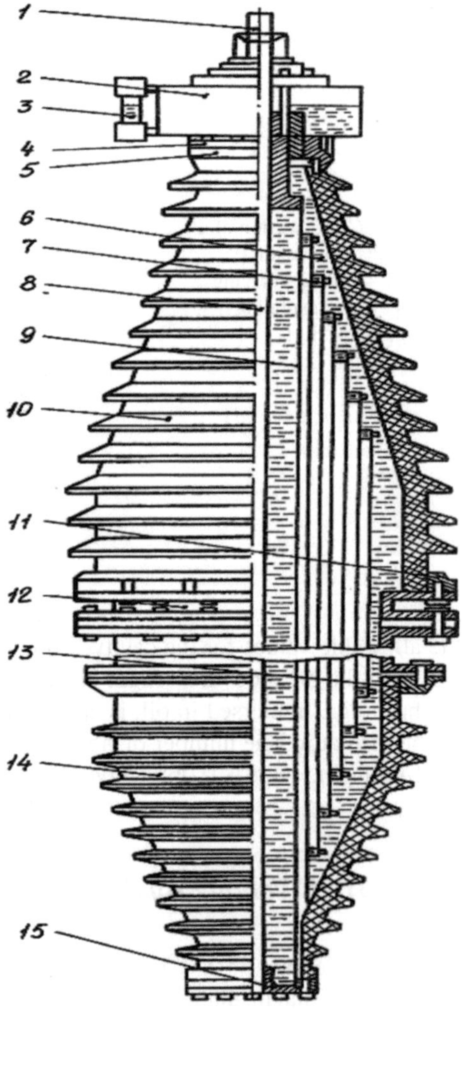

Fig. 2.38 Oil barrier bushing for voltage 220 kV: 1—current clamp, 2—expander, 3—oil level indicator, 4—sump, 5—flange, 6—transformer oil, 7—distance spacers, 8—current-carrying rod, 9—paper-bakelite cylinders, 10—top tire, 11—flange, 12—connecting sleeve, 13—flange, 14—bottom tire, 15—glass

The insulating cylinders are fastened with spacers made of wood or plastic. Oil, circulating in the annular gaps (2–5 cm thick) between the barriers, cools the current-carrying rod and promotes heat removal from the insulating partitions. The bushing is equipped with an expander and an oil level indicator.

Oil barrier insulation (OBI) are highly reliable, but have large dimensions due to the relatively weak regulation of the electric field (a small number of equalizing plates) and the relatively low electrical strength of the OBI. Depending on the number of oil channels (from 2–3 at Un = 110 kV to 8 at Un more than 220 kV) and their thickness, as well as the size of the capacitor plates, the permissible field strengths in oil should not exceed 40–65 kV/cm.

The diameter of the current-carrying rod (pipe) and its cross section are chosen so that the maximum electric field strength on its surface and the density of the transmitted current do not exceed the allowable values.

Capacitor-type bushings have a more advanced design that allows them to be used at voltages of 220–500 kV and higher. Bushings of this type include (a) paper-bakelite and (b) paper-oil capacitor bushings [1, 9, 10].

Bakelized paper 0.06 mm thick is used as the main insulation in paper-bakelite bushings, which is wound on a current-carrying rod or pipe in a hot state (160–185 °C). In order to equalize the electric field in the volume of insulation, equalizing plates of aluminum foil are laid between the layers of paper at certain intervals. The resulting bushing is baked, the ends of the core are sharpened into a cone, after which the surface of the insulating framework is covered with a moisture-resistant varnish. The advantages of such bushings are—ease of manufacture, compactness, high mechanical strength (500–800 kgF/cm^2); disadvantages—hygroscopicity of insulation, low tracking resistance, the possibility of thermal breakdown.

PBBs for outdoor installation are placed in porcelain cover, and the space between the cover and the insulating framework is filled with special mastic. If the lower part of the bushing runs in oil, then the lower porcelain cover is missing.

Currently, bushings with ***capacitor oil-paper insulation (COPI)*** are widely used. In this design of the input, cable paper is wound spirally on a current-carrying rod (pipe) or on a paper-bakelite cylinder with positive overlap. At certain intervals between the layers of paper, capacitor plates made of perforated aluminum foil with a thickness of about 0.01 mm are laid. More modern winding is also used—in two layers of paper: one is clean, and the other is with a printed metal coating.

The insulating core of the insulator is impregnated with oil poured into the cavity of the porcelain cover. On Fig. 2.39 shows a drawing of a capacitor paper-oil input for a voltage of 110 kV.

In the upper part of the bushing, an oil expander with an oil level indicator is attached to the flange, which can be equipped with a hydraulic seal to prevent oil moistening.

Modern designs of bushings are hermetic, which prevents moisture and oxidation of the oil by atmospheric air. Such bushings have remote pressure tanks (compensators), which are connected by a copper pipeline to the internal bushing cavity. Bellows are placed in the pressure tank (cylindrical containers filled with nitrogen at a pressure of 0.025 MPa (0.25 kgF/cm^2) at t = 20 °C. The bushing compensator for

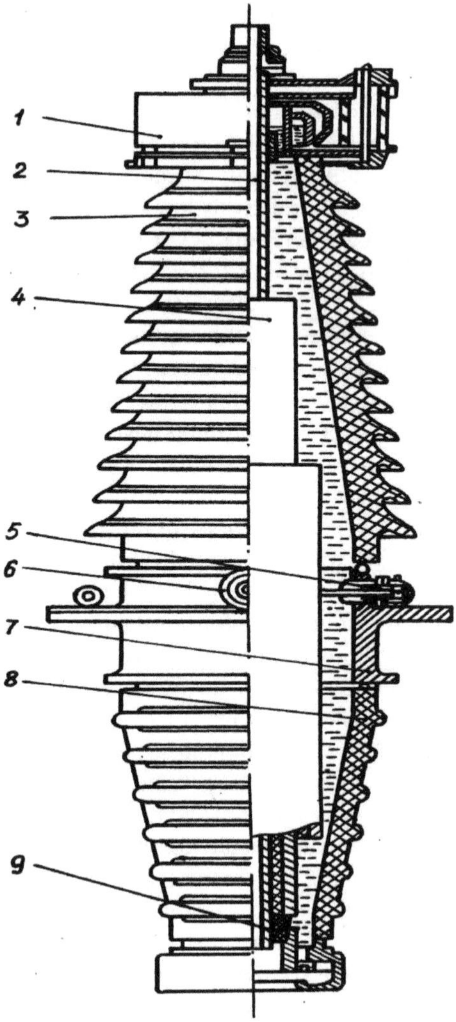

Fig. 2.39 Capacitor paper-oil bushing for voltage 110 kV: 1—expander with a water seal, 2—current-carrying pipe, 3—top cover, 4—insulating frame, 5—measuring input, 6—oil sump, 7—connecting sleeve, 8—bottom cover, 9—glass

500 kV has 76 bellows. When the oil expands due to heating, part of it moves through the pipeline to pressure tank (bellows will shrink). When the oil in the bushing cools, its volume will decrease, and the oil from the compensator under nitrogen pressure in expanding bellows will flow into the bushing.

Sealed bushings do not have expanders with hydraulic seals, because their role is played by compensators.

New insulating materials and improved design and manufacturing technologies have made it possible to reduce the size and simplify the design of high-voltage bushings, Fig. 2.40a–d.

Diagnostics of bushings for voltages from tens of kilovolts to 500 kV is discussed in Chap. 10.

Fig. 2.40 **a** Air—SF_6 bushing for switchgear 220 kV, for oil circuit for switchgear 220 kV, 2000–3150 A. **b** Air—oil bushing breakers 35–220 kV, 1000–3150 A. **c** Air-oil bushing for power transformers bushings for power and shunt reactors 12–1200 kV, 315–2500 A. **d** Air—oil detachable transformers 20–35 kV, 6–20 kA

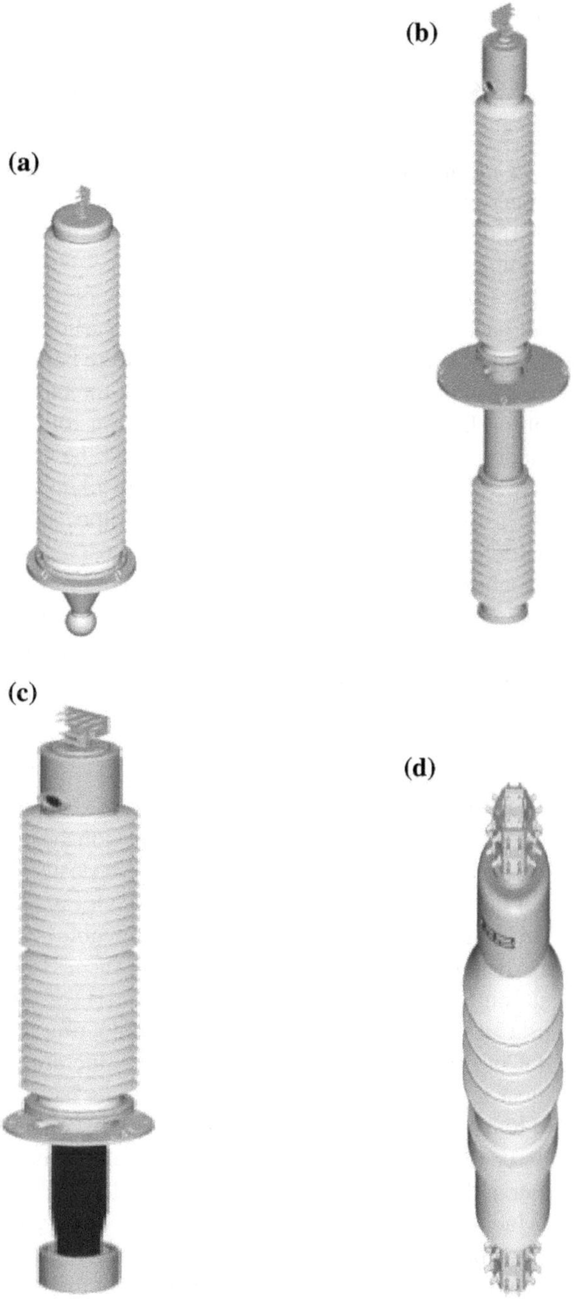

2.6 Cables

2.6.1 *Short History and Future of Electrical Cables*

The telegraph was the first device utilizing electrical energy to become of any commercial importance and its development necessarily required the use of wires. Underground construction was planned for most of the earliest commercial lines including underground power systems.

The first available information on the use of a cable to transmit an electric signal dates back to 1812. This is probably the earliest use of a continuously insulated conductor. While early telegraph systems were being developed, other experimenters were solving the problems connected with the commercial development of electric lighting. Edison planned his first installation for New York City, and decided that an underground system of distribution would be necessary. Edison designed a rigid, buried system consisting of copper rods insulated with a wrapping of jute. Two or three insulated rods were drawn into iron pipes and a heavy bituminous compound was forced in around them. The high first cost and heavy losses made such systems uneconomical for general distribution. Accordingly, they were used in special cases. For example, for the power supply of urban areas with a high density of electrical load. The first installations operated at a voltage of 1000 V. Over time, the areas of application of cable lines expanded, their length and, accordingly, the level of operating voltage increased. For example, in 1954, the Swedish State Power Board energized the 60-mile, 100-kV dc submarine cable utilizing U. Lamm's mercury arc valves at the sending and receiving ends of the world's first high-voltage direct current (HVDC) link connecting the Baltic island of Gotland and the Swedish mainland.

The development of dc power transmission systems is constrained by high price of high-voltage cables with a rather complex design necessary to overcome the large water barriers (oceans, seas, and large lakes), Fig. 2.41.

In future, the pace of development of DC cable systems of power transmission over long distances will largely depend on the success in the development of industrial-scale high-temperature superconductivity (HTS). In late 1986 a ceramic material LaBaCuO was discovered to be superconducting at 35 K and in 1987 the material YBaCuO was found to have a critical temperature of 92 K. Since that time the critical temperatures of these HTS materials have progressively increased to over 130 K. The enormous significance of these discoveries is that these materials will be superconducting in liquid nitrogen, which has a boiling point of 77 K and is much easier and cheaper to provide than helium [7].

Much work has been directed towards finding materials with higher T_c values but this has remained at 133 K for some time. There are many trials being made of the application of the HTS cables throughout the world including USA, Europe, and Japan. Detroit (USA) is being re-equipped with HTS power cable for its transmission system and copper cables weighing over 7 tons are being replaced with HTS cables of less than 0.12 tons.

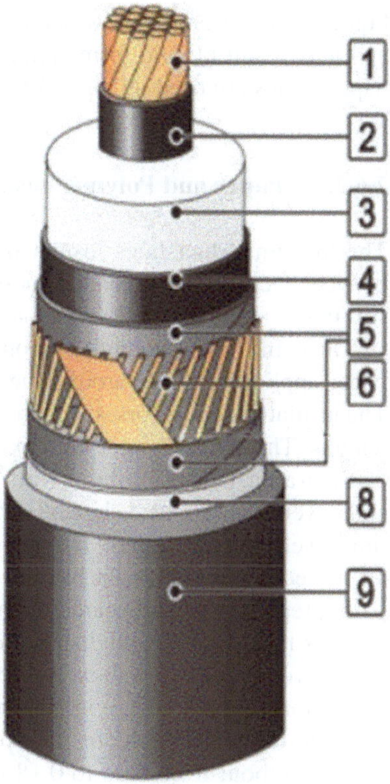

Fig. 2.41 The design of a single-core cable with XLPE insulation with a voltage of 10 kV: 1—conductive stranded core (conductor); 2—internal electrically conductive screen along the core; 3—insulation made of cross-linked polyethylene; 4—external electrically conductive insulation screen; 5—separating layer; 6—wire screen; 8—transverse sealing; 9—outer shell

2.6.2 Insulation and Design of Power Cables

As insulating materials in cable lines vulcanized rubber, PVC, oil impregnated polyester film and paper, cable oil (low viscosity mineral oil), high-pressure gases are used. On a small scale, as part of pilot projects, cryogenic power lines are used.

Vulcanized rubber, insulated cables are used for wiring of houses, buildings and factories for low power work.

PVC is inert to oxygen, oils, alkalis and acids and therefore, if the environmental conditions are such that these things are present in the atmosphere, PVC is more useful than rubber.

Polyethylene is used for high frequency cables. This has been used to a limited extent for power cables also. The thermal dissipation properties are better than those of impregnated proper. The maximum operating temperature of this cable under short circuits is 100 °C.

In case of impregnated paper, a suitable layer of the paper is lapped on the conductor depending upon the operating voltage. It is then dried by the combined application of heat and vacuum. The compound used in case of impregnated paper is semifluid and when the cables are laid on gradients the fluid tends to move from higher to lower gradients which reduces the compound content at higher gradients and may result in void formation at higher gradients. For this reason, impregnated paper cables are used up to 3.3 kV.

2.6.2.1 Paper and Polymer Insulated Cables

The first important lines insulated with paper were installed by Ferranti in 1890 between Deptford and London for operation at 10,000 V. Some of these mains were still in use at the original voltage after more than 50 years.

The cables consisted of two concentric conductors insulated with wide strips of paper applied helically around the conductor and saturated with a rosin based oil. The insulated conductors were forced into a lead pipe and installed in 20 foot (6.1 m) lengths. These mains were not flexible and were directly buried in the ground.

Soon after, cables insulated with narrow paper strips saturated in a rosin compound and covered with a lead sheath (very similar in design to those in use at the present time) were manufactured. These were the first flexible paper-insulated cables, and all subsequent progress has been made through improvements in the general design.

Impregnated paper became the most common form of insulation for cables used for bulk transmission and distribution of electrical power, particularly for operating voltages of 12.5 kV and above, where low dielectric loss, a low dissipation factor, and a high ionization level are important factors in determining cable life. Impregnated paper insulation consists of multiple layers of paper tapes, each tape from 2.5 to 7.5 mils (about from 0.06 to 0.18 mm) in thickness, wrapped helically around the conductor to be insulated. The total wall of paper tapes is then heated, vacuum dried, and impregnated with an insulating fluid. The quality of the impregnated paper insulation depends not only on the properties and characteristics of the paper and impregnating fluid, but also on the mechanical application of the paper tapes over the conductor, the thoroughness of the vacuum drying, and the control of the saturating and cooling cycles during the manufacturing.

Increased knowledge of the chemical treatment of the wood (in order to obtain pure cellulose by the adjustment of the fiber content and removal of lignin), the control of tear resistance, and the availability of long fiber stock resulted in the almost universal use of wood pulp paper in cables. The impregnating compound was changed from a rosin-based compound to a pure mineral oil circa 1925, or oil blended to obtain higher viscosity, until polybutene replaced oil about 1983.

Paper insulated, lead-covered cables were the predominant primary cables of all the large, metropolitan distribution systems in the world, throughout the twentieth century. Their reliability was excellent. It was, however, necessary to have a high degree of skill for proper splicing and terminating.

A shift towards extruded dielectric cables began about 1975 in those metropolitan areas, but the majority of the distribution cables of the large cities remain paper insulated, lead-covered cables as the century ends. Considerable research has been carried out by the utilities, technical organizations, and manufacturer's of cables to obtain improved paper and laminated PPP (polypropylene-paper-polypropylene, now used in transmission cables) tapes and insulating fluids able to withstand high, continuous operating temperatures, etc.

Impregnated paper insulation has excellent electrical properties, such as high dielectric strength, low dissipation factor, and dielectric loss. Because of these properties, the thickness of impregnated paper insulation was considerably less than for rubber or varnished cambric insulations for thc samc working voltages. Polyethylene and crosslinked polyethylene cables in the distribution classes are frequently made with the same wall thickness as today's impregnated paper cables.

2.6.2.2 Extruded Dielectric Power Cables

The development of polyethylene in 1941 triggered a dramatic change in the circa insulation of cables for the transmission and distribution of electrical energy.

There are two major types of extruded dielectric insulation in wide use today for medium voltage cables:

(a) Crosslinked polyethylene or tree-retardant crosslinked polyethylene
(b) Ethylene propylene rubber.

Thermoplastic polyethylene (PE), which was widely used through the 1970s, was introduced during World War I1 for high-frequency cable insulation. PE was furnished as 15 kV cable insulation by 1947. Large usage began with the advent of Underground Residential Distribution (URD) systems early in the 1960s.

The development of modem URD systems may be viewed as the result of drastically lowering first costs through technology. Post-war URD systems were basically the same as the earlier systems except that there were two directions of feed (the loop system) System voltages rose from 2400/4160 to 7620/13,200 V. The pre-1950 systems were very expensive because they utilized such items as paper insulated cables, vaults, and submersible transformers. Underground service was, therefore, limited to the most exclusive housing developments. But for three developments in the 1960s, the underground distribution systems that exist today might not be in place.

First, in 1958–1959, a large Midwestern utility inspired the development of the pad-mounted transformer; the vault was no longer necessary nor was the submersible transformer.

Second, the polyethylene cable with its concentric neutral did not require cable splicers, and the cable could be directly buried. While possibly not as revolutionary, the loadbreak elbow (separable connector) allowed the transformer to be built with a lower, more pleasing appearance. The booming economy of some countries and the environmental concerns of the nation made underground power systems the watchword of the Great Society. In a decade, URD had changed from a luxury to a necessity. The goal for the utility engineer was to design a URD system at about the same cost as the equivalent overhead system. There was little or no concern about costs over the system's life because that PE cable was expected to last 100 years!

During the early part of the 1970s, isolated reports of early cable failures on extruded dielectric systems began to be documented in many parts of the world. "Treeing" was reintroduced to the cable engineer's vocabulary. This time it did not have the same meaning as with paper insulated cables. By 1976 research confirmed the fact that thermoplastic polyethylene insulated cables were failing in service at a rapidly increasing rate.

Cross-linked polyethylene exhibited a much lower failure rate that was not escalating nearly as rapidly.

The realization of the magnitude and significance of the problem led to a series of changes and improvements to the cables:

- Research work was initiated to concentrate on solutions to the problem
- Utilities began replacing the poorest performing cables
- Suppliers of component materials improved their products
- Cable manufacturers improved their handling and processing techniques
- The use of inert gas at high pressure within the metal sheath and indirect contact with the dielectric.
- Because of the good thermal characteristics and high dielectric strength of the gas SF_6, it is used for insulating the cables also. SF_6 gas insulated cables can be matched to overhead lines and can be operated corresponding to their surge impedance loading. These cables can be used for transporting thousands of MVA even at UHV, whereas the conventional cables are limited to 1000 MVA and 500 kV.

2.6.2.3 Cables with Oil-Soaked Paper Insulation

At a voltage of 110 kV and above, cables with paper insulation impregnated with liquid (low viscosity) oil are used. Application of insulation thickness grading (field regulation) and oil circulation under excess pressure through the oil-conducting channel located in the center of the current-carrying core make it possible to reduce the likelihood of gas inclusions in the insulation and, accordingly, increase the maximum operating electric field strength in the insulation up to 9–12 kV/mm and stability of electrical characteristics. The diagram of the 110 kV cable insulation device is shown in Fig. 2.42.

Oil circulating through the cable line and penetrating from the oil-conducting channels into the thickness of the insulation not only prevents the formation of gas or vacuum cavities, but also improves cable cooling conditions, which can be used to increase the transmitted power.

In Russia, such cables are manufactured for a voltage of 110 kV of low (up to 1 atm) and medium (3–5 atm) pressure in a lead sheath. As a rule, cables are made in single-phase execution. In some countries, cables are manufactured in a three-phase version for a voltage of 35–60 kV with oil-conducting channels located in the space between the cores and having the shape of spiral tubes.

High-pressure cables (up to 15 atm) have been developed, in which additional channels on the inside of the lead sheath are used to more efficiently fill the insulation

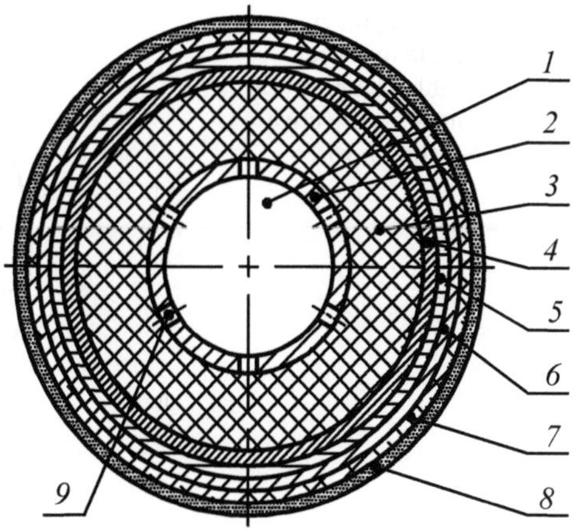

Fig. 2.42 Scheme of the cable insulation device 110 kV: 1—oil channel; 2—perforated current-carrying core; 3—paper-oil insulation; 4—semiconducting layer; 5—hermetic coating; 6—pillow; 7—armor; 8—anti-corrosion coating; 9—holes for the passage of oil into the insulation

with oil. In this case, a second lead sheath is used to increase the mechanical strength. In oil-filled cables, the impregnating composition must have a low tgδ to avoid the risk of thermal breakdown.

2.6.2.4 Cables with Paper Insulation Filled (Impregnated) with Pressurized Gas

By design, such cables are similar to oil-filled ones, but an increase in the dielectric strength of the insulation is achieved by the presence of high-pressure gas, which makes it possible to increase the ionization voltage in paper insulation with depleted impregnation. Depending on the gas pressure (nitrogen, freon, sulfur hexafluoride), low (0.7–1.5 atm), medium (1.7–3.0 atm) and high (11.0–15 atm) pressure cables are distinguished.

Gas-filled cables for voltages of 10, 35 and 60 kV are usually made in a common metal sheath with solid or sector sealed conductors with semi-conductive paper screens. A layer of metallized perforated paper and a copper perforated tape with a gap between the turns are applied over the insulation. The gas is supplied by means of flexible galvanized steel gas-permeable tubes laid in the space between the cores.

For voltages of 110 kV and above, single-phase cables with a central channel are used.

Permissible field strengths in the insulation of low pressure cables are 2.5 kV/mm, medium pressure—6.5 kV/mm.

Gas-filled cables are used when laying on steep slopes.

Fig. 2.43 High pressure oil-filled cable in a steel pipe: 1—cable phase; 2—oil; 3—pipeline; 4—protective covers; 5—sliding wires

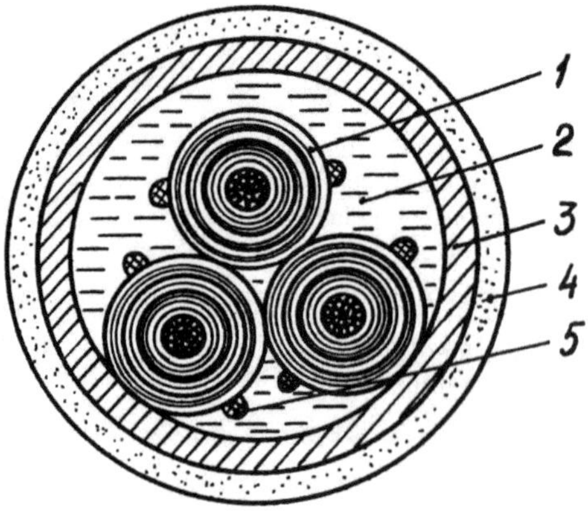

2.6.2.5 Cables in Pressurized Oil or Gas Insulated Steel Pipes

Such high pressure cables (up to 15 atm) are made in a steel pipeline. A sketch of an oil-filled high-pressure cable in a steel pipe is shown in Fig. 2.43.

For impregnation of high pressure oil-filled cable in a steel pipe, a more viscous oil is used than in cables of low and medium pressure. Since the pipe protects the cable from mechanical damage, the outer sheath is made of lightweight copper tape, over which slip armor is applied which ensures their sliding in the pipe (during installation).. Such cables, as a rule, are manufactured for higher rated voltages of 110–500 kV. Screens of metalized paper tapes are applied over the core and insulation. Permissible stresses in the cable insulation are 6–11 kV/mm, and the insulation thickness is from 12 to 29 mm.

For example, the outer diameter of one phase of a cable in a lead sheath for a voltage of 220 kV is 80 mm, the weight is 18.5 t/km, and the pipe diameter is 240 mm. Such a line as a whole is 20–30% cheaper than an oil-filled high-pressure cable. At a pressure of 12–15 atm, the allowable electric field strengths reach 10–12 kV/mm. The disadvantage of such cables is the worst heat dissipation conditions.

2.6.2.6 Cables in the Form of Gas-Filled Pipes Under High Pressure

The cable consists of an outer sheath (steel or plastic pipe) and a central conductor, which is installed along the axis of the pipe and fastened with centering insulators. Such a cable combines the features of overhead and cable lines. The main insulation is high pressure gas. The main problem is the creation of a spacer (centering insulator) with a high value of the flash voltage across the insulator surface. Such lines are capable of transmitting high power (up to 100 MW).

2.6.2.7 Cryogenic and Superconducting Transmission Lines

The idea of ***cryogenic transmission lines*** is based on the effect of reducing the resistance of metals (especially pure ones) with decreasing temperature. For example, cooling pure aluminum (99.99 Al) to a temperature of 20 K (−253 °C—the temperature of liquid hydrogen) reduces its resistance by about 500 times. When developing cryogenic lines, they usually focus on cooling the current-carrying core to a temperature below 80 K, i.e. up to liquid nitrogen temperature (−196 °C).

Such a power transmission line should be enclosed in a special hermetic sheath, inside which the current conductors are cooled along the entire length of the line. The lines are cooled using special cryogenic installations located along the power transmission line at a certain distance from each other. To ensure low temperatures close to the boiling points of helium or nitrogen, and to reduce the influx of heat from the external environment into the hollow with conductors, it is proposed to use a deep vacuum (10^{-2} Pa) as a heat-insulating medium, where heat exchange is carried out only by radiation.

The conductors of the cryogenic line are immersed in helium or liquid nitrogen and fixed with the help of centering insulators made of synthetic materials. End high-voltage bushings into the cooled zone on lines of high voltage classes are very responsible and structurally complex. Therefore, the key principle for creating such lines will always be the maximum use of their capabilities in the transmission of electricity by increasing the current, i.e. cryogenic power engineering is a multi-ampere power engineering.

Sections of cryogenic lines with a length of 10–15 m will be manufactured at the factory, and interconnected directly at the location of the transmission line. Compensating devices (bellows, etc.), materials with a low coefficient of thermal linear expansion are needed to avoid the occurrence of large thermal stresses and subsequent destruction of power transmission lines during their cooling.

The basis for the creation of ***superconducting power lines*** is the phenomenon of superconductivity (discovered in 1911)—a decrease in the electrical resistance of some pure metals to zero when cooled to a certain critical temperature. For aluminum, it is 1.2 K, for lead—7.2 K, for mercury—4.2 K. For a number of metals that have low resistance at ordinary temperatures (platinum, gold, silver, and copper) at temperatures of 0.05–0.1 K, the effect of superconductivity is absent.

The designs of superconducting and cryogenic cables are similar, with characteristic advantages and disadvantages. Such cables can transmit currents of 300–500 A at a voltage of 10–100 kV.

Superconducting and cryogenic transmission lines open up new technical possibilities in the power engineering. However, due to complexity, low maintainability, difficulties in ensuring high reliability and due to a number of other equally complex technical problems, these lines will not yet be able to compete with overhead and cable transmission lines that have existed for many decades in the near future.

Question and Tasks

1. What is the main purpose of high-voltage power transformers?
2. Give the classification of instrument transformers.
3. What gases are used as an insulating medium in power transformers?
4. What is the design feature of dry transformers?
5. Name the main reasons for the failure of power transformers.
6. What explains the increase in demand for cable transmission lines?
7. Draw and explain the structure of an oil-filled cable.
8. What are the advantages of cables with XLPE insulation?
9. How are HV circuit breakers classified according to the type of insulating medium?
10. Name the advantages of vacuum circuit breakers.
11. What are the types of high-voltage capacitors used in electric power systems?
12. What is the main purpose of cosine capacitors?
13. What types of cooling are used in high power rotating machines?
14. What is the disadvantage of oil barrier bushings?
15. What does the term "capacitor-type insulation" mean?
16. What is the purpose of aluminum foil linings in the bushing insulation?

References

1. Kurtenkov GE (1999) Fundamentals of designing insulation of high-voltage electrical equipment. NTL Publishing House, Tomsk, p 276 (in Russian)
2. Warne DF (2005) Newnes electrical power engineers' handbook, 2nd edn. Linarce House, Burlington, Amsterdam, p 477
3. Ushakov V (2004) Insulation of high voltage equipment. Springer, Berlin, Heidelberg, 2004, p 421
4. Haddad B, Warne D (eds) (2004) Advances in high voltage engineering. Textbook, IET Amazon, p 552
5. Stone GC, Boulter EA, Culbert I, Dhirani H (2004) Electrical insulation for rotating machines: design, evaluation, aging, testing, and repair. Wiley, p 389
6. Ushakov V (2001) High voltage engineering. Textbook. Tomsk, TPU Press, p 253
7. Thue WA (ed) (1999) Electrical cable engineering. Marcel Dekker Inc.
8. Fink DG, Beaty HW (2006) Standard handbook for electrical engineers, 15th edn. McGraw-Hill
9. Winders JJ (2002) Power transformers. Principles and applications. Marcel Dekker Inc., p 296
10. Flanagan WM (1993) Handbook of transformers design and applications, 2nd edn. McGraw Hill, p 496
11. Emadi A (2004) Energy-efficient electric motors, 3rd edn. CRC Publisher, p 424
12. Boldea I (2005) Synchronous generators. CRC Publisher, p 448

Chapter 3
Insulating Materials and Media Used in High-Voltage Elements of Electric Power Systems

Abstract The severity of the consequences from the impact on high-voltage equipment of numerous factors (Chapter 4) to a decisive extent depends on the type and quality of the dielectric materials and media used in it. It is quite natural that most methods for control the condition of equipment are based on monitoring the initial quality of insulation and changes in its characteristics during operation. The main characteristics of all insulating materials and media—solid, liquid, gaseous and vacuum—are considered. The requirements for electrical insulation are formulated taking into account the modes of operation and external conditions of operation of the equipment.

3.1 The Most Important Characteristics of Insulating Materials and Media

3.1.1 Resistivity of Dielectric Materials

The measured resistance R of insulation depends on the geometry of the specimen or system measured, which for a parallel-plate arrangement is:

$$R = \frac{r \times t}{A} Ohms, \tag{3.1}$$

where: t is the insulation thickness in centimeters, A is the area in square centimeters, and r is the dielectric resistivity in ohm-centimeters. If t and A vary from place to place, the effective "insulation resistance" will be determined by the effective integral of the t/A ratio over all the area under stress, on the assumption that the material resistivity p does not change. If the material is not homogeneous and materials of different resistivity appear in parallel, the system can be treated as parallel resistors. In this case, the lower-resistivity material usually controls the overall behavior. But, if materials of different resistivity appear in series in the electric field, the higher-resistivity material generally will control the current, and a majority of the voltage will appear across it, as in the case of series resistors. The resistance of dielectrics and

V. Y. Ushakov et al., *High-Voltage Equipment of Power Systems*, Power Systems,
https://doi.org/10.1007/978-3-031-38252-9_3

insulation is usually time-dependent and (for the same reason) frequency-dependent. The dc behavior of dielectrics under stress is an extension of the low-frequency behavior. The ac and dc resistance and permittivity can, in principle, be related for comparable times and frequencies.

Current flow in dielectrics can be divided into parts: (a) the true dc current, which is constant with time and would flow indefinitely, is associated with a transport of charge from one electrode into the dielectric, through the dielectric, and out into the other electrode, and (b) the polarization or absorption current, which involves, not charge flow through the interface between the dielectric and the electrode, but rather the displacement of charge within the dielectric.

Polarization current results from any of the various forms of limited charge displacement which can occur in the dielectric. The displacement occurring first (within less than nanoseconds) is the electronic and intramolecular charged atom displacement responsible for the very high frequency permittivity. The next slower displacement is the rotation of dipolar molecules and groups which are relatively free to move. The displacement most commonly observed in dc measurements, that is, currents changing in times of the order of seconds and minutes, is due to the very slow rotation of dipolar molecules and ions moving up to internal barriers in the material or at the conductor surfaces. When those slower displacement polarizations occur, the dielectric constant declines with increasing frequency and approaches the square of the optical refractive index h_2 at optical frequencies.

In composite dielectrics (material with relatively lower resistance intermingled with a material of relatively higher resistance), a large interfacial or Maxwell–Wagner type of polarization can occur. A circuit model of such a situation can be represented by placing two of the circuits in series and making the parallel resistance of one much lower than the other. To get the effect, it is necessary that the time constant R_pC be different for each material [1, 2].

A simple model of the polarization current predicts an exponential decline of the current with time: $Ip = Ae^{-t}$, similar to the charging of a capacitor through a resistor. Composite materials are likely to have many different time constants *1/ RC* superimposed. It is found empirically that the polarization or absorption current decreases inversely as a simple negative exponent of the time:

$$I = A \times t^{-n}. \tag{3.2}$$

The ratio of the current at 1 min to that at 10 min has been called the *polarization index* and is used to indicate the quality of composite machine insulation. A low polarization index associated with a low resistance sometimes indicates parallel current leakage paths through or over the surface of insulation (e.g., in adsorbed water films).

The level of the conduction current which flows essentially continuously through insulation is an indication of the level of the ionic concentration and mobility in the material. Frequently, as with salt in water, the ions are provided by dissolved, absorbed, or included impurity electrolytes in the material rather than by the material itself. Purifying the material will therefore often raise the resistivity. If it is liquid, purification can be done with adsorbent clays or ion-exchange resins.

The ionic conductivity also varies widely with material purity. Among the polymers and resins, nonpolar resins such as polyethylene are likely to have high resistivity, on the order of 10^{16} or greater, since they do not readily dissolve or dissociate ionic impurities. Harder or crystalline polar resins have higher resistivity than do similar softer resins of similar dielectric constant and purity. Resins and liquids of higher dielectric constant usually have higher conductivities because they dissolve ionic impurities better, and the impurities dissociate to ions much more readily in a higher dielectric constant medium. Ceramics and glasses have lower resistivity if they contain alkali ions (sodium and potassium), since these ions are highly mobile [3].

The change in dielectric properties with frequency is not considered in this textbook, because in the power system, high-voltage equipment operates mainly on alternating voltage of industrial frequency and on direct voltage. Lightning and switching overvoltages, due to their short duration, cannot significantly affect the properties of dielectrics through frequency dependence.

3.1.2 Variation of Dielectric Properties with Temperature

The trend in ac permittivity and conductivity, as measured by the dissipation factor, is controlled by the increasing ionic migrational and dipolar molecular rotational mobility with increasing temperature. This curve, which is indicated in Fig. 3.1, is in most respects a mirror image of the frequency trend since the two effects are interrelated.

The permittivity-dispersion and dissipation factor maximum region occurs below room temperature for viscous liquids and still lower for mobile liquids. In fact, mobile liquids may crystallize before they would show dispersion, except at high frequencies. With polymers, the dissipation factor maxima is likely to occur, at power frequencies, at a temperature close to a softening-point or internal second-order transition-point temperature. Dielectric dispersion and mechanical modulus dispersion usually can be correlated at the same temperature for comparable frequencies.

3.1.3 Potential Distribution in Dielectrics

The maximum potential gradient in dielectrics is of critical significance insofar as the breakdown is concerned, since breakdown or corona is usually initiated at the region of highest gradient. In a uniform-field arrangement of conductors or electrodes, the maximum gradient is simply the applied voltage divided by the minimum spacing. In divergent fields, the gradient must be obtained by calculation (which is possible for some simple arrangements) or by field mapping.

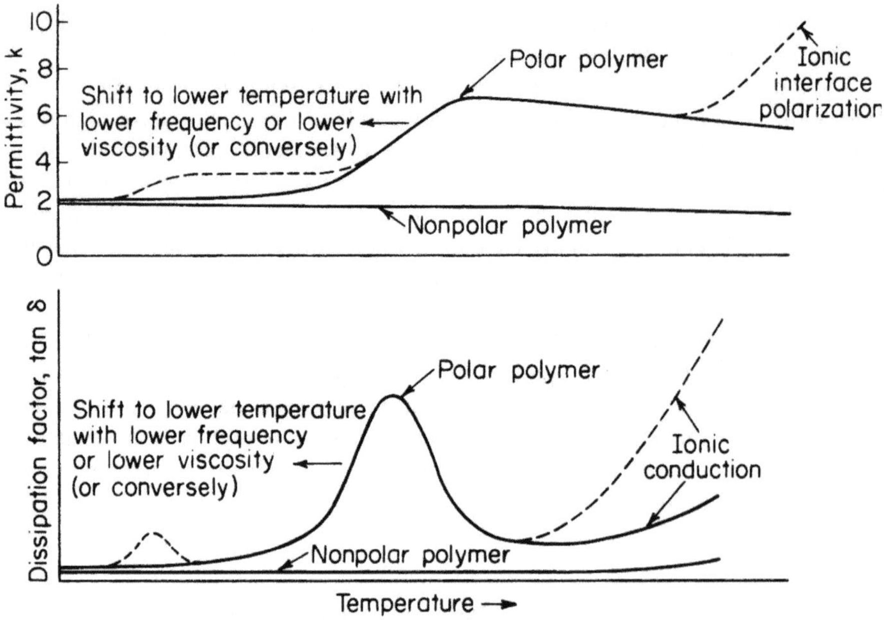

Fig. 3.1 Typical variation in dielectric properties with temperature

A common situation is the coaxial geometry with inner and outer radii R_1 and R_2. The gradient at radius r (centimeters) with voltage V applied is given by the equation:

$$E = \frac{V}{r \times \ln\left(\frac{R_1}{R_2}\right)}, \tag{3.3}$$

where E is in V/cm. The gradient is a maximum at $r = R_1$.

When different dielectrics appear in series, the greater stress with ac fields is on the material having the lower dielectric constant. This material will frequently break down first unless its dielectric strength is much higher [4].

The effect of the insulation thickness and dielectric constant (as well as the sharpness of the conductor edge) to create sufficient electric stress for local air breakdown (partial discharges). With dc fields, the stress distributes according to the resistivity of the materials, the higher stress being on the higher-resistivity material.

3.1.4 Dielectric Strength

This is the maximum potential gradient that the material can withstand without rupture. Practically, the strength is often reported as the breakdown voltage divided by the thickness between electrodes, regardless of electrode stress concentration.

Breakdown appears to require not only sufficient electric stress but also a certain minimum amount of energy. It is a property which varies with many factors such as thickness of the specimen, size and shape of electrodes used in applying stress, form or distribution of the field of electric stress in the material, frequency of the applied voltage, rate and duration of voltage application, fatigue with repeated voltage applications, temperature, moisture content, and possible chemical changes under stress.

The practical dielectric strength is decreased by defects in the material, such as cracks, and included conducting particles and gas cavities. The dielectric strength of liquids and gases is very adversely affected by conductive particles. To state the dielectric strength correctly, the size and shape of specimen, method of test, temperature, manner of applying voltage, and other attendant conditions should be particularized as definitely as possible.

ASTM standard methods of dielectric strength testing should be used for making comparison tests of materials, but the levels of dielectric strength measured in such tests should not be expected to apply in service for long times. It is best to test an insulation in the same configuration in which it would be used. Also, the possible decline in dielectric strength during long-time exposure to the service environment, thermal aging, and partial discharges (corona), if they exist at the applied service voltage, should be considered. ASTM has thermal life test methods for assessing the long-time endurance of some forms of insulation such as sheet insulation, wire enamel, and others. There are IEEE thermal life tests for some systems such as random wound motor coils.

The dielectric strength varies as the time and manner of voltage application. With unidirectional pulses of voltage, having rise times of less than a few microseconds, there is a time lag of breakdown, which results in an apparent higher strength for very short pulses. In testing sheet insulation in mineral oil, usually a higher strength for pulses of slow rise time and somewhat higher strength for dc voltages is observed.

The trend in breakdown voltage with time is typical of many solid insulation systems. With ac voltages, the apparent strength declines steadily with time as a result of partial discharges (in the ambient medium at the conductor or electrode edge). These penetrate the solid insulation. The discharges result from breakdown of the gas or liquid prior to the breakdown of the solid. Mica in particular, as well as other inorganic materials is more resistant to such discharges. Organic resins should be used with caution where the ac voltage gradient is high and partial discharges may be present. Since the presence of partial discharges on insulation is so important to the long- time voltage endurance, their detection and measurement have become very important quality control and design tools. If discharges continuously strike the insulation within internal cavities or on the surface, the time to failure usually varies inversely as the applied frequency, since the number of discharges per unit time increases almost in direct proportion to the frequency. But in some cases, ambient conditions prevent continuous discharges.

When organic resin insulation is fabricated to avoid partial discharges using conductors or electrodes intimately bonded to the insulation, as in extruded polyethylene cables with a plastic semi- conducting interface between the resin and the coaxial inner and outer metal conductors, respectively, the voltage endurance is greatly extended. Imperfections, however, in this "semicon"- resin interface, or at conducting particle inclusions in the resin, can lead to local discharges and the development of "electrical tree" growth. Vacuum impregnating and casting electrodes or conductors into resin also tend to avoid cavities and surface discharges and greatly improve the voltage endurance at high stresses.

The dc strength of solid insulation is usually higher and declines much less with time than the ac strength, since corona discharges are infrequent.

The dielectric strength is much higher where surface discharges are avoided and when the electric field is uniform. This can be achieved with solid materials by recessing spherical cavities into the material and using conducting paint electrodes.

The "intrinsic" electric strength of solid materials measured in uniform fields, avoiding surface discharges, ranges from levels on the order of 0.5–1 MV/cm for alkali halide crystals, which are about the lowest, upward to somewhat more than 10 MV/cm. Polymers and some inorganic materials, such as mica and aluminum oxide, have strengths of 2–20 MV/cm for thin films. The strength decreases with increasing thickness and with temperature above a critical temperature (which is usually from 1 to 100 °C), below which the strength has a level value or a moderate increase with increasing temperature. Below the critical temperature, the breakdown is believed to be strictly electronic in nature and is constant or increases slightly with temperature. Above this temperature, it declines owing to dielectric thermal heating.

The breakdown voltage of thin insulation materials containing defects, which give the minimum breakdown voltage, declines as the area under stress increases. The effect of area on the strength can be estimated from the standard deviation S of tests on smaller areas by applying minimum value statistics: $V1 - V2 = 1.497\ S \log(A1/A2)$, where $V1$ and $V2$ are the breakdown voltages of areas $A1$ and $A2$.

If the ac or dc conductivity of a dielectric is high or the frequency is high, breakdown can occur as a result of dielectric heating, which raises the temperature of the material sufficiently to cause melting or decomposition, formation of gas, etc. This effect can be detected by measuring the conductivity as a function of applied electric stress. If the conductivity rises with time, with constant voltage, and at constant ambient temperature, this is evidence of an internal dielectric heating. If the heat transfer to the electrodes and ambient surroundings is adequate, the internal temperature eventually may stabilize, but if this heat transfer is inadequate, the temperature will rise until breakdown occurs.

3.1.5 Influences Water Penetration and Ionizing Radiation

Water penetration into electrical insulation also degrades the dielectric strength by several mechanisms. The effect of water to increase the insulation conductivity contributes there by to a decreased dielectric strength, probably by a thermal breakdown mechanism. Another effect noticed recently, particularly in polyethylene cables, is the development of "water" or "electro- chemical trees." Water (and/or a similar high dielectric constant chemical) can diffuse through polyethylene and collect at tiny hygroscopic inclusion sites, where the water or chemical is adsorbed. Then the electric field causes an expansion and growth of the adsorbed water or chemical in the electric field direction. This may completely bridge the insulation or possibly increase the local electric stress at the site so as to produce an electric tree and eventual breakdown.

Ionizing radiation, as from nuclear sources, may degrade insulation dielectric strength and integrity by causing polymer chain scission, and cracking of some plastics, as well as gas bubbles in liquids. Also, the conductivity levels in solids and liquids are increased. Ionizing radiation in the electrical power engineering is a rather rare phenomenon (except for nuclear power plants), and therefore it is not considered in detail here.

3.1.6 Arc Tracking of Insulation

High-current arc discharges between conductors across the surface of organic resin insulation may carbonize the material and produce a conducting track. In the presence of surface water films, formed from rain or condensation, etc., small arc discharges form between interrupted parts of the water film, which is fairly conducting, and conducting tracks grow progressively across the surface, eventually bridging between conductors and causing complete breakdown. Materials vary widely in their resistance to tracking, and there are a variety of dry and wet tests for this property. With proper fillers, some organic resins can be made essentially nontracking. Some resins such as polymethyl methacrylate and polymethylene oxide burst into flame under arcing conditions.

3.1.7 Thermal Aging

Organic resinous insulating materials in particular are subject in varying degrees to deterioration due to thermal aging, which is a chemical process involving decomposition or modification of the material to such an extent that it may no longer function adequately as the intended insulation. The aging effects are usually accelerated by increased temperature, and this characteristic is used to make accelerated tests to failure or to an extent of deterioration considered dangerous. Such tests are made at appreciably higher than normal operating temperatures, if the expected life is to be several years or more, since useful accelerated tests reasonably should be completed in less than a year.

Frequently, other environmental factors influence the life in addition to the temperature. These include presence or absence of oxygen, moisture, and electrolysis. Mechanical and electrical stress may reduce the life by setting a required level of performance at which the insulation must perform. If this level is high, less deterioration of the insulation is required to reach this level.

Sometimes a complete apparatus is life-tested, as well as smaller specimens involving only one insulation material or a simple combination of these in a simple model. New tests are being devised continually, but there has been some standardization of tests by the IEEE and ASTM and internationally by the IEC.

It is important to note that frequently materials are assigned temperature ratings based on tests of the material alone. Often that material, combined with others in an apparatus or system, will perform satisfactorily at appreciably higher temperatures. Conversely, because of incompatibility with other materials, it may not perform at as high a temperature as it would alone. For this reason, it is considered desirable to make functional operating tests on complete systems. These can also be accelerated at elevated temperatures and environmental exposure conditions such as humidification, vibration, cold-temperature cycling, etc. introduced intermittently.

3.2 Solid Dielectrics

Solid insulating materials in high-voltage apparatus and devices simultaneously perform the functions of both insulating and structural materials, which is taken into account in the development and operation of such equipment.

The main requirements of the insulating materials used for power apparatus are:

1. High insulation resistance
2. High dielectric strength
3. Good mechanical properties i.e., tenacity and elasticity
4. It should not be affected by chemicals around it
5. It should be non-hygroscopic because the dielectric strength of any material goes very much down with moisture content
6. Resistance to high temperatures. The negative effect of elevated temperatures on solid dielectrics (mainly solid ones) is considered in the previous section. This requirement not universal, but common.

3.2.1 Organic Insulating Materials

Vulcanized rubber. Rubber in its natural form is highly insulating but it absorbs moisture readily and gets oxidized into a resinous material; thereby it loses insulating properties. When it is mixed with sulphur along with other carefully chosen ingredients and is subjected to a particular temperature, it changes into vulcanized rubber, which does not absorb moisture and has better insulating properties than even the pure rubber. It is elastic and resilient.

The electrical properties expected of rubber insulation are high breakdown strength and high insulation resistance. In fact, the insulation strength of the vulcanized rubber is so good that for lower voltages the radial thickness is limited due to mechanical consideration.

The physical properties expected of rubber insulation are that the cable should withstand normal hazards of installation and it should give trouble-free service.

Vulcanized rubber insulated cables are used for wiring of houses, buildings and factories for low-power work.

There are two main groups of synthetic rubber material: (i) general purpose synthetics which have rubber-like properties and (ii) special purpose synthetics which have better properties than the rubber e.g., fire resisting and oil resisting properties. The four main types are: (i) butyl rubber, (ii) silicon rubber, (iii) neoprene, and (iv) styrene rubber.

Butyl rubber. The processing of butyl rubber is similar to that of natural rubber but it is more difficult and its properties are comparable to those of natural rubber. The continuous temperature to which butyl rubber can be subjected is 85 °C whereas for natural rubber it is 60 °C. The current rating of butyl insulated cables is approximately same as those of paper or PVC insulated cables. Butyl rubber compound can be so manufactured that it has low water absorption and offers interesting possibilities for a non-metallic sheathed cable suitable for direct burial in the ground.

Silicone rubber. It is a mechanically weak material and needs external protection but it has high heat resistant properties. It can be operated at temperatures of the order of 150 °C. The raw materials used for the silicon rubber are sand, marsh gas, salt, coke and magnesium.

Neoprene. Neoprene is a polymerized chlorobutadiene. Chlorobutadiene is a colorless liquid which is polymerized into a solid varying from a pale yellow to a darkish brown color. Neoprene does not have good insulating properties and is used up to 660 V a.c. but it has very good fire resisting properties and therefore it is more useful as a sheathing material.

Styrene rubber. Styrene is used both for insulating and sheathing of cables. It has properties almost equal to the natural rubber.

Polyvinyl Chloride (PVC) is a polymer derived generally from acetylene and it can be produced in different grades depending upon the polymerization process. For use in cable industry the polymer must be compounded with a plasticizer which makes it plastic over a wide range of temperature. The grade of PVC depends upon the plasticizer. PVC is inferior to vulcanized in respect of elasticity and insulation resistance. PVC material has many grades. It is used both for sheathing and as an insulating material. In this compound monomeric plasticizers are used. It is to be noted that a V.R. insulated PVC sheathed cable is not good for use.

Hard grade PVC. These are manufactured with less amount of plasticizer as compared with general purpose type. Hard grade PVC are used for higher temperatures for short duration of time like in soldering and are better than the general purpose type. Hard grade can not be used for low continuous temperatures.

Heat resisting PVC. Because of the use of monomeric plasticizer which volatilizes at temperature 80–100 °C, general purpose type compounds become stiff. By using polymeric plasticizers it is possible to operate the material continuously around 100 °C. PVC compounds are normally costlier than the rubber compounds and the polymeric plasticized compounds are more expensive than the monomeric plasticized ones. PVC is inert to oxygen, oils, alkalis and acids and, therefore, if the environmental conditions are such that these things are present in the atmosphere, PVC is more useful than rubber.

Polythene can be used for high frequency cables. This has been used to a limited extent for power cables also. The thermal dissipation properties are better than those of impregnated paper and the impulse strength compares favorably with an impregnated paper-insulated device. The maximum operating temperature of this material under short circuits is 100 °C. The use of polythene for cables has been limited by its low melting point. By cross-linking the molecules, in roughly the same way as vulcanizing rubber, a new material is produced which does not melt but carbonizes at 250–300 °C. By using chemical process it has been made technically possible to cross-link polythene in conventional equipment for the manufacture of rubber. This is why the product is said to be ***"vulcanised" or "cross-linked" polythene***.

The polythene is inert to chemical reactions as it does not have double bonds and polar groups. Therefore, it was thought that polythene could be cross-linked only through special condition, e.g., by irradiating polythene with electrons, thereby it could be given properties of cross-linking such as change of tensile strength and better temperature stability. Many irradiation processes have been developed in the cable making industry even though large amounts of high energy radiations are required and the procedure is expensive.

Polythene can also be irradiated with ultraviolet light, after adding to it a small quantity of ultra- violet sensitive material such as benzophenone. Under the influence of ultraviolet light on benzophenone, a radical is formed of the same type as in the decomposition of peroxide by the radical mechanism. Organic peroxides have also been used successfully to crosslink the polythene.

It should be noted that polyethylene has proved to be an excellent material for the manufacture of large-sized insulating structures with a large thickness of the insulating layer for high-voltage electrophysical installations (Pulsed Power), Fig. 3.2, and some electric power equipment, Fig. 3.3, in turn, insulation parts (polyethylene rings) and vacuum diode are shown on Figs. 3.4 and 3.5.

In them, insulating structures (centering insulators in coaxial forming lines, bushings, diaphragm seals, and high-current electron gun housings) are not affected by most of the adverse factors that will be discussed in Chapter 4. But, on the other

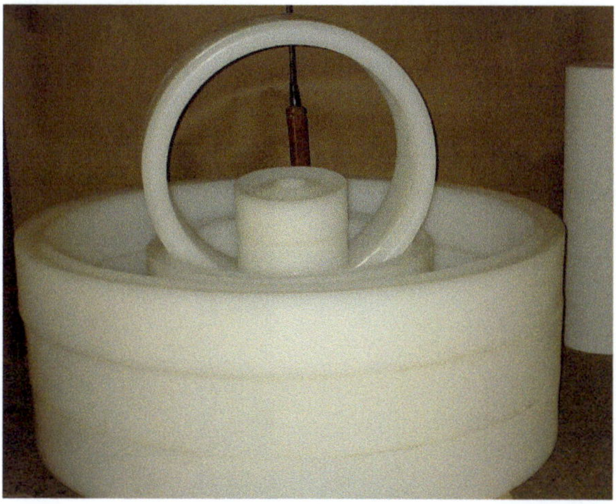

Fig. 3.2 Blanks made of high-density polyethylene (LDPE) of various shapes insulating structures for high-voltage installations (Outer diameter of horizontally lying rings—1.4 m)

Fig. 3.3 View and size of bushing

hand, they must operate at very high field strengths due to severe restrictions on the dimensions of high-voltage electrophysical installations. The manifestation of the effects of a strong electric field in an insulating material is enhanced by a high rate of change in field strength, since the operating voltage, as a rule, is pulses with nanosecond duration. Under these conditions, the reliability and service life of monolithic polymer insulation (for example, polyethylene) are determined mainly by the following processes:

Fig. 3.4 Polyethylene rings with an outer diameter of 3 m for a vacuum diode (electron gun) of a high-voltage electron accelerator

Fig. 3.5 View of the vacuum diode

- cold (autoelectronic) emission, supplying free electrons to the dielectric, which can cause ionization, initiate chemical reactions, and also distort the initial electric field,
- chemical transformations in polymers and changes in their supramolecular structure,
- development of submicro- and microcracks,
- the origin and development of dendrites.

To study the nature of these "fine" processes and their influence on electrical aging, it is necessary to apply special methods and techniques of physical experiment:

- sounding of space charge fields by albedo β-method and (or) acoustic probe method,

- study of the emergence and development of microinhomogeneities in a polymer by light scattering (for optically transparent dielectrics) and small-angle X-ray scattering,
- study of the processes of degradation (structural changes) of dielectrics in the final phases of aging—the formation and development of microcracks and dendritic channels—using optical and thermal imaging techniques.

In subsequent years, some of these methods found practical application for diagnostics, the use of others was postponed until the appearance of cheaper and more reliable equipment that meets the needs of practice.

These methods and the results obtained with their help are described in sufficient detail in [9].

Impregnated paper. A suitable layer of the paper is lapped on the conductor depending upon the operating voltage. It is then dried by the combined application of heat and vacuum. This is carried out in a hermetically sealed steam heated chamber. The temperature is 120–130 °C before vacuum is created. After the device is dried, an insulating compound having the same temperature as that of the chamber is forced into the chamber. All the pores of the paper are completely filled with this compound. After impregnation the device is allowed to cool under the compound so that the void formation due to compound shrinkage is minimized.

In case of pre-impregnated type the papers are dried and impregnated before they are applied on the conductor.

The compound used in case of impregnated paper is a semifluid and when the cables are laid on gradients the fluid tends to move from higher to lower gradient. This reduces the compound content at higher gradients and may result in void formation at higher gradients. This is very serious for cables operating at voltages higher than 3.3 kV. In many cases, the failures of the cables have been due to the void formation at the higher levels or due to the bursting of the sheath at the lower levels because of the excessive internal pressure of the head of compound.

Insulating press boards. If the thickness of paper is 0.8 mm or more, it is called paper board. When many layers of paper are laminated with an adhesive to get desired thickness, these are known as press boards and are used in bushings, transformers as insulating barriers or supporting materials. The electrical properties of press boards varies depending upon the resin content. The application of these press boards depends upon the thickness and density of paper used. For high frequency capacitors and cables usually low density paper (0.8 gm/cm^3) is used where medium density paper is used for power capacitors and high density papers are used in d.c. machines and energy storage capacitors. The electric strength of press board is higher than that of resins or porcelain. However, it is adversely affected by temperature above 20 °C. The loss angle tan also decreases with increase in temperature. The main advantage of this material is that it provides good mechanical support even at higher temperatures up to 120 °C.

Epoxy resins are low molecular but soluble thermosetting plastics which exhibit sufficient hardening quality in their molecules. The chemical cross-linking of epoxy resins is normally carried out at room temperatures either by a catalytic mechanism or by bridging across epoxy molecule through the epoxy or hydroxyl group.

Epoxy resins have high dielectric and mechanical strength. They can be cast into desired shapes even at room temperature. They are highly elastic and it is found that when it is subjected to a pressure of 175,000 psi, it returned to its original shape after the load is removed. The dielectric constant varies between 2.5 and 4.0. Epoxy resins basically being non-polar substances have high dc specific insulation resistance and low loss tan compared to polar materials like PVC. However, when the temperature exceeds 100 °C the specific insulation resistance begins to decrease considerably and tan δ in creases. Compared to porcelain the breakdown strength of epoxy resin is almost double at temperatures up to 100 °C but decreases rapidly at higher temperatures.

As filler materials, the inorganic substances like quartz powder (SiO_2) are used for casting applications. In SF_6 gas insulated systems having epoxy resin spacers, aluminium oxide and also dolomite are used as filler materials. These are found to be more compatible to the decomposed products of SF_6 by partial discharge and arcing discharges.

It is to be noted that the cast or encapsulation should not contain voids or humidity especially in high voltage applications and the material is desired to be homogeneous. It is, therefore, desirable to dry and degas the individual components of the mixture and casting is preferably carried out in vacuum.

The epoxy resins casts are inert to ether, alcohol and benzol. However, most of them are soluble in mineral oils at about 70 °C. It is for this reason that they are not found suitable for applications in filled transformers.

There are certain application which require insulating materials to operate between a high range of temperature e.g., −270 to 400 °C. Some of the applications are space shuttle solar arrays, capacitors, transformers high speed locomotive, microprocessor chip carriers, cryogenic cables and other applications at cryogenic temperatures. For this some thermoplastic polymer films are used which have unique combination of electrical, mechanical and physical quantities and these materials are able to retain these properties over a wide range of temperatures where other insulating materials may fail. Perfluoro carbon films have high dielectric strength very low dielectric constant of 2 and low dielectric loss of 2×10^{-4} at 100 Hz and 7.5×10^{-4} at 100 MHz. These films are used under extreme conditions of temperature and environment. These films are used for insulation on high temperature wires, cables, motor coils phase and ground insulation and for capacitors. This is also used as a substrate for flexible printed circuits and flexible cables.

Another insulating film in which has the best thermal properties in this category of insulating materials is polyimide film under the trade name of Kapton manufactured by DuPont of America. These films can be used between a very wide range of temperature variation varying between −270 °C and 350 °C. Its continuous temperature rating is 240 °C. It has high dielectric and tensile strength. The disadvantages of the film are:

i. high moisture absorption rate and
ii. it is affected by alkalies and strong inorganic acids.

Kepton films can be used capacitors, transformers formed coil insulation, motor state insulation and flexible printed circuits. The film is selectively costlier and is mainly used where its unique characteristics makes it the only suitable insulation. The use of this insulation for motors reduces the overall dimensions of the motors for the same ratings. It is, therefore, used in almost all situations whose space is a serious problcm and thc othcr naturc insulation result in bigger dimension.

Another recently developed resins is poly carbonate (PC) which is good heat resistant; it is flexible and has good dielectric characteristic. It is not affected by oils, fats and dilute acids but is adversely affected by alkalies, esters and aromatic hydrocarbons. The film being cost effective and fast resistant, it is used for coil insulation, slot insulation for motors and for capacitor insulation. This is known as the lexon polymer.

General Electric Co. of USA has developed a film under the trade name Ultem which is a polyetherimine (PEI) film which has dielectric strength comparable to that of polyimide film and has higher thermal conductivity and lower moisture absorption and is relatively less costlier. It is used as insulation for transformers and motors.

3.2.2 Inorganic Insulating Materials

Mica consists of crystalline mineral silicates of alumina and potash. It has high dielectric strength, low dielectric losses and good mechanical strength. All these properties make it useful for many electrical devices e.g., commutator segment separator, armature windings, electrical heating and cooling equipments and switchgear. Thin layers of mica are laminated with a suitable resin or varnish to make thick sheets of mica. Mica can be mixed with the required type of resin to obtain its application at different operating temperatures. Mica is used as a filler in insulating materials to improve their dielectric strength, reduce dielectric loss and improve heat resistance property.

Ceramics materials are produced from clay containing aluminium oxide and other inorganic materials. The thick parts of these substances is given the desired shape and form at room temperature and then baked at high temperature about (1450 °C) to provide a solid inelastic final structure. Ceramics also known as porcelain in one of its forms have high mechanical strength and low permittivity ($\varepsilon_r < 12$) are widely used for insulators and bushings. These have 40–50% of clay, 30– 20% of aluminium oxide and 30% of field-spar. The ceramics with higher permittivity ($\varepsilon_r > 12$) are used in capacitors and transducers.

The specific insulation resistance of ceramics is comparatively low. The tan δ of these materials is high and increases with increase in temperature resulting in higher dielectric loss. The breakdown strength of porecelain compared to other insulating material is low but it remains unaffected over a wide range of temperature variation. Porcelain is chemically insert to alkalies and acids and, therefore, corrosion resistant and does not get contaminated. Alumina (Al_2O_3) has replaced quartz because of its better thermal conductivity, insulating property and mechanical strength. It is used for the fabrication of high current vacuum circuit breakers.

Glass is a thermoplastic inorganic material consisting of silicondioxide (SiO_2), which is available in nature in the form of quartz. Different types of metal oxides could be used for producing different types of glasses but for use in electrical engineering only non-alkaline glasses are suitable having alkaline content less than 0.8%.

The dielectric constant of glass varies between 3.6 and 10.0 and the density varies between 2000 kg/m^3 and 6000 kg/m^3. The loss angle tan δ is less than 10^{-3} and losses are higher for lower frequencies. Its dielectric strength varies between 300 and 500 kV/mm and it decreases with increase in temperature. Glass is used for X-ray equipment, electronic valves, electric bulbs etc.

3.3 Liquid Dielectrics

3.3.1 General Consideration

The main functions of liquid insulation are: isolating the current-carrying parts of high-voltage equipment at different potentials, removing excess heat from operating equipment and extinguishing the arc in switches and other high-voltage switching equipment. In addition, liquid insulation helps to obtain information about the state of the internal insulation system as a whole and take timely measures to prevent equipment failures [3].

Thus, for the perfect performance of its functions, any liquid dielectric must have high electrical strength, good thermal conductivity, chemical and thermal stability, as little combustibility and toxicity as possible, suitable viscosity for cold conditions, and the ability to maintain its main properties over a long period of operation. At elevated temperatures and high electrical voltages. The list of basic requirements for liquid dielectrics is given in Fig. 3.6 [4].

The search for liquids that best meet the named list of properties began at the end of the nineteenth century and continues to the present, as shown below. Only research on polychlorinated biphenyls as insulation for high-voltage equipment was discontinued in 1970.

1884—Mineral oil
1930—polychlorinated biphenyl
1974—Silicone fluids
1985—Synthetic esters
1990—Natural Esters
2002—Mixed liquids
2008—Nanofluids

The most important properties of liquids as insulation for high-voltage equipment are dielectric strength, dielectric losses, dielectric constant, and volume resistivity.

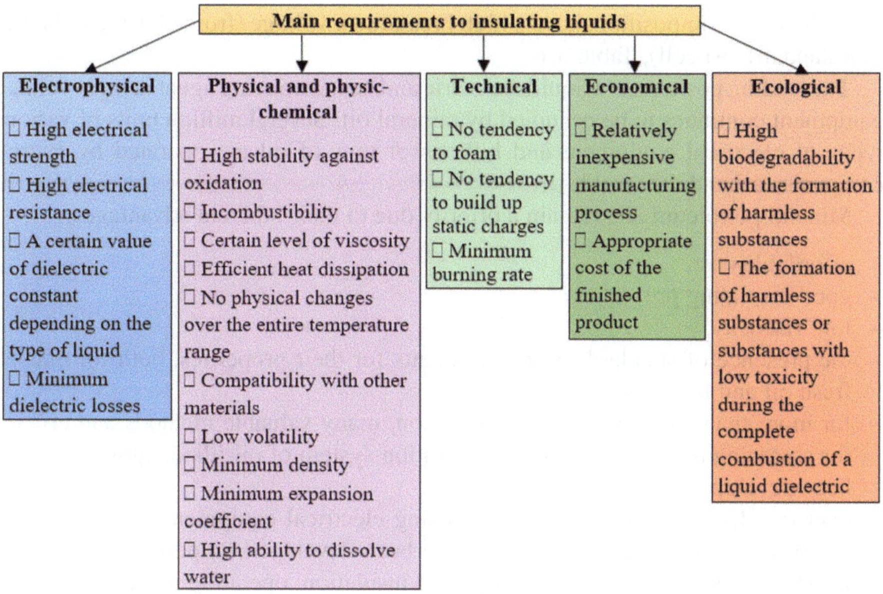

Fig. 3.6 Basic requirements for insulating liquids

The electrical properties of the most commonly used insulating liquids are shown in Table 3.1.

The most important properties of liquids as insulation for high-voltage equipment are dielectric strength, dielectric losses, dielectric constant, and volume resistivity. The electrical properties of the most commonly used insulating liquids are shown in Table 3.1.

For dielectric liquids, dielectric strength should be considered as an integral characteristic, since this indicator strongly depends on the degree of their contamination with moisture and conductive solid particles (cellulose fibers, carbon, dust, metal microparticles, colloidal compounds, etc.). Clean dielectric liquids prepared for pouring into the apparatus, free from water and other impurities, regardless of

Table 3.1 Electrical properties of insulating liquids [3, 4]

Indicator	BMIO	SF	SE	NE
Breakdown voltage, kV	30 ÷ 80	35 ÷ 60	50 ÷ 80	60 ÷ 85
Dielectric loss factor, 50/60 Hz, at 90 °C	<0.001	<0.002	<0.006	<0.040
Dielectric constant, 50/60 Hz, at 100 °C	2.1 ÷ 2.5	2.6 ÷ 2.9	3.0 ÷ 3.3	3.2 ÷ 3.3
Specific volume resistance, Ohm cm, at 23 °C	10^{13}	10^{15}	10^{13}	10^{14}

PBMIO—petroleum based mineral insulating oil
SF—silicone fluid (organosilicon or polydimethylsiloxane)
SE—synthetic ester
NE—natural ester

their chemical composition, have a high breakdown voltage (from 60 kV and higher in a standard test cell), Table 3.1.

The leading position as a liquid dielectric and heat-removing agent in high-voltage equipment continues to be occupied by mineral oil. Several million units of various types of electrical equipment and billions of tons of oil are operated by energy companies around the world.

Mineral petroleum oils remain a priority due to their inherent advantages:

- affordable cost,
- good insulating properties,
- low viscosity,
- the presence of standardized requirements for their properties, both for unused fresh oil and for oil in operation,
- for more than a century of their operation, many valuable methods and criteria for assessing the condition of the insulation system of oil-filled equipment have been developed,
- over the decades, methods for diagnosing electrical equipment, algorithms for restoring insulation characteristics have been developed, in particular, methods for regenerating mineral oil, drying solid insulation, operating in an oil medium, topping up, replacing or mixing petroleum oils of different brands, etc.

3.3.2 Mineral Insulating Oils

Mineral insulating oils are hydrocarbons (compounds of hydrogen and carbon) refined from crude petroleum deposits from the ground. About 10 types of oils are used only in the power engineering of Russia. The chemical composition of fresh (not exploited) oil is determined by the properties of the original oil feedstock and production technology [3–7].

They consist partly of aliphatic compounds with the general formula C_nH_{2n} comprising a mixture of straight- and branched-chain and cyclic or partially cyclic compounds. Many oils also contain a sizable fraction of aromatic compounds related to benzene, naphthalene, and derivatives of these with aliphatic side chains. The ratio of aromatic to aliphatic components depends on the source of the oil and its refining treatment. The percentage of aromatics is of importance to the gas-absorption or evaluation characteristics under electrical discharges and to the oxidation characteristics.

Many manufacturers now approve the use of any of several brands of mineral insulating oil in their apparatus provided that they meet their specifications which are similar to ASTM D1040. Low values of dielectric strength may indicate water or dirt contamination. A high neutralization number will indicate acidity, developed very possibly from oxidation, particularly if the oil has used been already. Presence of sulfur is likely to lead to corrosion of metals in the oil.

The solubility of gases and water in mineral oil is of importance about its function in apparatus. Solubility is proportional to the partial pressure of the gas above the oil where S is the amount dissolved at pressure p if the solubility is expressed as the amount S_0 dissolved at pressure p_0.

The solubility is frequently expressed in volume percent of the oil. Values for solubility of some common gases in transformer oil at atmospheric pressure (760 torr) and 25 °C are air 10.8%, nitrogen 9.0%, oxygen 14.5%, carbon dioxide 99.0%, hydrogen 7%, and methane 30% by volumc. Thc solubilities of all the gases, except CO_2, increase slightly with increasing temperature. Water is dissolved in new transformer oil to the extent of about 60 to 80 ppm at 100% relative humidity and 25 ^{0}C. The amount dissolved is proportional to the relative humidity. Solubility of water increases with oxidation of the oil and the addition of polar impurities, with which the water becomes associated. Larger quantities of water can be suspended in the oil as fine droplets.

3.3.2.1 Dielectric Properties of Mineral Oils

The permittivity of mineral insulating oils is low, since they are essentially nonpolar, containing only a few molecules with electric dipole moments. Some oils possess a minor fraction of polar constituents, which have not been identified. These contribute a dipolar character to the dielectric properties at low temperature and/or high frequency. A typical permittivity for American transformer oil at 60 Hz is 2.19 at 25 °C declining almost linearly to 2.11 at 100 °C. At low temperatures and high frequencies, values of permittivity as high as 2.85 have been noted in oils with a relatively high level of polar constituents. The dc conductivity levels of mineral oils range from about 10^{-15} *Ω per cm* for pure new oils up to 10^{-12} *Ω per cm* for contaminated used oils. This conductivity is due to dissociated impurity ions or ions developed by oil oxidation. It increases approximately exponentially with temperature about 1 decade in 80 °C.

Corona or partial breakdown can occur in mineral oil, as with any liquid or gas, when the electric stress is locally very high and complete breakdown is limited by a solid barrier or large oil gap (as with a needle point in a large gap). Such discharges produce hydrogen and methane gas, and sometimes carbon with larger discharges. Dissolved air is also sometimes released by the discharge. If the gas bubbles formed are not ejected away from the high field, they will reduce the subsequent discharge threshold voltage to as much as 80%. The resistance of insulating oils to partial discharges is measured by two ASTM gassing tests: D2298 (Merrill test) and D2300 (modified Pirelli test). These tests measure the amount of decomposition gas evolved under specified conditions of expo- sure to partial discharges. A minimum amount of gas is, of course, preferred, particularly in applications for cables or capacitors. In fact, conventional mineral oils are inadequate in this respect for application in modern 50 or 60 Hz power capacitor designs.

In power systems (mainly at substations), oil-filled equipment of the following types is used:

1. power transformers;
2. measuring current and potential transformers;
3. shunt reactors;
4. circuit breakers and other switching devices;
5. high-voltage bushings;
6. oil-filled cable lines.

Experts note that a significant proportion of oil-filled equipment is operated at the limit of its capabilities, i.e. over the standard operating life. Along with other parts of the equipment, oil is also subject to aging. Under the influence of electric and magnetic fields (PD, thermal destruction processes, etc.), its initial molecular composition changes. In addition, due to operation, a change in its volume is possible, which, in turn, can be dangerous both for the operation of the equipment and for the maintenance personnel. Therefore, correct and timely oil diagnostics is the key to reliable and safe operation of the equipment.

Oil analysis is carried out to determine not only its quality indicators, which must comply with the requirements of regulatory and technical documentation, but also to assess the condition of the equipment as a whole. This is due to the fact that the oil contains about 70% of information about the condition of the oil-filled equipment.

Table 3.2 shows the most commonly used quality indicators of transformer oil today and the types of violations detected with their help in oil-filled equipment.

Table 3.2 Quality indexes of transformer oil

Oil quality index	Detectable faults
Electrical strength	Interleaf shorts in the stator
Acid number	Locations of stray magnetic fields Evaluation of the efficiency of cooling systems
Content of water-soluble acids	Overheating of contacts of current-carrying tires, working and arcing chambers The state of tank insulation
Flash point in closed cup	Overheating of external and internal contact connections Deterioration of the internal insulation of the windings
The presence of mechanical impurities	Violation of the sealing of elements Breakage of shunt resistances
Oil color	Breakdown of element sections
Oil dielectric loss tangent	Overheating of contact connections
Total gas content in oil	Overheating of contact connections of switches, disconnectors, current transformers, cables, current-carrying tires, etc.
The presence of soluble sludge	Overheating of power cables, assessment of fire hazard of cables

3.3.2.2 Deterioration of Oil

Deterioration of oil in apparatus partially open or "breathing" is subject to air oxidation. This leads to acidity and sludge. There is no correlation between the amount of acid and the likelihood of a sludge or the amount of sludge. Sludge clogs the ducts, reduces the heat transfer, and accelerates the rate of deterioration. ASTM tests for oxidation of oils are D1904, D1934, D1313, and D1314. Copper and lead and certain other metals accelerate the oxidation of mineral oils. Oils are considerably more stable in nitrogen atmospheres.

Inhibitors are now commonly added both to new and to used oils to delay the oxidation. Ditertiary butyl paracresol (DBPC) is the inhibitor most commonly used at present.

3.3.2.3 Servicing, Filtering, and Treating

Oil in service is usually maintained by testing for acidity, dielectric strength, inhibitor content, interfacial tension, neutralization number, peroxide number, pour point, power factor, refractive index and specific optical dispersion, resistivity, saponification, sludge, corrosive sulfur, viscosity, and water content, as outlined in ASTM D117. These properties indicate various types of contamination or deterioration, which might affect the operation of the insulating oil.

Depending on the voltage rating of the apparatus, the oil is maintained above 16 to 22 kV (ASTM test D877). The usual contaminants are water, sludge, acids, and in circuit-breaker oils, carbon. The centrifuge is best suited for removing large quantities of water, heavier solid particles, etc. The blotter filter press is used for the removal of minute quantities of water, fine carbon, etc. In another method, after removing the larger particles, the oil is heated and sprayed into a vacuum chamber, where the water and volatile acids are removed. Sludge and very fine solids are then taken out by a blotter filter press. All units are assembled together so that the process is completed in a single pass. Some work has been done in reclaiming oil by treating it to reduce acidity. One process is similar to the later stages in refining. Another treatment uses activated alumina, Fuller's earth, or silica gel. The IEEE Guide for Maintenance of Insulating Oil is published as *IEEE Standards Publ.*

It has been found that analysis of the dissolved gas in oil or above the oil in oil-insulated transformers and cables is a good diagnostic tool to detect electrical faults, particularly, or deterioration, generally. For example, continuing or intermittent partial discharges produce hydrogen and low- molecular-weight hydrocarbons such as methane, ethane, and ethylene, which accumulate in the oil and can be measured accurately to assess the magnitude of the fault. Higher-current arc faults pro- duce acetylene in addition to H_2 and other low-molecular-weight hydrocarbons. Thermal deterioration of cellulosic or paper insulation is indicated by elevated concentrations of CO and CO_2 in the oil [5–7].

Experts note that a significant proportion of oil-filled equipment is operated at the limit of its capabilities, i.e. over the standard operating life. Along with other parts of the equipment, oil is also subject to aging. Under the influence of electric and magnetic fields (PD, thermal destruction processes, etc.), its initial molecular composition changes. In addition, due to operation, a change in its volume is possible, which, in turn, can be dangerous both for the operation of the equipment and for the maintenance personnel. Therefore, correct and timely oil diagnostics is the key to reliable and safe operation of the equipment [7–10].

Oil analysis is carried out to determine not only its quality indicators, which must comply with the requirements of regulatory and technical documentation, but also to assess the condition of the equipment as a whole. This is due to the fact that the oil contains about 70% of information about the condition of the oil-filled equipment. Under the pressure of tightening environmental regulations, manufacturers of high-voltage equipment with liquid insulation are forced to work not only to improve the culture of handling transformer oil, but also to find new insulating liquids that cause minimal environmental damage. The development of high-voltage apparatus building is closely related, among other things, to progress in the production of insulating materials, including insulating liquids based on esters.

Considering the specific properties of ester fluids (high viscosity and density, increased tendency to accumulate electrostatic charges, faster propagation of discharge channels in a non-uniform electric field compared to petroleum-based oils, etc.), energy companies are very cautious about using alternative dielectrics in electrical equipment. To date, synthetic and natural ethers are mainly used in electrical equipment installed in places where there are increased requirements for fire safety and environmental protection (densely populated areas, tunnels, subways, oil and gas platforms, river and marine fleets, underground substations, ecological zones and etc.), Table 3.3. The scope of biodegradable liquid dielectrics in high-voltage equipment is gradually expanding. Progress in the field of biogenic technologies allows us to successfully solve this problem. The use of biodegradable ester fluids in high-voltage equipment, in contrast to petroleum-based oils, fits this trend very well.

Transformers with *synthetic esters* as liquid insulation are mainly operated by European companies, while transformers with *natural esters* are used in the USA

Table 3.3 Use of insulating liquids in electrical equipment [3–6]

Type of electrical equipment	Mineral oil	Silicone fluid	Natural ether	Synthetic ether
Power transformers	++	-	+	+
Traction transformers	++	++	-	++
Distribution transformers	++	++	++	++
Current transformers	++	-	-	-
Voltage transformers	++	-	-	-
High voltage bushings	++	-	-	-
Regulation devices under voltage	++	++	-	+
High voltage circuit breakers	++	-	-	-

Note "++"—often used; "+"—used less often; "-"—not applicable and Brazil. In 1978, small power transformers filled with synthetic esters were tested. In 1992, mass production of new transformers began in Germany and the filling of old transformers with ethereal synthetic fluid began [4].

Unlike synthetic esters, natural ester fluids came into use much later, around the 1990s. In terms of the scope of application, they are still significantly inferior to oil. Basically, these are sealed transformers of small and medium power (with film protection, nitrogen cushion in the oil space or completely sealed). To date, according to expert estimates, the number of transformers and reactors filled with natural ethers is more than 2.5 million units. Data based on several years of operation of such transformers indicates that natural ester fluids can replace transformer oil in some types of high voltage equipment.

Due to the difference in physical, electrical, chemical and thermal properties of ester fluids from mineral oil, for their effective use in power transformers, it is necessary to modify some structural parts of the equipment.

3.3.3 Synthetic Liquid Insulation

Synthetic chlorinated diphenyl and chlorinated benzene liquids (askarels) have been used widely from the middle1930s up to the middle1970s and are still in service in many power capacitors and transformers, where they were adopted for their non-flammability as well as good electrical characteristics. Since the mid-1970s, their use has been banned in most countries due to their alleged toxicity and resistance to biodegradation in the environment. Now, when apparatus containing these fluids, which are commonly referred to as *PCBs*, are taken out of service, environmental regulations in the United States require that the fluid not be released into the environment. Waste fluid should be incinerated at high temperature with HCl reactive absorbent scrubbers in the stack, since this acid gas is a product of the combustion.

New synthetic fluids have been developed and are now widely applied in power capacitors where the electrical stresses are very high. These fluids include aromatic (containing benzene rings) hydro- carbons, some of which have excellent resistance to partial discharges. They are not fire-resistant, however.

Very high boiling, low-vapor-pressure, high-flash-point (about 300 °C) hydrocarbon oils are being tried for power transformers with some fire resistance. Methods for assessing the risk of fire with such liquids, as well as with silicones, are still being debated.

Perchlorethylene (tetrachloroethylene), a nonpolar liquid, is now in use in sealed medium-power transformers, where nonflammability is required. With a boiling point at atmospheric pressure of 121 °C, this fluid is completely nonflammable. It is also widely used in dry cleaning. Other important classes of synthetic insulating fluids are discussed in the following sections.

Fluorocarbon Liquids. A number of nonpolar nonflammable perfluorinated aliphatic compounds, in which the hydrogen has been completely replaced by fluorine, are available with different ranges of viscosity and boiling point from below room temperature to more than 200 °C. These compounds have low permittivities (near 2.0) and very low conductivity. They are inert chemically and have low solubilities for most other materials. The chemical formula for these compounds is one of the following: C_nF_{2n}, C_nF_{2n+2}, or $C_nF_{2n}O$. The presence of the oxygen in the latter formula does not seem to reduce the stability. These compounds have been used for filling electronic apparatus and large transformers to give high heat-transfer rates together with high dielectric strength. The vapors of these liquids also have high dielectric strengths.

Silicone Fluids. These fluids, chemically formed from Si—O chains with organic (usually methyl) side groups, have a high thermal stability, low temperature coefficient of viscosity, low dielectric losses, and high dielectric strength. They can be obtained with various levels of viscosity and correlated vapor pressures. Rated service temperatures extend from –65 to 200 °C, some having short-time capability up to 300 °C. Their permittivity is about 2.6–2.7, declining with increasing temperature. These fluids have a tendency to form heavier carbon tracks than other insulating liquids when breakdown occurs. They cannot be considered fireproof but will reduce the risk of fire due to their low vapor pressure.

Ester Fluids. There are a few applications, mostly for capacitors, where organic ester compounds are used [4]. These liquids have a somewhat higher permittivity, in the range of about 4 to 7, depending on the ratio of ester groups to hydrocarbon chain lengths. Their conductivities are generally somewhat higher than those of the other insulating liquids discussed here. The compounds are easily subject to hydrolysis with water to form acids and alcohols and should be kept dry, particularly if the temperature is raised. Their thermal stability is poor. Specifically, dibutyl sebacate has been used in high-frequency capacitors and castor oil in energy-storage capacitors.

3.4 Insulating Gases and Vacuum

3.4.1 General Properties of Gases

A gas is a highly compressible dielectric medium, usually of low conductivity and with a dielectric constant only a little greater than unity, except at high pressures. In high electric fields, the gas may become conducting because of impact ionization of the gas molecules by electrons accelerated by the field and by secondary processes, which produce partial breakdown / partial discharge (PD) or complete breakdown. Conditions which ionize the gas molecules, such as very high temperatures and ionizing radiation (ultraviolet rays, x-rays, gamma rays, high-velocity electrons, and ions such as alpha particles), will also produce some conduction in a gas.

The gas density d (grams per liter) increases with pressure p (torrs or millimeters of mercury) and gram-molecular weight M and decreases inversely with the absolute temperature T (degrees Celsius—273).

The use of air as *internal insulation* at pressures of 10–15 atm and above complicates the design of apparatuses, their inputs and seals and reduces the safety of maintenance and makes operation difficult. Therefore, for these purposes, gases with increased electrical strength are used much more often, which makes it possible to avoid the problems mentioned above [10–12].

The most important requirements for gases used as electrical insulation are high electrical strength and low liquefaction temperature, Table 3.4.

Gases must also be chemically inert, non-toxic, fire and explosion-proof, inexpensive, have sufficient thermal conductivity, and also be chemically resistant when exposed to an electric field and discharges. As follows from Table 3.4, SF_6 meets all the requirements most fully. In its pure form, it is inert, non-flammable, heat-resistant up to 800 °C, non-toxic, odorless. SF_6 molecules have a high electron affinity, which determines its high dielectric strength and good arc-quenching ability. It is applied in transformers, switches, capacitors, cables and some other high-voltage equipment.

In high-voltage structures nitrogen, carbon dioxide, sulfur hexafluoride (SF_6) and its mixture with nitrogen are mainly used as internal gas insulation. Their main characteristics are given in Table 3.5.

Nitrogen and carbon dioxide do not exceed atmospheric air in electrical strength, but they are chemically inert and therefore, in their environment, oxidative processes of thermal aging of structural elements made of solid materials are excluded or sharply slowed down. They, like SF_6, are fire and explosion-proof.

Table 3.4 The most important characteristics of gases used as electrical insulation or promising for such use

Gas	Air	N_2	SF_6	CCl_2F_2	CO_2	CCl_4	SeF_6
$U^G{}_{Br}/U^A{}_{Br}$	1.0	1.0	2.3–2.5	2.4–2.6	0.95	6.3	4.5
$T_{сж}$, °C	−192	−195.8	−62	−28	−34.6	76	49.2

Here $U^G{}_{Br}/U^A{}_{Br}$ is—the ratio of the electrical strength of a given gas and air, T_{liq}, °C—the temperature of liquefaction of air and gases

Table 3.5 Physical properties of gases used in high voltage equipment

Parameters	Nitrogen	Carbon dioxide	Sulfur hexafluoride
Molecular mass	28.013	44.011	146.05
Gas density, Kg/m^3, at 0 °C and pressure 0.1 MPa	1.25	1.98	6.39
Boiling point, K, at a pressure of 0.1 MPa	77.4	194.7	318.7
Heat capacity of gas at a pressure of 0.1 MPa and 0° C, kJ/kg.K	1.06	0.82	0.625
Thermal conductivity of gas at a pressure of 0.1 MPa and 0 °C, W/(m.K)	0.024	0.015	0.0127
Electrical strength of gas (relative to air strength)	1.0	0.9	2.6

The main dielectric "material" for creating the ***external insulation*** of high-voltage equipment is atmospheric air. Insulated current-carrying parts (buses of distribution devices, wires of overhead power transmission lines, external current-carrying parts of electrical apparatus) are located at certain distances from each other and from the ground (or grounded parts of the apparatus) and are strengthened in a given position with the help of insulators. Insulating distances in air are determined by the magnitude of the acting voltage and the dielectric strength of the air [13].

Under normal atmospheric conditions, the electric strength of air is relatively small and in intervals longer than 1 cm with a uniform field does not exceed 30 kV/cm. The electric strength of air in sharply inhomogeneous fields, which are predominant for high-voltage equipment, and at distances of 1–2 m is only 5 kV/cm, and at distances of 10–20 m it decreases, respectively, to 2.5–1.5 kV/cm. In this regard, the dimensions of the overhead transmission line and switchgears are largely determined by the electrical strength of the air and increase rapidly with an increase in the rated operating voltage. The discharge (breakdown) voltages of air gaps are influenced by pressure p, temperature T and absolute humidity Ɣ of the air, so the insulating distances through the air are chosen so that they have sufficient dielectric strength under adverse weather conditions. In particular, conventional electrical equipment is designed to operate at altitudes up to 1000 m above sea level and at ambient temperatures up to +40 °C. It should be borne in mind that every 100 m rise above sea level reduces the discharge voltage by about 1%. The same decrease gives an increase in temperature for every 3 °C above normal. As a normal temperature, 20 °C is taken, normal pressure (at sea level)—760 mm Hg. Art., normal humidity $Ɣ_0 = 11$ g/m^3. The given data characterizing the change in discharge voltages under the influence of atmospheric conditions refer to inter-electrode distances up to 1 m. At large distances, the effect of atmospheric conditions decreases as the distance increases. Rain has practically no effect on the discharge voltages of gaps with a non-uniform field.

If the gas is a vapor in equilibrium with a liquid or solid, the pressure will be the vapor pressure of the liquid or solid. The logarithm of the pressure varies as $-\Delta H/RT$, where ΔH is the heat of vaporization in calories per mole and R is the molar gas constant, 1.98 cal/(mol) (°C). This relation also applies to all common atmospheric gases at low temperatures, below the points where they liquefy.

Dielectric Constant. The dielectric constant k of gases is a function of the molecular electrical polarizability and the gas density. It is independent of magnetic and electric fields except when a significant number of ions is present.

Conduction. The conductivity of a pure molecular gas at moderate electric stress and moderate temperature can be assumed, in the absence of any ionizing effect such as ionizing radiation, to be practically zero. Ionizing radiation induces conduction in the gas to a significant extent, depending on the amount absorbed and the volume of gas under stress. The energy of the radiation must exceed, directly or indirectly, the ionization energy of the gas molecules and thus produce an ion pair (usually an electron and positive ion). The threshold ionization energy is on the order of 10–25 electron-volts (eV)/molecule for common gases (10.86 eV for methyl alcohol, 12.2 for oxygen, 15.5 for nitrogen, and 24.5 for helium). Only very short wavelength ultraviolet light is effective directly in photoionization, since 10 eV corresponds to a photon of ultraviolet with a wavelength of 1240 A. Since the photoelectric work function of metal surfaces is much lower (2–6 eV; e.g., copper about 4 eV), the longer-wavelength ultraviolet commonly present is effective in ejecting electrons from a negative conductors surface. Such cathode-ejected electrons give the gas apparent conductivity [14].

High-energy radiation from nuclear disintegration is a common source of ionization in gases. Nuclear sources usually produce gamma rays on the order of 10^6 eV energy. Only a small amount is absorbed in passing through a low- density gas. A flux of 1 R/h produces ion pairs corresponding to a saturation current (the flat part of the curve V_1-V_2 in the Fig. 3.7) of 0.925×10^{-13} A/cm^3 of air at 1 atm. pressure if all the ions formed are collected at the electrodes. The effect is proportional to the flux and the gas density. The part of curve V_2-V_3 corresponds to end of saturation part the beginning of the electron growth, so as electric field is strong enough for this. Rapid increasing of Volt-Ampere characteristic from the V_4 implies a shift to strong field area. The penetration to this region of Volt-Ampere characteristic means extremely fast growth of current carriers and next breakdown in any case. At a voltage stress below about 100 V/cm, some of the ions formed will recombine before being collected, and the current will be correspondingly less. Higher stresses do not increase the current if all the ions formed are collected. A very small current, on the order of 10^{-21} A/cm^3 of air, is attributable to cosmic rays and residual natural radioactivity. Electrons (beta rays) produce much more ionization per path length than gamma rays, because they are slowed down by collisions and lose their energy more quickly.

Correspondingly, the slower alpha particles (positive helium nuclei) produce a very dense ionization in air over a short range. For example, a 3-million-eV (MeV)

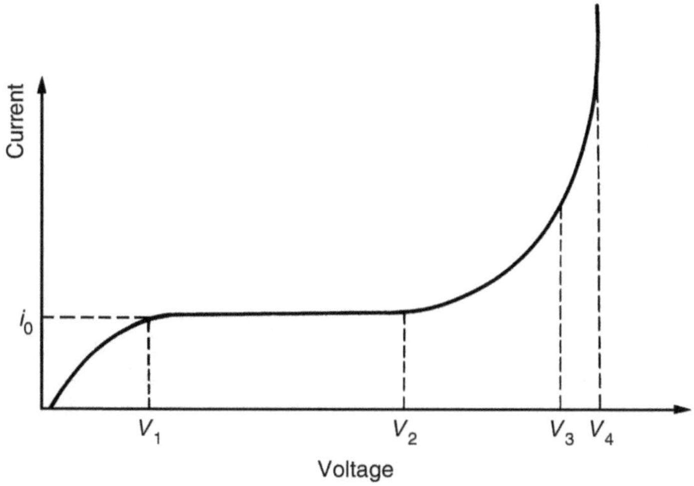

Fig. 3.7 Volt-Ampere characteristic

alpha particle has a range in air of 1.7 cm and creates a total of 6.8×10^5 ion pairs. A beta particle (an electron) of the same energy creates only 40 ion pairs per centimeter and has a range of 13 m in air.

It should be noted that ionizing radiation of significant levels has only a small effect on gas dielectric strength. For example, the ionization current produced by a corona discharge from a needle point is typically much higher than that produced by a radiation flux of significant level, 10^{11} gamma photons per square centimeter.

At temperatures increasing above 600 °C, it has been shown that thermionic electron emissions from negative conductor surfaces produce significant currents compared with levels typical of electrical insulation.

Since the rate of production of ions by the various sources mentioned above is limited, the current in the gas does not follow Ohm's law, unless the rate of collection of the ions at the electrodes is small compared with the rate of production of these ions.

3.4.2 Breakdown of Gases

Uniform Fields. The dielectric breakdown of gases is a result of an exponential multiplication of free electrons induced by the field. It is generally assumed that the initiation of breakdown requires only one electron. However, if only a few electrons are present prior to breakdown, it is not easily possible to measure the trend of current shown in Fig. 3.7. If the breakdown is completed between metal electrodes, the spark develops extremely rapidly into an arc, involving copious emission of electrons from the cathode metal and, if the necessary current flow is permitted, vaporization of metal from the electrodes.

In uniform electric fields, breakdown occurs at a critical voltage which is a function of the product of the pressure p and spacing d (Paschen's law), Fig. 3.8.

It would be more accurate to consider the gas density-spacing product, since the dielectric strength varies with the temperature only as the latter affects the gas density. It will be noted that the electric field at breakdown decreases as the spacing

Fig. 3.8 Paschen's curve

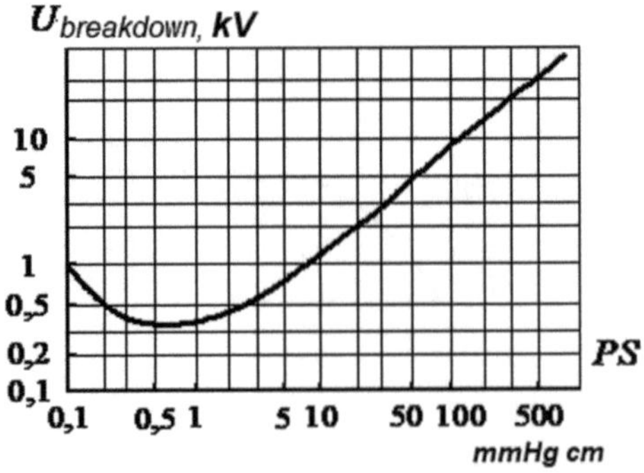

increases. This is typical of all gases and is due to the fact that a minimum amount of multiplication of electrons must occur before breakdown occurs. A single electron accelerated by the field creates an avalanche which grows exponentially as x.eα, where x is the distance and α is the Townsend ionization coefficient (electrons formed by collision per centimeter), which increases rapidly with electric field. At small spacing, α and the field must be higher for sufficient multiplication. In divergent electric fields or large spacing, it has been found that when the integral increases to about 18.4 (10^8 electrons), sufficient space charge develops to produce a streamer type of breakdown. It seems to be apparent that the final step in gas breakdown before arc development is the development of a branched filamentary streamer which proceeds more easily from the positive electrode toward the negative electrode.

The relative dielectric strength, with few exceptions, tends upward with increasing molecular weight. There are a number of factors other than molecular or atomic size which influence the retarding effect on electrons. These include ability to absorb electron energy on collision and trap electrons to form negative ions. The noble atomic gases (helium, argon, neon, etc.) are poorest in these respects and have the lowest dielectric strengths.

The relative strengths vary with the *p. d* value, as well as gap geometry, and particularly in divergent fields where corona begins before breakdown. It is best to consult specific references with regard to divergent field breakdown values.

In nonuniform fields, when the ratio of spacing to conductor radius of curvature is about 3 or less, breakdown occurs without prior corona. The breakdown voltage is controlled by the integral of the Townsend ionization coefficient α across the gap. At larger ratios of spacing to radius of curvature, corona discharge occurs at voltage levels below complete gap breakdown.

Corona in air at atmospheric pressure occurs before breakdown when the ratio of outer to inner radius of coaxial electrodes exceeds 2.72 or where the ratio of gap to sphere radius between spheres exceeds 2.04. These discharges project some distance from the small-radii conductor but do not continue out into the weaker electric field region until a higher voltage level is reached. Such partial breakdowns are often characterized by rapid pulses of current and radio noise. With some conductors at intermediate voltages between onset and complete breakdown, they blend into a pulseless glow discharge around the conductor. When corona occurs before breakdown, it creates an ion space charge around the conductor, which modifies the electric field, reducing the stress at sharp conductor points in the intermediate voltage range. At higher voltages, streamers break out of the space-charge region and cross the gap. The surface voltage stress at which corona begins increases above that for uniform field break- down stress, since the field to initiate breakdown must extend over a finite distance. An empirical relation developed by Peek is useful for expressing the maximum surface stress for corona onset in air for several geometries of radius *r*.

It has been shown by a number of investigators that the discharge-threshold voltage stresses on or between insulator surfaces is the same as between metal electrodes. Thus, the threshold voltage for such discharges can be calculated from the series dielectric capacitance relation for internal gaps of simple shapes, such as plane and coaxial gaps, insulated conductor surfaces, and hollow spherical cavities.

The corona-initiating voltage at a conductor edge on a solid barrier depends on the electric stress concentration and generally on the ratio of the barrier thickness to its dielectric constant, except with low surface resistance. Any absorbed water or conducting film raises the corona threshold voltage by reducing stress concentration at the conductor edge on the surface.

It is sometimes possible to overvoltage such gaps considerably prior to the first discharge, and the offset voltage may be below the proper voltage due to surface-charge concentration. With ac voltages, pulse discharges occur regularly back and forth each half cycle, but with dc voltage, the first discharge deposits a surface charge on the insulator surface which must leak away before another discharge can occur. Thus, corona on or between insulator surfaces is very intermittent with steady dc voltages, but discharges occur when the voltage is raised or lowered.

As has been mentioned in the previous section on partial discharges, the breakdown in gases is influenced by the presence of solid insulation between conductors. This insulation increases the electric stress in the gas. A particular case of this is the complete breakdown between conductors across or around solid insulator surfaces. This can occur when the conductors are on the same side of the insulation or on opposite sides. A significant reduction in flashover voltage can occur whenever a significant part of the electric field passes through the insulation. The reduction is influenced by the percentage of electric flux which passes through the solid insulation and the dielectric constant of the insulation.

3.4.3 Relative Dielectric Strengths of Gases

The relative dielectric strength, with few exceptions, tends upward with increasing molecular weight. There are a number of factors other than molecular or atomic size which influence the retarding effect on electrons. These include ability to absorb electron energy on collision and trap electrons to form negative ions. The noble atomic gases (helium, argon, neon, etc.) are poorest in these respects and have the lowest dielectric strengths.

The relative strengths vary with the $p.\ d$ value, as well as gap geometry, and particularly in divergent fields where corona begins before breakdown. It is best to consult specific references with regard to divergent field breakdown values.

3.4.4 Corona and Breakdown in Inhomogeneous Fields

In non-uniform fields, when the ratio of spacing to conductor radius of curvature is about 3 or less, breakdown occurs without prior corona. The breakdown voltage is controlled by the integral of the Townsend ionization coefficient α across the gap. At larger ratios of spacing to radius of curvature, corona discharge occurs at voltage levels below complete gap breakdown.

Corona in air at atmospheric pressure occurs before breakdown when the ratio of outer to inner radius of coaxial electrodes exceeds 2.72 or where the ratio of gap to sphere radius between spheres exceeds 2.04. These discharges project some distance from the small-radii conductor but do not continue out into the weaker electric field region until a higher voltage level is reached.

Such partial breakdowns arc often characterized by rapid pulses of current and radio noise. With some conductors at intermediate voltages between onset and complete breakdown, they blend into a pulseless glow discharge around the conductor. When corona occurs before breakdown, it creates an ion space charge around the conductor, which modifies the electric field, reducing the stress at sharp conductor points in the intermediate voltage range. At higher voltages, streamers break out of the space-charge region and cross the gap.

Figure 3.9 illustrates the case of a strongly non–uniform field in a positive point—plane gap. For the special case of a coaxial cylindrical geometry in air, an empirical relation based on many measurements of the critical field strength *Ec* (corona inception) for different diameters of the inner conductor *(2r)* and relative air density δ was developed by Peek of the form:

$$\frac{E_C}{\delta} = 31.53 + \frac{9.63}{\sqrt{\delta r}},$$

where E_C is in *kV/cm*, *r* in *cm*.

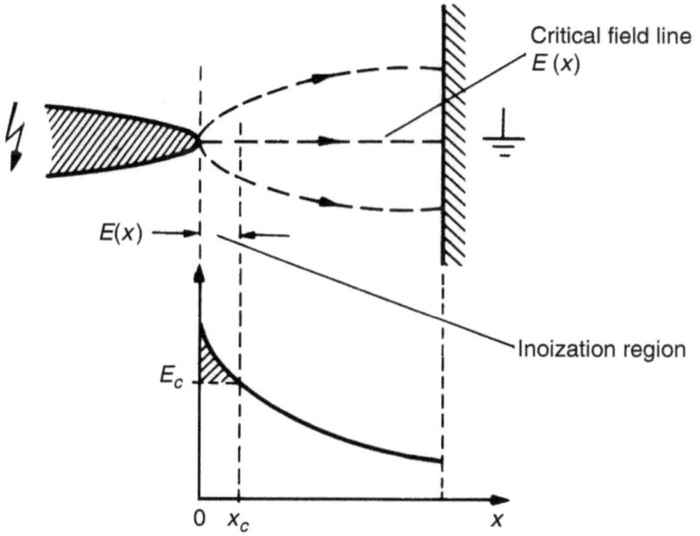

Fig. 3.9 Electric field distribution in a non-uniform field gap

In uniform field and quasi–uniform field gaps the onset of measurable ionization usually leads to complete breakdown of the gap. In non-uniform fields various manifestations of luminous and audible discharges are observed long before the complete breakdown occurs. These discharges may be transient or steady state and are known as ***corona***. The phenomenon is of particular importance in high voltage engineering where non–uniform fields are unavoidable. It is responsible for considerable power losses from high voltage transmission lines and often leads to deterioration of insulation by the combined action of the discharge ions bombarding the surface and the action of chemical compounds that are formed by the discharge. It may give rise to interference in communication systems. On the other hand, it has various industrial applications such as high–speed printing devices, electrostatic precipitators, paint sprayers, Geiger counters, etc.

The voltage gradient at the surface of the conductor in air required to produce a visual a.c. corona in air is given approximately by the Peek's expression. There is a distinct difference in the visual appearance of a corona at wires under different polarity of the applied voltage. Under positive voltage, a corona appears in the form of a uniform bluish–white sheath over the entire surface of the wire. On negative wires the corona appears as reddish glowing spots distributed along the wire. The number of spots increases with the current. Stroboscopic studies show that with alternating voltages a corona has about the same appearance as with direct voltages. Because of the distinctly different properties of coronas under the different voltage polarities it is convenient to discuss separately positive and negative coronas.

3.4.5 Corona Discharges on Insulator Surfaces

It has been shown by a number of investigators that the discharge-threshold voltage stresses on or between insulator surfaces is the same as between metal electrodes. Thus, the threshold voltage for such discharges can be calculated from the series dielectric- capacitance relation for internal gaps of simple shapes, such as plane and coaxial gaps, insulated conductor surfaces, and hollow spherical cavities.

The corona-initiating voltage at a conductor edge on a solid barrier depends on the electric stress concentration and generally on the ratio of the barrier thickness to its dielectric constant, except with low surface resistance. Any absorbed water or conducting film raises the corona threshold voltage by reducing stress concentration at the conductor edge on the surface.

It is sometimes possible to overvoltage such gaps considerably prior to the first discharge, and the offset voltage may be below the proper voltage due to surface-charge concentration. With ac volt- ages, pulse discharges occur regularly back and forth each half cycle, but with dc voltage, the first discharge deposits a surface charge on the insulator surface which must leak away before another discharge can occur. Thus, corona on or between insulator surfaces is very intermittent with steady dc voltages, but discharges occur when the voltage is raised or lowered.

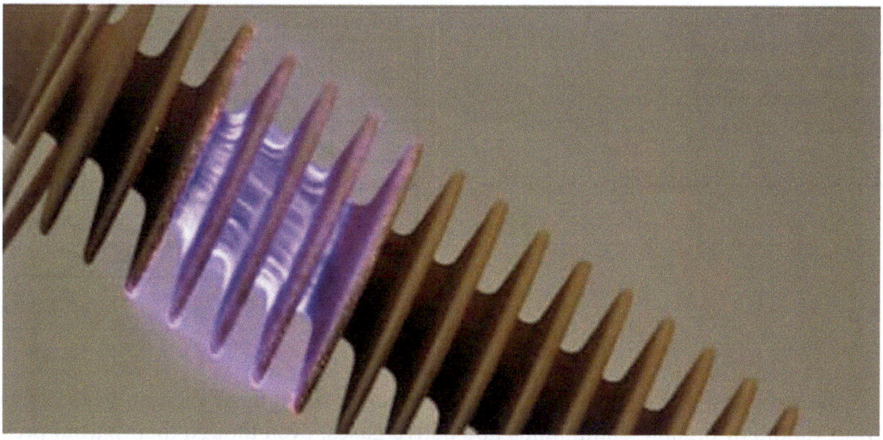

Fig. 3.10 Corona discharge over the surface of an insulator

Figure 3.10 shows a corona discharge over the surface of an insulator in atmospheric air at AC voltage.

3.4.6 Flashover on Solid Surfaces in Gases

As has been mentioned in the previous section on partial discharges, the breakdown in gases is influenced by the presence of solid insulation between conductors. This insulation increases the electric stress in the gas. A particular case of this is the complete breakdown between conductors across or around solid insulator surfaces. This can occur when the conductors are on the same side of the insulation or on opposite sides. A significant reduction in flashover voltage can occur whenever a significant part of the electric field passes through the insulation. The reduction is influenced by the percentage of electric flux which passes through the solid insulation and the dielectric constant of the insulation.

3.4.7 Vacuum

When going along the left branch of the Paschen curve toward low pressure of about 10^{-3} Pa and then to vacuum, the breakdown voltage first increases and then remains essentially unchanged. The pressure decrease is accompanied by a simultaneous decrease in the breakdown voltage spread. This is typical of fairly short gaps (less then 2 cm). Starting with p about 1–10^{-1} Pa or less, the behavior of $U = f(p)$ depends on many factors including the interelectrode gap length, residual gases, rate of increase in voltage across the gap (the voltage duration, and so on.

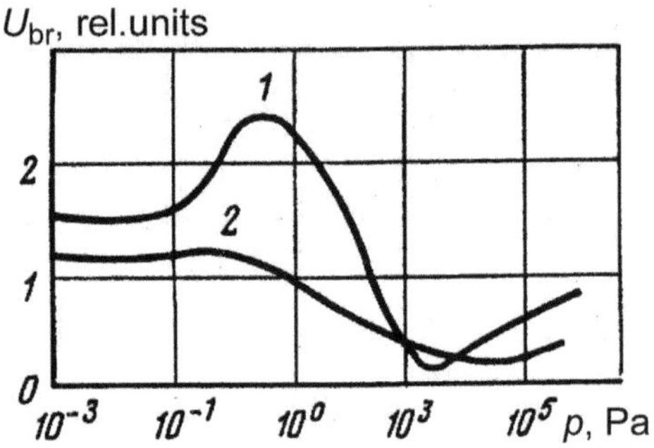

Fig. 3.11 Dependence of breakdown voltage on residual gas pressure for direct current, with d > 2 cm (1) and d < 2 cm (2)

The strongest negative influence on the electric strength of vacuum gaps comes from vapors of organic compounds and active gases, even though their partial pressure are much lower than the pressure of other residual gases and do not exceed 0.01–0.1 μPa. Impurities and active gases especially significantly reduce the long-term withstand voltage in the presence of dark currents, which facilitate cracking of hydrocarbon vapor at the surface and the formation of carbon-bearing film. For voltage pulses, their influence is much weaker.

For large interelectrode gap lengths at 10^{-1}–1 Pa, the electric strength for dc voltage and pulses of tens of microseconds or longer is greater than for higher vacuum, Fig. 3.11.

This excess can vary from several percent to double the electric strength as a function of the degree of evacuation, gap parameters, applied voltage, and electrodes material. For voltage pulses shorter than 1 μs, the residual gas pressure has essentially no effect on the breakdown voltage. Much attention has been paid to investigations into the dynamics of gas layer formation near electrodes and surface of design elements, and its effect on the electric strength of vacuum-insulated systems.

The electrical strength of the vacuum gap is largely affected by the material of the anode, primarily its refractoriness. For example, the electrical strength increases when the material of the electrodes is replaced in the following sequence: copper, iron, tungsten. High purity metals or alloys must be used in vacuum circuit breakers. As the distance between the electrodes increases, the average breakdown strength decreases, as in the case of other dielectric materials and media, due to the distortion of the macroscopic field and the field near microinhomogeneities on the surface of the electrodes.

In installations with vacuum insulation, as well as with gas insulation, the electrical strength of the gap is very often determined by the discharge voltage on the surface of solid insulators, which are used to fasten various components of the installation. To increase and stabilize the discharge voltage on the surface of a solid dielectric, a training is carried out, which is an exposure of the gap under voltage. Vacuum insulation is mainly used in electrophysical equipment (accelerators, space engines, electrostatic separators, etc.). In electric power systems, it is used in 20–50 kV capacitors, switches, vacuum arresters and relays. The use of vacuum insulation in switches is of interest due to the rapid restoration of the electrical strength of the gap after a breakdown (10^{-3}–10^{-4} s); the use of vacuum insulation in spark relays makes it possible to obtain good timing—the instability of the operation time is less than 10 ns. The disadvantage of vacuum insulation is the design and technological difficulties of retention in a high vacuum apparatus.

Questions and Tasks

1. What are the three main groups of causes of premature failure of high-voltage equipment.
2. List the factors that affect the equipment and cause it to age and fail.
3. Which kinds of circumstances could cause equipment failure?
4. Formulate requirements to high voltage insulation.
5. Which are the most important characteristics of insulating materials and media?
6. Which are the main requirements of the insulating materials used for power apparatus?
7. Which are the characteristic of organic insulating materials?
8. Which are the main insulating properties of polythene?
9. Call the reasons, which make a polythene main prospect insulation of high voltage cables?
10. Which are main features of epoxy resins?
11. Explain, which are good quality properties of mica?
12. Name the main parameters of transformer oil and their numerical values.
13. What are the main barriers to the wider use of ester fluids as high voltage insulating media?
14. Explain, general properties of vacuum as isolation?
15. Formulate main law, which is responsible for electrical strength of gas insulation.

References

1. Rafiq M, Shafique M, Azam A, Ateeq M, Khan IA, Hussain A (2020) Sustainable, renewable and environmental-friendly insulation systems for high voltages applications. Molecules 25(3901):1–43
2. Fofana I (2013) 50 years in the development to insulating liquids. IEEE Electr Insul Mag 29(5):13–25
3. CIRGE Brochure 436 (2010) Experiences in service with new insulating liquids. Working Group A2.35, p 95
4. Lyutikova M, Korobeinikov S, Vohan Rao U, Fofana I (2022) Mixed insulating liquids with mineral oil for high voltage transformer applications: a review. IEEE Trans Dielectr Insul 29(2):1–10. Paper number TDEI – 9938-2022

5. Ushakov VYa, Klimkin VF, Korobeynikov SM (2007) Impulse breakdown of liquids (ed: Ushakov VYa). Springer-Verlag, Berlin Heidrlberg, p 397
6. Ushakov VYa (2004) Insulation of high-voltage equipment. Springer, Verlag Berlin Heidelberg, p 422
7. Beyer M, Boeck W, Moller K, Zaengl W (1986) Hochspannungstechnik. Springer Verlag, Theoretische und Praktische Grundlagen
8. Bogatenkov IM, Bocharov YN, Gumerova NI, et al (2003) High voltage engineering: textbook for universities (ed: Kuchinsky GS). St. Petersburg: Energoatomizdat. St. Petersburg Branch, p 608 (in Russian)
9. Ushakov VYA (1988) Electrical aging and resource of monolithic polymer insulation. Energoatomizdat, Moscow, p 152 (in Russian)
10. Warne DF (2005) Newnes electrical power engineers's handbook. 2nd edn. Linarce House, Burlington, Amsterdam, p 477
11. Fink DG, Beaty HW (2006) Standard handbook for electrical engineers, 15th edn. McGraw-Hill
12. Wadhwa CL (2007) High voltage engineering, 2nd edn. New Age International Publishers, p 312
13. Arakelian VG, Fofana I (2007) Water in oil-filled high-voltage equipment part I: states, solubility and equilibrium in insulating materials. IEEE Electr Insul Mag 6(4): 15–27
14. Heathcote MJ (2007) Basic materials. In: J and P transformer book—a practical technology of the power transformer, 13th Edn. Elsevier

Chapter 4
Effects on Equipment Causing Insulation Aging and Failure

Abstract This chapter considers the influence on high-voltage equipment (primarily on its insulation) of the main factors that cause its aging and failure. All factors are divided into two groups. The first group includes factors that are constantly acting or easily predictable, the second—those that can be attributed to the consequences of force majeure. The materials of this chapter will allow students to understand what factors and what consequences of their impact are objects of control and diagnostics of power equipment, as well as measures that need to be taken to prevent them or reduce their negative consequences.

4.1 Effects of Electrical Voltage

Numerous causes of premature failure of high-voltage equipment can be combined (rather conditionally) into three groups:

1. natural aging of equipment elements, primarily insulating materials and structures, due to the influence of a number of predictable (expected) factors (4.1–4.6),
2. violation of the normal operation of the equipment or its complete failure under the influence of force majeure (unpredictable or difficult to predict) factors (4.7),
3. violation of technological standards by the manufacturer, causing the release of products with defects that were not detected in the process of acceptance tests (see Chapter 5).

It will not be a big exaggeration to say that insulating materials and media play a crucial role in ensuring the reliability of high-voltage equipment and determine its technical and economic performance. Ensuring the requirements for insulation predetermines the concept of its implementation, the design and shape of a particular insulating device, as well as the substance used. When performing (implementing) technical isolation, it is rarely possible to find an optimal solution from all points of view. Therefore, it is often necessary to make trade-offs that ensure the correct operation and the required lifetime of the equipment with the most economical solution.

V. Y. Ushakov et al., *High-Voltage Equipment of Power Systems*, Power Systems,
https://doi.org/10.1007/978-3-031-38252-9_4

Electrical insulation is a medium or a material which, when placed between conductors at different potentials, permits only a small or negligible current in phase with the applied voltage to flow through it. The term *dielectric* is almost synonymous with electrical insulation, which can be considered the applied dielectric. A perfect dielectric passes no conduction current but only capacitive charging current between conductors. Only a vacuum at low stresses between uncontaminated metal surfaces satisfies this condition.

The range of resistivity of substances which can be considered insulators is from greater than 10^{20} Ω*cm downward to the vicinity of 10^{6} Ω*cm, depending on the application and voltage stress. There is no sharp boundary defined between low-resistance insulators and semiconductors. If the voltage stress is low and there is little concern about the level of current flow (other than that which would heat and destroy the insulation), relatively low-resistance insulation can be tolerated.

The ability of fluid insulating materials (liquids and gases) to remove heat from current-carrying parts to the environment is very important for the normal operation of many types of high-voltage equipment [1].

In applying an insulating material, it is necessary to consider not only the electrical requirements but also the mechanical and environmental conditions of the application. Mechanical failure often leads to electrical failure, and mechanical failure is frequently the primary cause for failure of an aged insulation.

The initial properties of insulation are frequently more than adequate for the application, but the effects of aging and environment may degrade the insulation rapidly to the point of failure. Thus, the thermal and environmental stability should be considered of equal importance. The effects of moisture and surface dirt contamination should be particularly considered, if these are likely to occur.

While describing the dielectric and other properties of various insulating materials, the following paragraphs also mention their use in various electrical devices. However, the reverse process is also desirable, that is, which insulating materials are used for a particular apparatus, depending on its nominal characteristics and the environmental conditions in which the apparatus must operate. In this aspect, the problem of isolation was briefly discussed in Chapter 3.

The operating conditions of electrical insulation are determined primarily by the acting voltages, characterized by amplitude, duration, shape. There are three groups of acting voltages: (1) operating voltage, (2) internal (switching) surges, (3) atmospheric (lightning) surges. (Characteristics of internal and atmospheric surges will be brief discussed in Sect. 12.2).

The operating voltage acts on the insulation for the entire lifetime of the high voltage structure (typically 20–30 years). The value of this voltage is set by regulatory documents (in the USSR and Russia—by the corresponding GOSTs) in the form of a rated voltage of the electrical network and may differ from the latter, depending on the transmission mode, up or down. This deviation decreases as the voltage class increases and is 20% for the equipment class 3–10 kV, 15% for 15–220 kV, 10% for 330 kV, 5% for 500–750 kV.

At the same time, processes that worsen its electrical and mechanical characteristics should not be observed in the insulation of new (not working) equipment.

In case of violation of the normal mode of operation of power transmission lines (a sharp increase in voltage), as well as a deterioration in the properties of insulation, due to changes in environmental conditions (especially important for the insulation of overhead transmission lines and open substations), undesirable phenomena such as corona, sliding (surface) discharges, partial discharges, trees (in polymer insulation), reducing the reliability and durability of high-voltage installations.

4.2 Mechanical Effects

Mechanical forces in the elements of high-voltage equipment occur both during normal operation (forces during scheduled switching, electrodynamic forces between conductors of different phases, wind and ice loads, etc.), and in emergency modes (increased electrodynamic forces between current-carrying parts during short circuits). The significantly lower mechanical strength of insulating parts and structures compared to metal ones causes a greater danger of mechanical impacts on them.

4.3 Thermal Effects

Thermal loads in the insulation arise due to the heat released in the current-carrying parts during the long-term flow of the rated current, as well as due to dielectric losses in the insulation when exposed to an electric field.

When short-circuit currents flow in emergency mode, the insulation experiences a short-term overheating, the so-called "thermal shock".

For example, when a transformer becomes hot, the insulation on the windings slowly breaks down and becomes brittle over time. The rate of thermal breakdown approximately doubles for every 10 °C. 10 °C is referred to as the Montsinger Factor and is a rule describing the Arrhenius theory of electrolytic dissociation (V. M. Montsinger [1930] formulated the main relationships between winding temperature, load and service life of power transformers).

4.4 Atmospheric Influences

When used outdoors, the insulation is exposed to rain, fog, dew, snow, ice, natural and man-made pollution, fluctuations in temperature and pressure, etc. As a rule, all these effects lead to a decrease in the dielectric strength of the insulation, and the appearance of additional mechanical loads.

4.5 Time Factor

Over time, even under normal operating conditions, the dielectric and mechanical properties of the insulation gradually deteriorate, and the "aging" of the insulation occurs. At the same time, a combination of various unfavorable factors has a great influence on the rate of "aging".

4.6 Exposure to Aggressive Media

During the operation of high-voltage equipment, aggressive gases and liquids can affect its insulation, leading to deterioration of its properties and premature failure.

When choosing insulation for this equipment, it must be taken into account that the negative factors listed above can act in various combinations and further worsen the electrical, mechanical and thermal properties of the insulation.

In some cases, when choosing dielectric materials and designing insulation, it is necessary to take into account the specific conditions of its operation. For example, when operating equipment in the tropics, insulation is affected by high humidity, temperature, solar radiation, as well as microorganisms and animals.

4.7 Force Majeure Factors Causing Equipment Failure

To get a clearer picture of the reliability problems of the power system, it is necessary to know the causes of faults and failures its main elements, which manifest themselves in force majeure circumstances. Let us briefly consider the impact of such factors on (a) underground cable lines, (b) transformers and (c) overhead transmission lines. These factors usually include the following:

1. Earthworks
2. Lightning
3. Tree Contact
4. Birds
5. Squirrels
6. Snakes
7. Insects
8. Bears, Bison, and Cattle
9. Mice, Rats, and Gophers
10. Vandalism

The influence of these factors is different for the three main elements of the energy system named above. Most of them are especially dangerous for overhead transmission lines. The first one applies only to cable lines.

1. **Earthworks**

a. **Underground cable lines**

The major reliability concern for underground cable are earthworks/*diging-in* (factor 1). This is when earthmoving equipment cuts one or more cables. To prevent excavation, utilities should encourage contractors to identify cable routes before excavation begins. In extreme cases where high reliability is required, utilities can place cable in concrete-encased duct banks.

Treeing in insulation and defects *by joints and terminations* are important reasons for reducing the reliability of underground cable lines. They can hardly be attributed to force majeure circumstances.

A major reliability concern pertaining to underground cables is **electrochemical treeing**. Treeing occurs when moisture penetration in the presence of an electric field reduces the dielectric strength of cable insulation. When the dielectric strength is degraded sufficiently, transients caused by lightning or switching can result in dielectric breakdown. Electrochemical treeing usually affects extruded dielectric cable such as cross-linked polyethylene (XLPE) and ethylene-propylene rubber (EPR), and is largely attributed to insulation impurities and bad manufacturing. Typical underground cable fault is shown on Fig. 4.1.

Not all underground cable system failures are due to cable insulation. A significant percentage is accounted for by joints and terminations. The main reasons are related to water ingress and poor workmanship. Heat-shrinkable coatings can be used to waterproof these joints and improve reliability.

b. **Transformers**

Short circuits cause extreme physical stress on transformer windings and are a major cause of transformer failure. ***Overloads*** rarely result in transformer failures, but do cause thermal aging of winding insulation. When a transformer becomes

Fig. 4.1 Underground cable fault

hot, the insulation on the windings slowly breaks down and becomes brittle over time. The rate of thermal breakdown approximately doubles for every 10 °C. Every 10 °C is referred to as the "Montsinger Factor" and is a rule of thumb describing the Arrhenius theory of electrolytic dissociation [2].

Because of this exponential relationship, transformer overloads can result in rapid transformer aging. When thermal aging has caused insulation to become sufficiently brittle, the next fault current that passes through the transformer will mechanically shake the windings, a crack will form in the insulation, and an internal transformer fault will result.

Extreme hot-spot temperatures in liquid-filled transformers can also result in failure. This is because the hot spot can cause free bubbles that reduce the dielectric strength of the liquid. Even if free bubbles are not formed, high temperatures will increase internal tank pressure and may result in overflow or tank rupture.

Many transformers are fitted with load tap changers (LTCs) for voltage regulation. These mechanically moving devices have historically been prone to failure and can substantially reduce the reliability of a transformer. Manufacturers have addressed this problem and new LTC models using vacuum technology have succeeded in reducing failure rates.

c. **Overhead Transmission Lines**

Overhead transmission lines (OTL) are the most vulnerable element of the power system to unpredictable external influences. This is due to their location in open space and huge size (length, height above the earth's surface). The vast majority of the factors listed above that affect the reliability of the power system relate specifically to overhead lines.

2. **Lightning**

A lightning strike occurs when the voltage generated between a cloud and the ground exceeds the dielectric strength of the air. This results in a massive current stroke that usually exceeds 30,000 amps. To make matters worse most strokes consist of multiple discharges within a fraction of a second. Lightning is a major reliability issue for utilities located in areas with high thunderstorm activity. An global map of thunderstorm activity is shown in Fig. 4.2.

Lightning can affect power systems through direct strikes (the stroke contacts the power system) or through indirect strikes (the stroke contacts something in close proximity and induces a traveling voltage wave on the power system). Lightning can be protected against by:

1. Having a high system basic insulation level,
2. Using shield wires,
3. Using surge arrestors to clamp voltages across equipment,
4. Having a low impedance ground.

Direct strikes are virtually impossible to protect against on a distribution system (see Sect. 7.2 for more information).

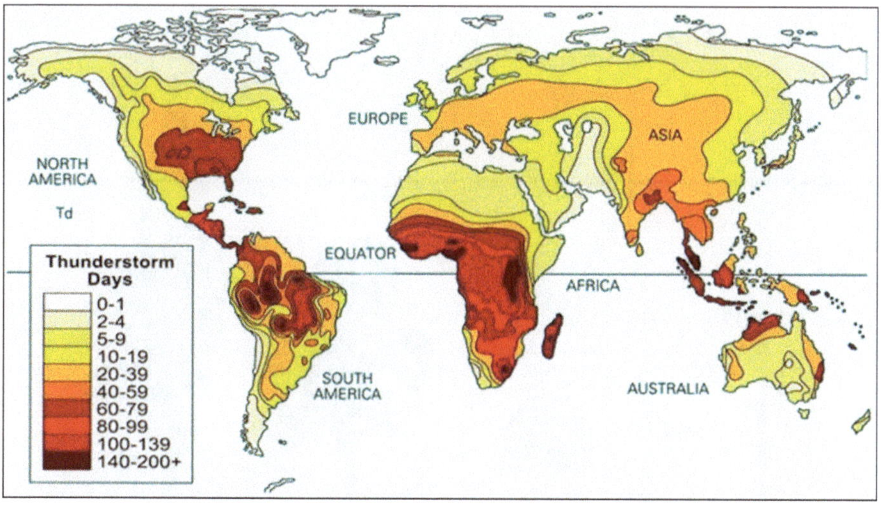

Fig. 4.2 Number of thunderstorm days per year

3. **Tree Contact**

Trees continuously grow, can fall onto conductors, can drop branches onto conductors, can push conductors together, and can serve as gateway for animals. This is why many utilities spend more on tree trimming than on any other preventative maintenance activity. When a tree branch bridges two conductors, a fault does not occur immediately. This is because a moist tree branch has a substantial resistance. A small current begins to flow and starts to dry out the wood fibers. After several minutes, the cellulose will carbonize, resistance will be greatly reduced, and a short circuit will occur. Branches brushing against a single-phase conductor typically do not result in system faults. Tree distances from overhead lines is shown on Fig. 4.3.

Faults due to tree contact can be reduced by using wire with an insulated jacket similar to cable. Thise wire can be effective, but faults tend to result in conductor burndown since they will not motor (move themselves along the conductor) like faults on bare conductor.

4. **Birds**

Birds are the most common cause of animal faults on both transmission systems and air insulated substations. Different types of birds cause different types of problems, but they can generally be classified as nesting birds, roosting birds, raptors, and woodpeckers.

Nesting birds commonly build their homes on transmission towers and in substations. Nesting materials can cause faults, and bird excrement can contaminate insulators. Nesting birds also attract predators such as raccoons, snakes, and cats. These predators can be a worse reliability problem than the birds themselves. Process of Seattle City Light workers remove an osprey bird nest near the Duwamish river that caused a short circuit is shown on Fig. 4.4.

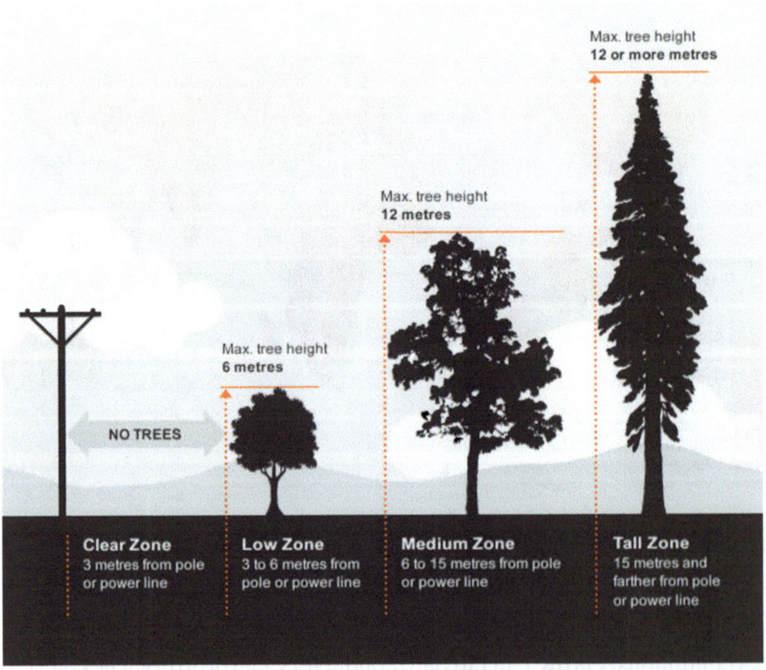

Fig. 4.3 Tree distances from overhead lines

Fig. 4.4 Seattle City Light workers remove an osprey bird nest near the Duwamish River that caused a short circuit

Roosting birds use electrical equipment to rest on or to search for prey. They can be electrocuted by bridging conductors with their wings, and their excrement can contaminate insulators. To prevent birds from roosting, anti-roosting devices can

be placed on attractive sites. For locations that cater to thousands of roosting birds, more extreme deterrent methods such as pyrotechnics can be used.

Raptors are birds of prey such as eagles, hawks, ospreys, owls, and vultures. Reliability problems are similar to other roosting and nesting birds, but special consideration may be required since most raptors are protected by the federal government.

Woodpeckers peck holes in wood with their beaks as they search for insects. This does not harm trees (the bark regenerates), but can cause devastating damage to utility poles This can be prevented by using steel poles, by using repellent, or by tricking a woodpecker into believing that there is already a resident woodpecker (woodpeckers are quite territorial).

5. **Squirrels**

Squirrels are a reliability concern for all overhead distribution systems near wooded areas. Squirrels will not typically climb utility poles, but will leap onto them from nearby trees. Problem which could cause by a squirrel is shown on Fig. 4.5.

They cause faults by bridging grounded equipment with phase conductors. Squirrel problems can be mitigated by cutting down nearby access trees or by installing animal guards on insulators.

6. **Snakes**

Snakes are major reliability concerns in both substations and underground systems. They can squeeze through very small openings, can climb almost anything, and have the length to easily span phase conductors. Snakes are usually searching for food (birds in substations and mice in underground systems), and removing the food supply can often remove the snake problem. Special "snake fences" are also available.

Fig. 4.5 A squirrel chewed into a power line in Trumbull, Connecticut, where the Nasdaq's computer center is located, shutting down trading for 34 min

Fig. 4.6 Fire ants nesting inside transformer

7. **Insects**

It is becoming more common for fire ants to build nests in pad mounted equipment. Fire ants nesting inside transformer are shown on Fig. 4.6.

Their nesting materials can cause short circuits, the ants can eat away at conductor insulation, and they make equipment maintenance a challenge.

8. **Bears, Bison, and Cattle**

These large animals do not typically cause short circuits, but degrade the structural integrity of poles by rubbing on guy wires. Bears can also destroy wooden poles by using them as scratching posts, and black bears can climb.

9. **Mice, Rats, and Gophers**

These rodents cause faults by gnawing through the insulation of underground cable. They are the most common cause of animal-related outages on underground equipment. To make matters worse, they will attract snakes (also a reliability problem). Equipment cabinets should be tightly sealed to prevent these small animals from entering. **Ultrasonic devices** can also be used to keep rodents away (ultrasonic devices will not keep snakes away).

10. **Vandalism**

Vandalism can take many different forms, from people shooting insulators with rifles to professional thieves stealing conductor wire for scrap metal. Addressing these reliability problems will vary greatly from situation to situation [3].

Question and Tasks

1. What are the three main groups of causes of premature failure of high-voltage equipment?

2. List the factors that affect the equipment and cause it to age and fail.
3. Which kinds of circumstances could cause equipment failure?
4. Formulate the requirements for the electrical strength of high-voltage insulation.
5. Which are the important influences to stable electric energy system work?
6. Which are the main requirements of the insulating materials used for power apparatus?
7. How snakes and bears could cause electric energy system damage?
8. How squirrels and woodpeckers could cause electric energy system damage?
9. Which result could take place due to fire ants nesting inside transformer?
10. List examples of vandalism in electric power system work.

References

1. Khalyasmaa AI, et al (2015) Diagnostics of electrical equipment of power stations and substations: textbook. Ural Publishing House, Ekaterinburg, p 64 (in Russian)
2. Hickey RB (2003) Electrical engineer's portable handbook, 2nd edn. McGraw-Hill, Publich, p 575
3. Electric systems issues at the turn of the 21st century. CIGRE ELECTRA (Special issue), 2000

Chapter 5
Defects in High Voltage Equipment: Types and Content of Tests

Abstract The word *defect* refers to each individual non-compliance of equipment with regulatory requirements. Defects in equipment can occur at different points in its life cycle: (1) during manufacture, installation, configuration (identified as a result of testing), (2) during operation (identified as a result of monitoring), (3) during testing and repair (identified as a result of diagnosing). They are varied and their consequences can also be different. This chapter briefly discusses the various stages of development of these defects and in sufficient detail—the types and methods of testing for their diagnosis.

5.1 Classification of Defects

It is customary to distinguish the following four main degrees of defect development:

1. There are no defects (the normal condition of the equipment).
2. Defect in the initial stage of development. The presence of such a defect does not significantly affect the operation of the equipment.
3. Strongly developed defect. The presence of such a defect limits serviceability of the equipment or shorten its service life.
4. Defect in the emergency stage of development. The presence of such a defect makes the operation of the equipment unacceptable.

Depending on the degree of development of defects, it is necessary to implement the following measures to eliminate them:

- replace the equipment, its part or element;
- to repair the equipment or its element, after which conduct an additional survey to assess the quality of the repair;
- leave the equipment in operation, but strengthen control over its condition;
- carry out other additional tests.

If defects are found and decisions are made regarding the further operation of the equipment, it is necessary to ensure reliability of information about its condition obtained by non-destructive testing (NDT). Measurement results may contain errors, so there is always the possibility of obtaining a false control result:

V. Y. Ushakov et al., *High-Voltage Equipment of Power Systems*, Power Systems,
https://doi.org/10.1007/978-3-031-38252-9_5

- a serviceable object will be recognized as unusable (false defect or error of the first kind);
- a faulty object will be recognized as fit (undetected defect or type II error).

Mistakes in NDT lead to various consequences. If mistakes of the first kind only increase the volume of restoration work, then mistakes of the second kind entail an emergency breakdown of equipment.

For any type of NDT, a number of factors can be identified that affect the results of measurements and analysis of the data obtained. They can be conditionally divided into three main groups:

1. environment;
2. human factor;
3. technical aspect.

The first group includes factors such as weather conditions (air temperature, humidity, wind strength, etc.), time of day.

The second group includes the qualifications of the personnel–professional knowledge of the equipment and competent conduct of the control itself.

The third group includes an information base about the diagnosed equipment (material, passport data, year of manufacture, etc.).

In real conditions, there are much more factors that affect the result of applying NDT methods than those listed above. But this topic is so vast that it is beyond the scope of this textbook.

It is precisely because of the possibility of making mistakes in the application of each type of NDT that there is its own regulatory documentation that regulates the purpose of NDT methods, the procedure for NDT, NDT tools, analysis of NDT results, possible types of defects identified during NDT, recommendations for their elimination, etc. For example, only in Russia the use of non-destructive testing methods is regulated by 19 GOSTs and a large number of industry regulations.

5.2 Test Types

Tests allow you to check the performance of electrical equipment, identify and eliminate deficiencies and ensure the safety of its operation. Industry orders and rules for the operation of electrical installations provide for the norms and frequency of inspections (tests), as well as the procedure for performing these operations.

There are five types of such tests:

1. type (normalized),
2. proof (control),
3. acceptance,
4. periodic (operational),
5. special.

5.2.1 Type (Normalized) Tests

Each new type of electrical equipment is subjected to type tests for compliance with the dielectric strength of its insulation with all the requirements of the relevant standards. Tests are subjected to the head sample or a sample from the first batch produced. They are also carried out when changing the design of insulation or technology. This type of verification is carried out at enterprises—manufacturers of electrical equipment. They are necessary to check the technical characteristics and safety before launching into mass production.

5.2.2 Proof (Control) Tests

These tests of the finished equipment before its transfer (sale) to the consumer are carried out at the manufacturer. Compared to type tests, proof tests are less complex and require less time. However, measurements must be included in the test program to verify compliance with current safety standards.

5.2.3 Acceptance Tests

Each sample of manufactured high-voltage equipment is subjected to acceptance tests in order to avoid the release of products with reduced dielectric strength of their insulation.

Regulatory documents establish the following types and levels of voltage during acceptance tests:

- *lightning impulses*; they test equipment for all voltage classes;
- *switching impulses*; they only test equipment of classes 330 kV and above, with the exception of power transformers and shunt reactors, whose internal insulation is subjected to these tests at voltage classes of 220 kV and above);
- *short-term voltage of industrial frequency* (one minute and with a smooth rise) and long-term voltage of industrial frequency.

When testing with lightning impulses, positive and negative full (1.2/50 μs) and cut off (1.2/5 μs) impulses are used according to the 3-shock method (for internal) and according to the 15-shock method (for external insulation and internal insulation of gas-filled equipment).

When testing with *switching impulses (waves)* of external and internal insulation of such devices as: current and voltage transformers, coupling capacitors, insulators and some others, aperiodic impulses of 250/500 μs are used. For internal insulation of power transformers—an oscillatory pulse with parameters of 100/1000 μs, shunt reactors—50/500 μs. Pulses of both polarities are applied using the 3-shock method (for internal) or the 15-shock method (for external insulation) with time intervals between voltage applications of 1 min.

When testing the internal insulation of electrical equipment with a *short-term voltage of industrial frequency*, a single application of a one-minute test voltage is used, and when testing external insulation, a threefold application of voltage with a smooth rise. The test continuous voltage is a voltage with a frequency of 50 Hz or increased frequency, but not higher 400 Hz. The exposure time of the test voltage does not depend on its frequency. When testing with continuous voltage of industrial frequency, the PD intensity should be measured in accordance with the state standard and standards for certain types of equipment.

The insulation of equipment in which thermal breakdown is possible must also be tested for resistance to thermal breakdown by applying a voltage of 50 Hz to the insulation relative to earth.

For voltage classes from 3 to 35 kV $U_{test} = 1.1\ U_n$.

When testing equipment of voltage classes of 110 kV and above, $U_{test} = 1.1\ U_n/\sqrt{3}$, where U_n is the highest operating linear voltage.

The test voltages of external insulation in the regulatory documents are given for normal conditions: atmospheric pressure 760 mm Hg, air temperature 20 °C, absolute air humidity 11 g/m^3. Under other test conditions, the values of the test voltages must be reduced to the corresponding standardized conditions—the installation height of the equipment is not more than 100 m at the upper value of the ambient temperature is not higher than 45 °C.

The test voltages of insulation (dry) intended for operation at an altitude of H from 1000 to 3500 m above sea level are obtained by multiplying the tabular values by the coefficient $K_H = 1/1, 1 - H/10000.$

When testing external insulation in rain, the test voltage is determined by multiplying the normalized values by the coefficient $K_I = 1 + 0.75\ (K_H - 1)$.

The normalized amplitudes of lightning and switching impulses and the magnitude of short-term voltage of industrial frequency for testing high-voltage equipment of all voltage classes can be found in the relevant regulatory documents and manuals.

5.2.4 Periodic (Operational) Tests

The equipment of each manufactured type must be subjected to periodic tests after the period specified in the standards for the corresponding electrical equipment, for compliance with the dielectric strength of its insulation with all the requirements of the standard, with the fulfillment of all instructions for type tests. They can also be carried out for preventive purposes. In the latter case, the frequency of inspections is indicated in the technical documentation.

5.2.5 *Special Tests*

Tests of this kind are carried out in research and test laboratories to obtain the results that scientists or engineers need. For example, with the aim of the definition of limit operating modes, etc.

5.3 Norms and Test Methods

The procedure for testing electrical equipment, as well as the regulatory documents that are used in this case, depend on the type of equipment, the place of its use, purpose and other parameters. They are indicated in the accompanying documentation, operating instructions (State standards, in Russia—GOSTs).

Different methods are used for different types of tests:

- Control and type tests are carried out within the framework of the requirements established by GOST. They may vary depending on the type of device being tested.
- Acceptance checks are carried out on the basis of the requirements specified in the Rules for the arrangement of electrical installations (RAEI).
- Operational tests, methods to check the equipment during exploitation, the timing of testing are regulated by a large number of documents. In Russia, the main ones are the "Electrical Equipment Testing Standards", as well as the RAEI. However, other standards may apply, up to internal ones, for example, those introduced at a given enterprise. (Other countries follow national level regulations such as those of the European Union.) In addition, the technical documentation, which is compiled and provided by the manufacturer, is taken into account. Equipment tests are carried out by specialists who have the necessary knowledge and experience, and also know what standards should be followed and which test methods should be used.

Since the 1960s, a system of *planned preventive maintenance (PPM)* has been successfully used to prevent and timely detect defects in power equipment. However, in modern conditions, this methodology needs to be updated, since it does not take into account the increase in the number of elements of the electric power system that have exhausted their regulatory resource. In addition, the problem is complicated by the need for a sufficient economic justification of the standards used and the difficulty of taking into account a number of important factors, such as:

- current technical condition and residual life of the equipment in operation;
- change in the nature of loads;
- use in the development of new equipment of more modern components, with other technical characteristics and properties. The use of new materials and technologies has made it possible to significantly increase the reliability of the units and, thus, to extend the overhaul period of their operation;

- the documents regulating the PPM provided for an excess safety stock, quite often fully functional equipment was taken out for scheduled repairs; premature maintenance caused unnecessary downtime and increased the cost of maintaining the facility.

Thus, in modern realities, the use of the PPM system is unjustified and is clearly costly. The best way out of this situation was the introduction of systems for periodic or continuous monitoring of the state of high-voltage equipment, thanks to which it is possible to carry out maintenance and repair not according to a calendar schedule, but according to a plan developed taking into account the actual technical condition (TC) of the facility. In addition, this approach makes it possible to extend the life of the equipment in excess of the standard, up to the complete exhaustion of the resource provided by the manufacturer. The concept of "control" includes two procedures—*diagnostics* and *monitoring*.

5.4 Diagnostics and Monitoring

Diagnostics and *monitoring*, together with *testing*, form a triad on which the control of the state of high-voltage equipment is based. Since the vast majority of failures in such equipment are due to insulation, the main components of this triad, shown in Fig. 5.1 [1, 2] relate to monitoring the state of the insulation. In turn, individual components may contain original methods and their hardware implementation (see Chapters 6–9).

5.4.1 Diagnostics

Diagnostics or *technical diagnostics* (from Greek—recognition) is a set of measures that allows you to study and establish signs of a malfunction (malfunction) of equipment, establish methods and means by which a conclusion (diagnosis) is given (diagnosed) about the presence (absence) of a malfunction (defect). In other words, technical diagnostics allows you to assess the state of the object under study. Such diagnostics is mainly aimed at finding and analyzing the internal causes of equipment failure. There are two types of diagnostics according to the depth (detailedness) of the studies carried out in this case: "*parametric diagnostics*" and "*fault diagnostics*". In the first case, the process is potentially safe for the object, and in the second case, it can worsen its condition. For their designation, two terms "*parametric diagnostics*" and "*fault diagnostics*" are used.

Modern diagnostic systems make it possible to fully and reliably assess the condition of equipment. The main goals of diagnostics are:

- determination (assessment) of the condition of the diagnosed equipment;

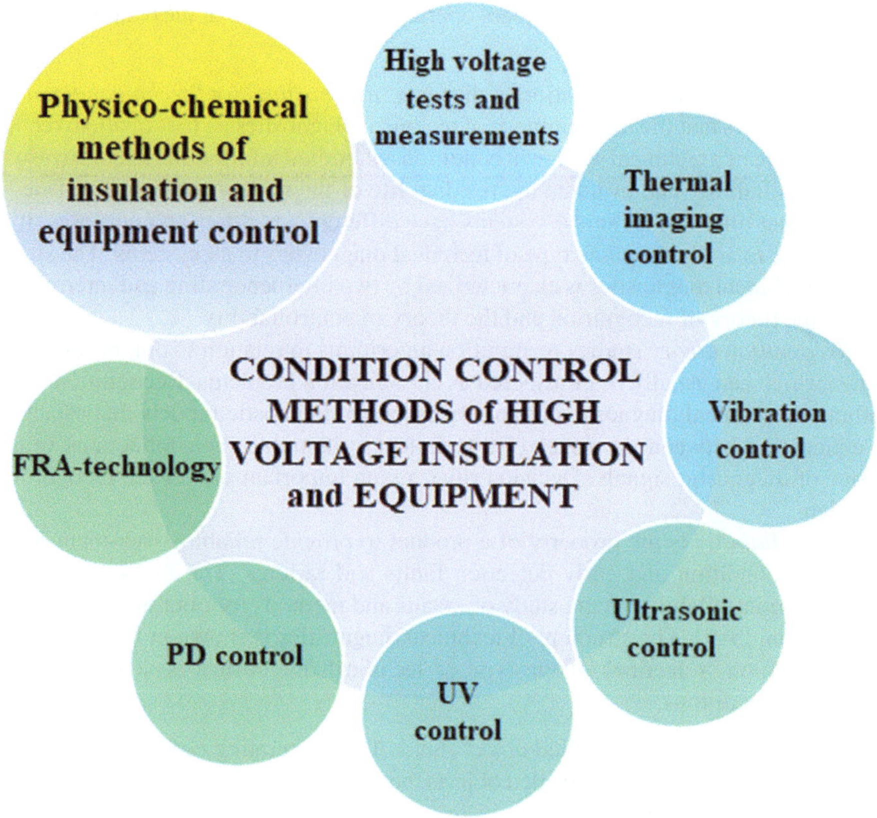

Fig. 5.1 Methods for monitoring the state of insulation of electrical equipment

- identification of the type of defect, its scale, location, causes of occurrence. This serves as the basis for deciding on the next operation of the equipment (putting it out for repair, additional examination, continued operation, etc.) or complete replacement of equipment;
- forecast of the period of subsequent operation—assessment of the residual service life of electrical equipment.

Modern diagnostic systems make it possible to fully and reliably assess the condition of equipment, including high-voltage equipment. The main goals of diagnostics are:

- determination (assessment) of the condition of the diagnosed equipment;
- identification of the type of defect, its scale, location, causes of occurrence. This serves as the basis for deciding on the next operation of the equipment (putting it out for repair, additional examination, continued operation, etc.) or complete replacement of equipment;

– forecast of the period of subsequent operation—assessment of the residual service life of electrical equipment.

Already from the interpretation of the term "*diagnostics*" (or "*technical diagnostics*") it follows that the main purpose of technical diagnostics is, first of all, to recognize the state of a technical system in conditions of limited information and, as a result, increase reliability and evaluate the residual life of the system (equipment). Due to the fact that different technical systems have different structures and purposes, it is impossible to apply the same type of technical diagnostics to all systems. The structure of technical diagnostics is characterized by two interpenetrating and interrelated areas: the theory of recognition and the theory of controllability.

Recognition theory studies recognition algorithms in relation to diagnostic problems, which can usually be considered as classification problems. Recognition algorithms in technical diagnostics are partly based on diagnostic models that establish a connection between the states of a technical system and their reflections in the space of diagnostic signals. Decision rules are an important part of the recognition problem.

Controllability is the property of a product to provide reliable assessment of its technical condition and early detection faults and failures. The main task of the theory of controllability is the study of means and methods for obtaining diagnostic information [3–5]. The structure of technical diagnostics is shown in Fig. 5.2.

Application (selection) of the type of technical diagnostics is determined the following conditions:

1. the purpose of the controlled object (field of use, operating conditions, etc.);
2. the complexity of the controlled object (the complexity of the design, the number of controlled parameters, etc.);
3. economic feasibility;
4. the degree of danger of the development of an emergency situation and the consequences of the failure of the controlled object.

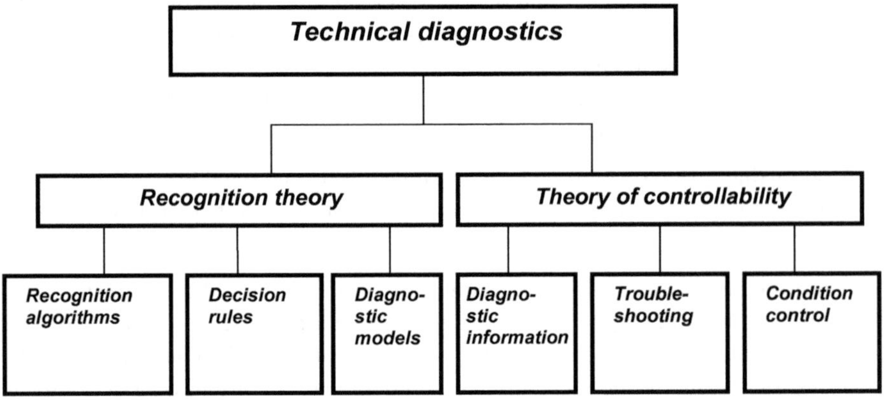

Fig. 5.2 Structure of technical diagnostics

The state of the system is described by a set of defining it parameters (signs), when diagnosing a system, they are called *diagnostic parameters*, and the procedure itself—*parametric diagnostics (control by limit values of parameters)*—is the control of normalized equipment parameters, detection and identification of their dangerous changes. It is used for emergency protection and equipment control, and diagnostic information is contained in the aggregate of deviations of these parameters from the nominal values.

When choosing diagnostic parameters, priority is given to those that meet the requirements for the reliability and sufficiency of information about the technical state of the system in real operating conditions. In practice, several diagnostic parameters are usually used simultaneously.

Diagnostic parameters can be parameters of working processes (power, voltage, current, etc.), accompanying processes (vibration, noise, temperature, etc.) and geometric quantities (clearance, backlash, beat, etc.). The number of measured diagnostic parameters also depends on the types of instruments used to obtain data and the degree of development of diagnostic methods. So, for example, the number of measured diagnostic parameters for power transformers and shunt reactors can reach 38, oil circuit breakers—29, SF_6 circuit breakers—25, surge suppressors and arresters—10, disconnectors (with a drive)—14, oil-filled measuring transformers and coupling capacitors—9.

In turn, diagnostic parameters should have the following properties:

1. high sensitivity;
2. uniqueness;
3. stability;
4. latitude of change;
5. informativeness;
6. frequency of registration;
7. availability and convenience of measurement.

The sensitivity of a diagnostic parameter is the degree of change in the diagnostic parameter when the functional parameter is varied, i.e. the larger the value of this value, the more sensitive the diagnostic parameter is to the change in the functional parameter.

The uniqueness of a diagnostic parameter is determined by its monotonically increasing or decreasing dependence on the functional parameter in the range from initial to limit change functional parameter, i.e. each value of the functional parameter corresponds to a single value of the diagnostic parameter, and each value of the diagnostic parameter, in turn, corresponds to a single value of the functional parameter.

Stability sets the possible deviation of the diagnostic parameter from its average value at multiple measurements under constant conditions.

Latitude of change—the range of change in the diagnostic parameter corresponding to the set value of the change of the functional parameter; thus, the greater the range of change of the diagnostic parameter, the higher its information content.

Informatively is a property of a diagnostic parameter that a swarm with insufficiency or redundancy can reduce efficiency the diagnostic process itself (diagnosis reliability).

The frequency of registration of a diagnostic parameter is determined based on the requirements of technical operation and manufacturer's instructions, and depends on the rate of possible formation and development of a defect.

Availability and convenience of measuring a diagnostic parameter directly depend on the design of the object being diagnosed and the diagnostic tool (instrument).

There are different classifications of diagnostic parameters. In our case, for the diagnostics of high-voltage equipment, we will adhere to the types of diagnostic parameters presented in [5, 6].

According to their purpose, diagnostic parameters are divided into three types:

1. representing the characteristics of the object (information type);
2. representing the current technical specification about the elements of the object;
3. which are derivatives of several parameters.

Information view diagnostic options include:

1. type of object;
2. commissioning time and duration of operation;
3. repair work carried out at the facility;
4. technical characteristics of the object obtained during testing at the factory and/or during commissioning.

Diagnostic parameters representing the current technical characteristics of the elements of the object, most often are the parameters of working (sometimes related) processes.

To diagnostic parameters that are derivatives several parameters, primarily include such as:

1. the maximum temperature of the hottest point of the transformer at any load;
2. dynamic characteristics or their derivatives.

In many ways, the choice of diagnostic parameters depends on the specific type of equipment and the diagnostic method used for this equipment.

5.4.2 Monitoring

Monitoring (or ***preventive diagnostics***) is the detection of all potentially dangerous defects at an early stage of development, monitoring their development and, on this basis, a long-term forecast of the state of the equipment [5]. In other words, monitoring is a continuous observation of an object without active interference in its work (as if from outside), which includes the periodic registration of parameters of interest, their registration, storage and analysis. Monitoring can be called *the* *control of the current values of the parameters of the observed object.* A set of measures is always extended in time and allows you to study the properties of an object in dynamics, understand the direction of its development, and identify hidden problems. Monitoring allows you to understand the cause of a defect (fault) and to trace its dynamics. The monitoring system is primarily focused on solving two problems:

- real-time determination of the current state of high-voltage equipment, identification of destructive processes at the early stages of their development and assessment of the residual life;
- solution of technical and economic problems.

Unlike diagnostics, this type of control is carried out using modern monitoring systems and makes it possible to obtain reliable, timely information about the real state of electrical equipment. Its advantages are as follows:

1. High-quality diagnostics of the high-voltage equipment allows timely detection and localization of defects. In this case, the study is carried out using non-destructive control methods (see Chapter 6).
2. Periodic monitoring of the high-voltage equipment makes it possible to track the dynamics of the development of damage, evaluate the residual operational life, predict the time and scope of the necessary maintenance and/or repair of defective equipment.
3. Assessment of the state of equipment using a monitoring system does not require stopping the equipment, which allows you to quickly identify rapidly developing defects and, thereby, prevent the occurrence of emergencies in the intervals between scheduled inspections. In order to achieve the required degree of detail, it is necessary to install many different sensors and data collection devices, organize information transfer channels to centers (servers) for collecting, analyzing and storing research results, which require certain investments. However, this approach is justified, since in this case the decision on the need for maintenance or repair of this or that equipment is made taking into account its actual condition. Maintenance and repair work is carried out in optimal time, since a set of characteristics is used to determine the current state of electrical equipment and its residual life. In this case, the equipment is examined during operation (without stopping the production process).

Diagnostic and monitoring systems allow you to control the current state of any element of the power system, regardless of the age of the equipment. With their help, you can achieve the following:

1. extending the service life of electrical equipment;
2. savings on planned preventive maintenance (PPM);
3. prevention of emergencies;
4. reduction of injuries of operating and maintenance personnel as a result of damage to electrical equipment;

5. minimizing the volume of investments in the unreasonable renewal of fixed assets;
6. reducing the labor costs of operational personnel through the introduction of automated monitoring and diagnostic systems;
7. reduction of operating costs and losses with an individual approach to the planning of repair work;
8. reducing the duration of scheduled repairs with stopping electrical equipment due to timely and targeted preparation for repair work;
9. reduction of cases of undersupply of electric energy and claims from consumers for the damage caused;
10. increasing the terms of cost-effective operation of electrical equipment in conditions of reduced loads.

5.5 Expert Systems

Modern systems for assessing the technical condition (ATC) high-voltage equipment are automated expert systems designed to solve several important tasks, including:

1. determination of the current technical condition;
2. identification of signs of a faulty state of electrical equipment;
3. planning the timing and scope of the upcoming repairs;
4. assessment of the effectiveness of the repair work performed.

There are two methods by which expert diagnostic systems for power equipment are created, namely:

1. *A specialized system* designed for a certain type of equipment. It is created on the basis of the knowledge of a practical expert or a group of such experts.
2. *A universal self-adjusting system*, formed on the basis of neural networks or created on the basis of artificial intelligence. It is based on a universal software product. Practical experts are involved in the work within their competence only at the stage of adaptation or self-learning of the expert system.

The problem arises of choosing the more efficient option out of the two.

The neural network is aimed at finding and diagnosing defects that have already been fixed earlier. Its functionality cannot detect and register the manifestation of conditions that were not previously diagnosed, since it does not take into account the features of physical processes in controlled devices.

The fixed expert system is based on the deep knowledge of an experienced consultant (or a group of consultants) about the processes occurring in the equipment. Such a system can fully take into account the occurrence of repetitive and modified defects. In addition, it has great potential in terms of modernizing diagnostic algorithms.

The structure of all modern ATC systems is approximately the same and consists of four main components:

1. database (DB)—the initial data on the basis of which the ATC of the equipment is performed;
2. knowledge base (KB)—a set of knowledge in the form of structured rules data processing, including all kinds of expert experience;
3. the mathematical apparatus with which it is described mechanism of operation of the ATC system;
4. results. This component usually includes formalized or non-formalized assessments of the ATC of equipment and control actions based on the assessments received—recommendations on the further operation of the equipment being evaluated.

As input parameters (DB), data obtained during the use of various non-destructive testing methods, equipment tests, or data obtained from various monitoring systems, sensors, etc. are usually used.

Various rules can be used as a knowledge base, such as presented in the working documentation (WD) and other regulatory documents, and in the form of complex mathematical rules and functional dependencies.

The results usually differ only in the type of estimates (indices) of the state of the equipment, possible interpretations of the classifications of defects and control actions. But the main difference between ATC systems from each other is the use of different mathematical tools (models), on which the reliability and correctness of the system itself and its operation as a whole depend to a greater extent.

Despite all the undoubted advantages of existing ATC systems, in modern conditions, they have a number of significant drawbacks:

- focused on solving a specific problem of a specific equipment operator (for specific schemes, specific equipment, etc.) and, as a rule, cannot be used at other similar facilities without major modifications;
- use different-scale information with different levels of accuracy, which can lead to unreliability of the assessment;
- do not allow to take into account the dynamics of changes in the equipment ATC criteria, i.e. systems are not trained.

These shortcomings deprive modern ATC systems their versatility, which is why the current situation in the electric power industry a number of countries forces to improve existing or look for new methods for modeling ATC systems. Modern ATC systems should have the properties of data analysis (introspection), search for patterns, forecasting and, ultimately, learning (self-learning). Such opportunities are provided by artificial intelligence methods. Today, the use of artificial intelligence is not only a generally recognized direction of scientific research, but also quite successful implementation of the actual application of these methods for various energy facilities [7–9].

Question and Tasks

1. Classify defects according to the stages of their development.

2. Name the types of testing of high-voltage equipment.
3. Explain the term "Diagnosis". Tasks and methods for performing this procedure.
4. Name the tasks and methods of equipment monitoring.
5. List the problems that can be solved with the help of diagnostics and monitoring

References

1. Lyutikova M, Korobeinikov S, Konovalov A (2021) New insulating fluids and diagnostic techniques for paper-oil insulated equipment. Power Technol Eng 55:136–142
2. Lyutikova MN, Korobeinikov SM, Konovalov AA, Kozlov VK, Garifullin MSh (2019) Control of the antioxidant additive (ionol) content in liquid insulation of high-voltage equipment in electric grid companies using modern instrumental methods. Power Technol Eng 53(1):118–125
3. Khalyasmaa AI et al (2015) Diagnostics of electrical equipment of power stations and substations: textbook. Ural Publishing House, Ekaterinburg, p 64 (in Russian)
4. Bubnicki Z (2002) Modern control theory. Springer Verlag, Berlin, p 423
5. Ignatov S (2020) Diagnostics and monitoring of high-voltage equipment. Modern technical and software solutions. https://marketelectro.ru/node/diagnostika-i-monitoring-vysokovoltnogo-oborudovaniya (in Russian)
6. Kurtenkov GE (1999) Fundamentals of insulation design of high-voltage electrical equipment: textbook. NTL Publishing House, Tomsk, p 276 (in Russian)
7. Fink DG, Beaty HW (2006) Standard handbook for electrical engineers, 15th edn. McGraw-Hill, New York
8. Kundur P (2007) Power system stability and control. CRC, Boca Raton, p 360
9. Webster J (2003) Electrical measurement signal processing and displays. CRC, Boca Raton, p 768

Chapter 6
Non-electrical Diagnostic Methods

Abstract Already from the titles of the paragraphs of this chapter, we can conclude that the progress of recent decades in the field of non-destructive testing of a wide range of products and materials has given power engineers a large number of various reliable methods and means of monitoring the state of high-voltage equipment of electric power systems. Some of them have been used by power engineers for many decades and continue to improve them, others have been developed relatively recently, and specialists strive to fully use their potential. For the diagnostics of oil-filled equipment, the leading place belongs to the methods of chemical control of the state of insulating liquids, to control the state of the insulating structures of overhead power lines and open substations—methods based on the registration of electromagnetic radiation of various frequency ranges.

6.1 Chemical Control Methods

Liquid dielectrics used in high-voltage equipment perform not only the functions of an insulating material, a heat-removing and arc-extinguishing medium, but also an indicator of changes in the state of the internal insulation of the apparatus during its operation in order to take timely measures to ensure its reliable operation. (For example, more than 38% of power transformer failures are due to defects in the insulation system.)

To date, many tools and methods have been developed for monitoring the state of electrical equipment during its service life.

With regard to oil-filled equipment, chemical and physico-chemical methods of control are in the lead among them, since they allow working directly with samples of internal insulation taken from the equipment, which means assessing the state of the insulation system from the inside according to certain characteristics and marker substances. In addition, unlike many methods based on high-voltage measurements of some insulation parameters (for example, insulation tests with increased voltage of industrial frequency of 50 Hz or tests with increased direct voltage), physicochemical tests of insulating materials are non-destructive methods for monitoring the condition of an insulation system that do not reduce device service life.

V. Y. Ushakov et al., *High-Voltage Equipment of Power Systems*, Power Systems,
https://doi.org/10.1007/978-3-031-38252-9_6

During the operation of insulating materials, profound changes occur, both at the physical and chemical levels (with a change in the molecular composition). Such processes can be designated by one concept of "aging". In general, aging is accompanied by a change in the insulating properties of the liquid and paper insulation. The main factors causing the aging of liquid dielectrics in operating high-voltage equipment are elevated temperature, dissolved oxygen in the air, as well as electric and magnetic fields. The aging rate of dielectrics is influenced by high-energy radiation, structural metal elements (especially those made of metals of variable valence: copper, iron, nickel, cobalt, etc.), as well as water, salts of organic acids and other polar impurities dissolved in the dielectric liquid. As a result of electrochemical and thermochemical oxidation of numerous components of a dielectric liquid hazardous polar substances appear, such as alcohols, aldehydes, ketones, carboxylic acids, macromolecular and colloidal compounds. Colloidal compounds combine into large agglomerates and form deposits in the insulating gaps and on the windings. Gradually, the precipitate thickens and becomes insoluble, which prevents efficient heat removal from the heated parts of the transformer.

The traditional physical and chemical indicators characterizing the dynamics of aging or the degree of wear of insulating materials in operating electrical equipment, first of all, should include such as: acid number, content of water-soluble acids and alkalis, dielectric loss tangent at 90 °C, optical turbidity, surface tension at the interface between water and oil, the content of the antioxidant additive (ionol) and furan compounds. Other characteristics (breakdown voltage, total concentration of different forms of water in a dielectric liquid, flash point in a closed crucible, etc.), as a rule, depend, among other things, on the values of the above indicators.

The oil contains about 70% of information about the condition of the equipment, so monitoring the condition of the oil is important for taking timely measures to ensure the reliable operation of high-voltage equipment. Table 6.1 shows the most commonly used quality indicators of transformer oil and the types of defects detected with their help.

Below is a brief explanation of the content of these indicators.

The acid number is the amount of caustic potassium (ACP), measured in milligrams, that is required to neutralize free acids in 1 g of oil. An increase in the acid number indicates the oxidation of the oil, which may cause corrosion of structure elements, emergence of soaps with metal ions, and development of colloidal-dispersed processes leading to a decrease in the electrical strength of the oil. Acids, due to their polarity, can also help increase the water absorption of paper insulation.

The content of water-soluble acids and alkalis that can form during oil manufacturing or as a result of oil oxidation in the course of operation. These ingredients are quite corrosive and contribute to the development of corrosion and the aging of paper insulation.

The moisture content, as an indicator of solid insulation and oil conditions, which is periodically monitored during the operation of transformers. Atmospheric moisture penetrates into the oil or emerges in it due to the following reasons: (1) absence or malfunction of dryers in transformers with free breathing, (2) suction of moist air or rain water into the oil in transformers with a forced cooling system when the latter

Table 6.1 Admissible concentrations of gases in oil for block[a] and grid transformers

Gas	Block transformers	Grid transformers
Hydrogen (H_2)	260	500
Methane (CH_4)	250	200
Ethane (C_2H_6)	160	170
Ethylene (C_2H_4)	250	260
Acetylene (C_2H_2)	20	70
Propane (C_3H_8)	40	30
Propylene (C_3 H_6)	40	40
Carbon monoxide (CO)	280	260
Carbon dioxide (CO_2)	3500[a]	4000

Note [a] A block transformer is a transformer rigidly connected by busbars to a synchronous generator. Its power is equal to the power of the generator

is leaking, (3) suction of humid air through other leaks, (4) formation of moisture as a result of aging processes in insulation itself.

The gas content of the oil is monitored in operating transformers with a film protection of the oil from oxidation in order to estimate transformer tightness. An increase in the gas content (including the air) promotes more intensive oil oxidation and a decrease in the dielectric strength of insulation in the active part of a transformer.

The above physical and chemical indicators have been used for many years in traditional methods for diagnostics of power transformer conditions, elaborated in the vast literature available to English-speaking readers [1–4]. Therefore, such indicators are not considered in more detail in this textbook.

6.1.1 Chromatographic Analysis of Dissolved Gases

Chromatographic analysis of dissolved gases (CADG) is the most common type of monitoring, used in almost every country. All major electrical companies and transformer manufacturing enterprises widely use the analysis of gases dissolved in oil, combined with diverse systems for evaluating and identifying the type of transformer defects.

This method is attractive by its high sensitivity to electrical discharges in insulation (PD, dendrites, creeping discharges) and to local overheating, responsible for decomposition of paper insulation and oil. Special attention is paid to the content of furan derivatives, which may be indicative to the destruction of paper insulation. Thermolysis, oxidation and hydrolysis of insulation, causing partial destruction of cellulose macromolecules, lead to the emergence of furan series components released into transformer oil.

Regulatory documentation used since the 1980–1990s in all countries with a developed electric power industry provides for the CADG method to estimate the paper insulation condition for power transformers in operation. Since that time, a fairly large experience in using the method has been acquired in relation to power transformers with a voltage of 110–750 kV. This experience makes it possible to select a number of indicators that have a relatively high diagnostic value, and to determine the type and nature of defects identified for making decisions on further operation of a transformer.

The CADG method can be used to identify two groups of defects in power transformers:

- overheating of current-carrying connections and frame structure elements;
- electrical discharges in oil.

The concentration of the following seven gases is then determined: hydrogen (H_2), methane (CH_4), acetylene (C_2H_2), ethylene (C_2H_4), ethane (C_2H_6), carbon monoxide (CO) and carbon dioxide (CO_2).

Defects in power transformers are identified using a division of gases into basic (key) and accompanying (characteristic) gases.

In the case of overheating in current-carrying connections and transformer frame elements, the basic gases are C_2H_4 (with oil-paper insulation and oil being heated above 500 °C) and C_2H_2 (during an arc discharge). The accompanying gases in both cases are H_2, CH_4 and C_2H_6.

In the case of partial discharges in oil, the main gas is H_2, while the accompanying gases with a low content are CH_4 and C_2H_2.

In spark and arc discharges, the basic gases are H_2 or C_2H_2, while the accompanying gases with any content are CH_4 and C_2H_4.

In the case of solid insulation overheating, the main gas is CO_2. It should also be noted that a concomitant indicator of destruction for cellulose insulation in a transformer is an increase in the content of oxide and carbon dioxide dissolved in transformer oil. At the same time, in accordance with the recommendations of The International Council on Large Electric Systems (CIGRE), the presence of a total CO and CO_2 concentration of more than 1% may indicate the degradation of cellulose insulation [1].

The basic gas is determined by relative concentrations of hydrogen and hydrocarbons, taking into account the corresponding boundary concentrations according to the formula

$$a_i = A_i / A_{\Gamma pi}, \tag{6.1}$$

where a_i is the relative concentration of a selected gas; A_i is the measured value of the selected gas concentration, % vol.; $A_{\Gamma pi}$ is the boundary concentration of the selected gas in volume percent, % vol.

According to calculated relative concentrations, the maximum value of a_i corresponds to the basic gas, except for CO_2, being a basic gas if $A_{CO_2} >$ % vol. In this case,

$a_i > 1$, characteristic gas with a high content;
$0.1 < a_i < 1$, characteristic gas with a low content;
$a_i < 0.1$, uncharacteristic gas.

It should be pointed out that in analyzing the composition and concentration of gases dissolved in oil, as well as in diagnosing the operational condition of power transformers, it is necessary to take into account the factors that cause transformers to experience alterations.

The operational factors causing an increase in the concentration of gases dissolved in transformer oil include: residual gas concentrations from a previous defect eliminated during repair (the oil has not been degassed); increasing the load of a transformer; topping up with used oil containing dissolved gases; performing welding work on the tank; overflow of gases from the expander tank of an OLVR contactor to the transformer tank. The operational factors causing a decrease in the concentration of gases dissolved in transformer oil include: a decrease in the load of a transformer; oil degassing; topping up with degassed oil; replacement of silica gel.

To diagnose defects developing in power transformers, the following main criteria are used:

- criterion of boundary concentrations;
- criterion of gas growth rate;
- criterion of pair ratios for characteristic gases.

According to regulatory documentation, the values of parameters exceeding some predetermined limits (limit values) should be regarded as indicators of defects that may lead to equipment failure. In this regard, the CADG method is different from the above ideology, since normative boundary gas concentrations are those whose occurrence indicates a mere chance for defects to develop in a transformer. Such transformers are taken under control with increased oil samplings and CADG procedures.

The criterion of boundary concentrations permits selecting transformers with possible defects in development from the total number of units in a transformer stock, while the hazard degree for a defect to develop is determined by the relative growth rate of gas (gases). The amount of accumulated experience shows that once the relative growth rate of gas (gases) exceeds 10% a month, this indicates the presence of a rapidly developing transformer defect.

The nature of defects in development, according to CADG results, is determined by a criterion of concentration ratios for various gas pairs. It is customary to distinguish between thermal and electrical defects. Defects of thermal nature include the occurrence of short-circuited circuits, increased heating of insulation and contacts of on-load voltage regulators (OLVR) and those of switching without excitation (SWE), taps, pins, and other frame and tank metal structures of a transformer. Defects of electrical nature include discharges in oil of various intensities. Naturally, the development of transformer failures may be of mixed character. Analysis of methods available for estimating the nature of defects in development (thermal or electrical ones), based on the results of CADG, shows significant differences in both types and gas pair ratios used. These ratios have been established in a number of techniques: Dörnenburg's method, Müller's method, CEGB/Rogers Ratios, Schliesinger's method, *normogram* method, Duval's method, IEC 60599.

Methods of interpreting CADG results for transformers used by companies in different countries are noticeably different. Thus, ABB has adopted the Dörnenburg's method using three basic gas ratios, according to IEC, with a set of four types of defects (thermal, partial discharges of low and high energy, powerful discharges) given a simple graphic interpretation. Limits of gas concentration have been recommended for transformers without defects, regardless of the time of their operation.

Spain has adopted the ASINEL system applying seven digital codes assigned to six gas ratios and hydrogen concentrations, as well as to a sum of the C_1 and C_2 hydrocarbons. Three of the codes correspond to IEC 60599 Codes and have the same meanings; one of the gas ratios corresponds to the Rogers Code, and another to CO_2/CO, adopted for estimating the degradation of paper insulation. The ratio C_2H_2/H_2 is used to estimate the location of a failure: once it is higher than 2, there is an OLVR defect, and once lower, the defect is located in the main tank of a transformer.

In Canada, the diagnosis is made using the Duval triangle based on relative concentrations of the H_2, C_2H_2, C_2H_4, C_2H_6 gases for six main types of defects. The criteria for gas concentrations under normal, acceptable and hazardous conditions are significantly varied for different gases and, in particular, for different values of voltage and service life of transformers. To obtain a confident diagnosis, CADG data are entered into an expert system for estimating the condition of a transformer.

A detailed analysis of CADG efficiency for transformers has been made by the Canadian IREQ Institute. It is based on 25,000 samplings taken by the energy company Hydro Quebec.

The system by the LABELEC Laboratory (Portugal) is based on the IEC 599 gas ratios, as well as on the CO_2/CO and C_2H_2/H_2 ratios; however, it provides for a first-priority assessment of absolute values for gas concentrations. If the volume of at least one of the gases goes beyond the limit value, then five gas ratios are calculated. Additionally, the Schlesinger method is used to confirm the presence of more than one type of simultaneous defects.

RWE Energy (Germany) has adopted a system using five gas ratios and four limit concentrations. Depending on the value of a ratio, the severity of a defect is identified and compared with the concentration of gases. Significant basic gas ratios and concentrations are regarded as indicators of a significant stage of insulation degradation. The system distinguishes between five types of defects. A similar system (by KEMA Transformer) is adopted in The Netherlands.

Belgium has adopted a system based on the Laborelec Code Table. The codes represent concentrations of the H_2, $C_2\,H_2$, CO gases and the sums of C_1 and C_2 hydrocarbons. In addition, the CH_4/H_2 ratio is taken into account. The system includes nine ways of identifying defects being distinct in severity. Depending on concentrations and gas ratios, 27 different types of diagnosis are provided, indicated by codes. The normal condition of a transformer corresponds to a hydrogen concentration of up to $200 \cdot 10^{-6}$ rel. units, with the sum of C_1 and C_2 hydrocarbons being $300 \cdot 10^{-6}$ rel. units, and CO being $800 \cdot 10^{-6}$ rel. units.

France has adopted the system by the LCIE Laboratory, based on comparing gas concentrations with a standard, taking into account the voltage, type of design (presence or absence of OLVR), terms and conditions of transformer operation. Once concentrations go beyond the standard values, gas ratios are analyzed to identify the type of defects. The rate of increase in gas concentrations is taken into account as well. The LCIE Laboratory also uses C_3 hydrocarbons to interpret the results of CADG.

The National Grid energy company (Great Britain) traditionally uses the Rogers Ratios method. It is believed that the C_2H_6/CH_4 gas ratio, excluded from the IEC 60599 system, is efficient at identifying an increase in temperature as compared to an operating temperature. The method also takes into account an increase in gas emission over time, and a comparison of gas concentrations with the admissible one. The method continues to improve.

Siemens Trafo Union (Germany) uses concentrations of characteristic gases and their ratios as criteria. The concentration of acetylene, hydrogen and that of a sum of C_1 and C_2 hydrocarbons is analyzed for propylene, whereas the sum of CO and CO_2 is used for propane.

In Poland, gas concentrations and ratios of characteristic gases are also used as criteria taking into account increase rates for gas content in oil. It is noteworthy that admissible concentrations of gases in oil, determined from practical experience, are assumed to be different for block and grid transformers, being 10^{-6} units, see Table 6.1.

Energopomiar, implementing CADG, annually conducts over 650 oil samplings from large transformers. CADG is also widely implemented by Elta Manufacturing, being the main transformer enterprise in the country.

According to the results of CADG performed in Russia in the late 90s, it was found that the previously recommended IEC 60599 method has the highest diagnostic ability to identify the nature of developing defects.

The practice of Russian power systems relies on gas chromatography as a basic method for estimating the technical condition of transformer equipment. Criteria are distinguished by the assignment of equipment to frequent control (limit concentration criterion), the type of defects in development (concentration ratio criterion), and the danger of defects (rise rate criterion). To identify thermal defects at early stages of their development, oil analysis is carried out to determine the presence (additional to the seven basic gases) of C_3 and C_4 hydrocarbons. For this analysis, the gas chromatographic complex Tsvet 500-TM is recommended, as well as the complex manufactured by OOO NPF "Elektra".

The experience of running CADG on transformers operated by various enterprises shows that it is advisory to take into account different factors when interpreting the results of analysis. Taking into account changes in gas concentration when comparing the results of two samplings conducted at different operation periods makes it possible to identify defects in development and estimate their danger. It should be noted that comparison of absolute gas release values for different transformers necessitates a reduction to one and the same oil volume. In particular, this technique is applied to study the on-load dependence of gas release.

Positive results are ensured by analyzing thermodynamic processes in a transformer prior to its damage: at the location of defect development there is a zone where the temperature is sufficient for releasing the primary decomposition products of complex oil molecules. These products penetrate into the cooling zone, where equilibrium is achieved. The relative concentration of constituent gases can be calculated as a temperature function and then applied to estimate the change.

In many systems, the presence or absence of OLVR is duly taken into account; the sampling of oil is made on a regular basis; the rate of increase in gas concentration is accounted for, and the use of CADG at the initial period of transformer operation is duly distinguished. In determining the limit values of gas concentration, one takes into account the voltage class, the load intensity, and the possibility of gases penetrating from the OLVR tank into the main transformer tank.

To interpret the results of analysis, other criteria are also used, developed using statistical and practical operating data. For instance, such criteria are given by an increase in gas concentration following 100 h of transformer operation, and by a difference between grid and block transformers.

The lack of a unified methodology for interpreting the results of CADG in power transformers makes it difficult to compare the condition of transformers operated and controlled by different enterprises. It is difficult to coordinate most of the criteria for assessment of transformer conditions and put to use the experience of other power industry participants.

Based on efficiency analysis for monitoring transformer conditions by different methods using CADG, the CIGRE Working Group 15.01 has come up with an innovative method for interpreting CADG results, *CIGRE WG 15.01*. For large network and block transformers, the method normalizes the concentration of basic (key) gases, see Table 6.2.

The new method provides the following key gas ratios:

Ratio No. 1—C_2H_2/C_2H_6 (acetylene/ethane). Ratio No. 1 is considered to be crucial in determining the presence of electrical discharges; a value larger than a unity indicates a defect.

Table 6.2 Normal concentrations of basic gases according to the innovative method CIGRE WG 15.01

Basic gas, 10^{-6} rel. units	Excess concentration	Possible defect
C_2H_2	20	Intense discharges
H_2	100	Partial discharges
C_XH_Y sum:		
C_1, C_2, C_3 gases	1,000	Thermal defects
C_1 and C_2 gases	500	Thermal defects
CO_2 and CO gases	10,000	Cellulose degradation

Ratio No. 2—H_2/CH_4 (hydrogen/methane). The presence of partial discharges is determined in relation to No. 2. Usually, the value is more than ten. (IEC 60599 uses a methane/hydrogen ratio).
Ratio No. 3—C_2H_4/C_2H_6 (ethylene/ethane). Ratio No. 3 is the ratio of unsaturated hydrocarbons to saturated hydrocarbons and reveals thermal effects. Usually, the value should be larger than a unity. Unsaturated hydrocarbons are formed mainly in overheated oil.
Ratio No. 4—CO_2/CO (carbon dioxide/monoxide). Ratio No. 4 determines the degree of cellulose degradation. If the value is more than ten, then cellulose overheating takes place. If the value is less than three, then it is usually regarded as evidence of cellulose degradation under the influence of electrical defects. A furan analysis according to IEC 61198 is recommended to confirm the diagnosis.
Ratio No. 5—C_2H_2/H_2 (acetylene/hydrogen). Ratio No. 5 is used to determine the penetration of gases into the common tank from the OLVR tank. In this case, the ratio usually equals two or more units, and the C_2H_2 concentration is no less than $30 \cdot 10^{-6}$ rel. units. Since hydrogen is less soluble in transformer oil as compared to acetylene, the latter diffuses from the OLVR tank more rapidly; only a small amount of hydrogen diffuses into the main tank. As a result, the amount of acetylene in the transformer oil exceeds that of hydrogen.

CADG Result Interpretation Procedure

- ratios and concentrations of key gases dissolved in oil are identified;
- if all concentrations are below normal ones, then the result is indicated by K1. If at least one of the concentrations has exceeded the normal range values, then code K2 is assigned;
- if all key ratios are below their specified limits (inside the limits, in the case of CO and CO_2), the result is indicated by code E1;
- if any of the ratios exceeds the specified limits (reaches outside the limits, in the case of CO and CO_2), the result is indicated by code B2.

Combined results:

a. K1 and E1: no measures are taken; the transformer most likely has no defects;
b. K2 and K1: the transformer is likely to have a defect; further analysis is required;
c. K1 and K2: defects in development are possible; further analysis is required;
d. K2 and B2: more than one type of defect is likely; further analysis is required.

For the reliability of a diagnosis, sampling is repeated to calculate the above combinations of codes, with the exception of the first one. Further analysis and other examinations are carried out to find the nature of defects in occurrence. The obtained data are compared with statistical distributions of the probability of a transformer defect—as dependent on the content of H_2, CH_4, C_2H_2, C_2H_4, C_2H_6, N_2, O_2 gases dissolved in oil—available in the databases of various organizations for different voltage classes, workload and service life of transformers. These diagrams show, for different voltage classes and workload levels of a transformer, that the hydrogen concentration does not affect the distribution pattern. For the distribution of a sum of hydrocarbons (C_XH_Y), such dependence does actually exist. The distribution of CO_2 is influenced by voltage classes; CO_2 concentrations are much higher for block transformers, normally operated at base load, than they are for grid transformers.

Research is in progress to determine the "normal" values for various classes of transformers. The innovative method CIGRE WG 15.01 is characterized by a step-by-step assessment of transformer conditions: first, one uses simple methods to determine a possible presence of defects and a need for further analysis of the transformer, and then the nature and severity of defects are identified by a more detailed analysis of CADG data. The new method not only complies with the IEC 60599 recommendations, but also develops them in a way suitable for practice.

According to the Transformers Gas Analyst™ method by Delta-X Research, division of Hydra-Centaurs Technologies Inc (Canada), results of analysis and trends of changes are displayed, and control data are documented. The database also contains monitoring rules and instructions, and thereby makes it possible to reflect the experience of equipment operation and operating conditions.

Analysis of gas from gas relays is regarded by some foreign enterprises as a particularly effective direction for diagnosing the condition of a transformer. Sampling is made when a gas relay is triggered by a signal; then diagnostic methods are used to determine the amount, location and nature of the defect.

Taken together, the results of CADG are indications for extraordinary measurements of winding insulation resistance, dielectric loss angle tangent, winding resistance to direct current, idling losses, thermal imaging control of transformer tank and cooling system surfaces, as well as for execution of CADG in contactor tank oil.

Based on the amount of measurements and their results, a number of questions can be answered:

- the amount and duration of overloading the transformer;
- the temperature in the upper layers of oil and air;
- the possibility of a decrease in the oil level relative to the top of the radiator;
- the operability of the transformer fans and radiators;
- the orientation of the transformer in relation to the sunrays;
- the fact of the transformer being closed on three sides without blowing.

Based on the results, one of the following decisions, or a number of them, must be taken:

- retain the transformer in operation under enhanced CADG control;
- verify the operation of the oil pump engines;
- verify the presence of gas overflow from the contactor tank to the transformer tank;
- degas the oil;
- remove the transformer out for repair.

Each factor of this nature may cause an intense emission of methane. Discrete hotspots cause an increase in the concentration of ethane or ethylene, but if the basic gas is methane then this is caused by a cooling system defect or by personnel overloading the transformer. Numerous problems can be solved by adding heat sinks and fans or by changing transformer load.

6.1.2 Chemical and Physical Indicators for Estimation the Condition of Paper Insulation for Oil—Filled Apparatus in Operation

During the operation of a power transformer, the cellulose insulation of the windings is subject to destruction and dehydration processes aggravating its physical and chemical properties. First of all, this is reflected by deterioration of mechanical strength, oxidation, formation of pores, and chemisorptions of acidic products due to the aging of transformer oil and metallic compounds of variable valence.

The electrical strength of oil-impregnated paper remains largely unaffected by aging, since the damaged areas of cellulose insulation are rapidly filled with oil, and the electrical indicators (insulation resistance and dielectric loss angle tangent) neither change significantly nor can serve as indicators of aging. Assessment of wear/aging for the winding insulation of any particular transformer should include a direct analysis of the physicochemical condition of cellulose insulation and of the related indicators for the degree of degradation in development. In addition, one needs a series of diagnostic features that allow one to make an objective conclusion on the amount of insulation wear and make a decision on the possibility and expediency of further transformer operation.

Due to a rather complex network of parallel and sequential chemical reactions leading to degradation, and the multiplicity of factors affecting the kinetics of their development, it is not possible to predict with the required accuracy the degree of deterioration of the winding insulation by analyzing the effects of operating factors. An assessment of deterioration for the winding insulation of any particular transformer should include a direct analysis for the physicochemical condition of cellulose insulation and for the related indications as to the amount of degradation development. It is evident that an acquisition of numerous diagnostic attributes for estimating the wear of winding insulation should be based on a profound study of physicochemical processes occurring in cellulose insulation under the influence of operational factors, among which the most significant ones are electric field, temperature, air (oxygen), chemically active impurities (aging products) and moisture.

The main physical and chemical phenomena causing the degradation of winding insulation during the operation of power transformers are as follows:

- thermal destruction and dehydration,
- hydrolysis of cellulose insulation,
- oxidative destruction when exposed to acidic products of oil aging and oxidants contained in oil,

– catalytic acid alcoholysis (alcoholysis is a general term for a group of exchange reactions between various organic compounds: epoxides, anhydrides and halides of carboxylic acids).

Electric field enhances the effect of most physicochemical factors and also promotes the adsorption (on the surface of cellulose insulation) of the aging products of transformer oil and construction materials. It also accelerates another important process of cellulose degradation, being catalytic acid alcoholysis under the action of hydroxyl-containing hydrocarbons (alcohols) in the presence of low molecular weight organic acids and other products formed in the oil during aging. The hydrolysis of cellulose insulation, which proceeds in parallel with the process of acid alcoholysis, makes, as compared to the latter, a substantially smaller contribution to overall degradation, which, in particular, is due to a rather low content of moisture in the insulation during normal transformer operation.

An important factor in the aging of cellulose insulation is its thermolysis, caused by elevated temperatures. Under the influence of high temperatures (more than 90 °C) in cellulose insulation, apart from an acceleration of the above-mentioned processes, thermal degradation is also activated, being destruction and dehydration in amorphous and mesomorphic regions with a formation of furfural and furan compounds. In addition, along with the indicated degradation processes, during transformer operation cellulose insulation is subject to oxidative destruction once exposed to acidic products of oil aging and to oxidants contained in these products. This process leads to the formation of oxidized (mainly carboxyl) groups and structure disturbances, as well as to the chemisorptions of low molecular weight degradation products and acidic products of oil aging, copper and iron ions formed during the corrosion of metal elements of a transformer in the course of its operation. This process is accompanied by the release of carbon oxide and dioxide into the oil; a visual sign of catalytic thermal oxidative destruction of cellulose insulation of windings is its dark brown color.

Currently, preference is given to two methods for assessing the condition of paper-oil insulation:

– according to the presence of furan compounds in oil;
– according to the polymerization degree of insulation samples.

It should be noted that the destruction of cellulose insulation during the operation of a transformer can be accompanied by a release of furan compounds into the transformer oil, the most of which should be regarded as furfural and hydroxymethylfurfural. Around 80% of furfural dissolves in insulating oil, and hydroxymethylfurfural is adsorbed mainly in paper insulation.

A permissible content of furan compounds (limiting the normal condition region) is set at no more than 0.0015% by weight (with furfural being no more than 0.001% by weight). For a number of reasons this indicator does not reflect the actual degree of cellulose destruction, namely, the degradation dynamics for cellulose insulation. An objective indicator that makes it possible to assess the wear of winding insulation is the ***degree of polymerization***, which directly characterizes the depth of its physicochemical destruction. At the same time, a decrease in the amount of polymerization has a monotonous dependence and reflects a monotonous decrease in the mechanical strength of insulation, which determines a deterministic diagnostic value of using this indicator. The resource of winding paper insulation is regarded to be exhausted when the degree of polymerization is reduced to 250 units, since in this case there is at least a fourfold decrease in the mechanical strength of insulation in comparison with the original strength. This, in turn, dramatically increases the risk of short-circuited turns and transformer damage when mechanical forces make appearance, primarily in the case of through SC currents.

For an objective assessment of the wear sustained by transformer winding insulation, it is necessary to measure the degree of polymerization of a sample of turn insulation, selected in one of the upper coils. The sampling of coil insulation can be carried out using a disconnected transformer both during overhaul and when partially draining the oil through the hatches.

Moisture contained (dissolved or bound) in transformer oil is one of the most important factors affecting the insulating properties of oil-paper insulation. The immediate reason for a decrease in the electrical density of transformer oil is the presence of moisture dissolved in it. However, bound moisture readily transforms into dissolved moisture, so that it is believed expedient to determine the total amount of moisture in transformer oil. Moisture content up to $200 \cdot 10^{-6}$ rel. units hardly affects oil conductivity and dielectric strength. Once this quantity of moisture is exceeded, treelike structures of increased conductivity start to emerge, and then free water inclusions (drops) begin to appear, sharply reducing the value of electrical strength. Electrical effects and thermal aging of cellulosic material in electrical equipment lead to the appearance of carbon oxides and water, and therefore paper in electrical equipment is not only an adsorbent and principal moisture carrier, but also the main source of water.

Monitoring the process of ***dehydration*** due to polymerization is highly effective, since it allows one to examine the amount of wear sustained by the paper insulation of windings. Output of water released by paper with a polymerization degree of more than 300 rel. units is around $10^{-3} \ldots 10^{-2}\%$ of the mass and has no significant influence on the performance of insulation. Upon reaching the degree of polymerization below 250 rel. units, the release of water due to dehydration may be more than 6% of the mass, which leads to a decrease in the dielectric strength of insulation.

Measurements of polymerization degree carried out to obtain a reliable assessment of the wear sustained by the winding insulation of power transformers should be made by determining the viscosity characteristics of cellulose insulation solutions in a cadmium-ethylenediamine complex. This makes it possible to ensure the absence of significant destructive changes in cellulose samples under examination, including oxidized ones. The use of other solvents tends to cause a chemical degradation of cellulose. Analyzing the polymerization of insulation by converting the latter into ethers may lead to overestimated values of the indicator in question, resulting from dissolution sustained by the low molecular weight fraction, with subsequent erroneous conclusions.

Mechanical impurities contained in oil dramatically reduce the dielectric strength of insulating gaps.

The electrical strength of the oil depends on the size of the impurity particles, since particles of larger size have a larger effect on increasing the field inhomogeneity in the oil, especially if they possess an increased conductivity, in particular due to the presence of moisture. Initially, transformer oil contains very small particles, caused by impurities in crude oil or formed during oil processing. During the manufacture and assembly of a transformer, its oil may be penetrated by cellulose fibers, dust, resin and metal particles. During the operation of a transformer, the concentration of particles, such as cellulose fibers, as well as metal and resin particles, increases as the materials age. The particles are carried throughout the entire transformer bulk by forced circulation of the oil. Local overheating and partial discharges also increase the concentration of carbon particles.

Having determined the nature of contaminants and aging products contained in transformer oil, one may proceed to choose optimal methods for regeneration of the used oils: for some oils a simple cleaning from mechanical impurities is sufficient, while for others deep processing is required, sometimes using chemical reagents. Used oil recovery methods can be divided into physical, physicochemical, chemical and combined ones. In practice, combined methods are normally used to obtain high quality recovered oils.

One particular case of oil contamination is the formation of sparingly soluble colloidal particles of copper and iron naphthenates. To detect colloidal particles contained in oil, turbidity measurements are used to determine the decrease in intensity due to the scattering by the particles of light passing through an oil cuvette.

The above makes it clear that the basic regeneration methods for used oils cannot be applied separately; in practice, it is often necessary to resort to various combinations of methods in order to ensure that the maximum cleaning effect is achieved. When choosing a regeneration method or a combination of methods, it is necessary to take into account the nature of aging products contained in used oils and the requirements for regenerated oils, as well as the amount of collected oils. One also needs to be aware of environmental consequences inherent in certain regeneration methods, so as to choose the most appropriate ones under the circumstances. Having these data available, one can determine the physicochemical properties of oil that need be corrected, and therefore one can choose an appropriate method of recovery.

The problem of removing moisture and impurities is effectively solved by the adsorption purification of oil using thermosyphon filters with an adsorbent, silica gel, due to its good adsorption properties and large specific surface area, being 350–450 m^2/g.

Dehumidified air needed for the operation of a transformer is obtained using air-drying cartridges filled with adsorbent (silica gel) and a small amount of silica gel indicator. The latter helps to determine the degree of adsorbent humidification and the need for its replacement or regeneration.

In the course of transformer overhaul, silica gel reduces the acid number of used transformer oils and gives them a second life. These processes of adsorption purification are extremely important in power engineering and can be achieved using sorbents of high and proven quality.

A new highly effective method for diagnosing the state of insulation of oil-filled equipment was proposed by a team of authors [5, 6]. The method is based on the analysis of phenolic compounds, including ionol and related compounds using chromato-mass spectrometry. (The second name of the method is gas chromatography with a mass selective detector.)

A common method for assessing the condition of oil insulation is to compare the actually measured values of oil quality indicators with the limit values specified in regulatory documents.

6.2 Vibration Control

Over the past decades, vibration diagnostics has become the basis for monitoring and predicting the state of rotating equipment/machines. The reason for its rapid development is the huge amount of diagnostic information contained in the oscillatory forces and vibrations of machines operating in both nominal and abnormal modes. Every machine exhibits dynamic forces. These forces are sources not only of noise and vibration, but also of defects that change the properties of the forces and, accordingly, the characteristics of noise and vibration. Functional diagnostics of machines without changing their mode of operation is the study of dynamic forces, and not vibration or noise itself. The latter simply contain information about dynamic forces, but in the process of converting forces into vibration or noise, some of the information is lost. Even more information is lost when the forces and the work they do are converted into heat energy. That is why, of the two types of signals (temperature and vibration), vibration is preferred in diagnostics.

The technical support of vibration diagnostics is high-precision means of measuring vibration and digital signal processing, the capabilities of which are constantly growing, and the cost is decreasing.

Currently, diagnostic information about the state of rotating machines is extracted from the parameters not only of vibration, but also of other processes, including working and secondary ones, occurring in machines. Naturally, the development of diagnostic systems follows the path of expanding the information received, not only due to the complication of signal analysis methods, but also due to the expansion of the number of controlled processes.

Vibration diagnostics, like any other diagnostics, includes three main areas:

- parametric diagnostics,
- fault diagnostics,
- preventive diagnostics.

Parametric diagnostics is used for emergency protection and equipment control. In this case, diagnostic information is contained in the totality of deviations of the values of these parameters from the nominal values. Parametric diagnostic systems usually include several channels for monitoring various processes, including vibration and temperature of individual equipment components. The amount of vibration information used in such systems is limited, i.e., each vibration channel controls two parameters, namely the value of the normalized low-frequency vibration and the rate of its increase. Usually vibration is normalized in the standard frequency band from 2 (10) Hz to 1000 (2000) Hz. The magnitude of the controlled low-frequency vibration does not always determine the actual state of the equipment, but in a pre-emergency situation, when chains of rapidly developing defects appear, their relationship increases significantly. This allows you to effectively use the means of emergency protection of equipment in terms of low-frequency vibration.

Simplified vibration alarm systems are most widely used. These systems are most often used for timely detection of errors of the personnel operating the equipment.

Fault diagnosis in this case is a vibration maintenance of rotating equipment, called vibration adjustment, which is performed based on the results of monitoring its vibration, primarily to ensure safe vibration levels of high-speed critical machines with a rotation speed of ~3000 rpm and above. Exactly in high-speed machines, increased vibration at rotational speed and multiple frequencies significantly reduces the life of the machine, on the one hand, and on the other hand, it is most often the result of defects in the machine or foundation. Identification of a dangerous increase in machine vibration in steady or transient (start-up) modes of operation, followed by identification and elimination of the causes of this increase, is the main task of vibration adjustment. As part of the vibration adjustment, after detecting the causes of vibration growth, a number of service works are performed, such as centering, balancing, changing the vibrational properties (detuning from resonances) of the machine, as well as replacing the lubricant and eliminating those defects in the machine components or foundation structures that led to dangerous growth vibrations.

Vibration preventive diagnostics (VPD) of machinery and equipment is the detection of all potentially dangerous defects at an early stage of development, monitoring their development and, on this basis, a long-term forecast of the condition of the equipment. VPD of machines, as an independent direction in diagnostics, began to form only at the end of the 80s of the last century. The main task of VPD is not only detection, but also identification of incipient defects. Knowledge of the type of each of the detected defects allows a sharp increase in the reliability of the forecast, since each type of defect has its own rate of development.

Preventive diagnostic systems consist of measuring instruments the most informative processes occurring in the machine, means or software for the analysis of measured signals and software for recognition and long-term prediction of the state of the machine. The most informative processes usually include vibration of the machine and its thermal radiation, as well as the current consumed electric motor used as an electric drive, and the composition of the lubricant. To date, only the most informative processes have not been determined, which make it possible to determine and predict the state of electrical insulation in electrical machines with high reliability.

Preventive diagnostics based on the analysis of one of the signals, for example, vibration, has the right to exist only in those cases when it allows you to detect the absolute (more than 90%) number of potentially dangerous types of defects at an early stage of development and give a forecast of trouble-free operation of the machine for a period sufficient to prepare for current repairs. This possibility can currently not be implemented for all types of machines and not for all industries. The greatest successes of VPD are connected with the forecast of the state of low-speed loaded equipment used, for example, in metallurgy, paper and printing industries. In such equipment, vibration does not have a decisive influence on its reliability, i.e., special measures to reduce vibration are used extremely rarely. In this situation, the vibration parameters most fully reflect the condition of the equipment units, and taking into account the availability of these units for periodic vibration measurement, preventive diagnostics gives the maximum effect at minimum cost.

The most complex problems of VPD are solved for reciprocating machines and high-speed gas turbine engines. In the first case, the useful vibration signal is repeatedly blocked by vibration from shock pulses that occur when the direction of movement of the inertial elements changes, and in the second case, by the flow noise, which creates strong vibration interference at those control points that are available for periodic vibration measurement. The success of VPD machines with a speed of rotation from ~300 to ~3000 rpm also depends on the type of diagnosed machines and the features of their work in different industries. The problems of monitoring and predicting the state of common pumping and ventilation equipment are most simply solved, especially if it uses rolling bearings and an asynchronous electric drive. Such equipment is used in almost all industries and in public utilities.

Diagnostics of electrical equipment of power plants and substations and its transfer for maintenance and repair according to the actual state does not require large financial and time costs.

The working body is a source of a large number of vibrations of various nature. As a result of improvement, vibration control has become a very effective method for monitoring the condition of not only rotating machines, but also power transformers. For these devices, the scale of application of the VPD is less than for rotary machines, but, nevertheless, their use as an auxiliary diagnostic method is justified.

The weakening of winding pressing can be regarded as the initial stage of damaged condition, causing the development of more serious and dangerous winding defects, such as coil descent and buckling, or turn-to-turn short circuits. Vibration control technology is used to monitor the degree of winding pressing. The purpose of transformer vibration diagnostics is to estimate the condition of a mechanical system, as well as to identify and eliminate defects in both external equipment (damage to pipelines due to resonant vibrations, wear of bearings of oil pumps and fans) and internal systems (unpressing of windings and magnetic conductor, displacement of magnetic shunts due to vibration).

Transformer vibrations are polyharmonic vibrations with frequencies being multiples of 100 Hz. The source of transformer vibrations is a magnetic conductor. The cause of this phenomenon is magnetostriction. Electric motors of oil pumps and fans are vibration sources by themselves, but their energy is much smaller. The vibration frequency of attached equipment is due to the rotational speed of electric motors, being approximately 720…1440 rpm. Vibrations are transmitted from sources to other units and elements of a transformer. An examination begins with measuring the vibration of a transformer tank. The most important vibration characteristics are as follows:

- vibration velocity, characterizing the vibration energy whose values are used to estimate the condition of the transformer tank, as well as to assess the impact of the transformer on the foundation;
- vibration acceleration, characterizing the inertial forces acting on the transformer tank as a result of the movement of its internal elements;
- vibration displacement, characterizing the vibration loads crucial for the condition of the transformer tank, welds and other elements.

The frequency spectrum of vibration velocity allows one to identify the vibration sources. Measurements are carried out in the frequency range up to 1000 Hz, since more than 90% of the total vibration energy of a transformer is concentrated in this range.

For a general assessment of transformer condition, a need for additional examinations arises at the following parameter values:

- vibration acceleration being more than 10 m/s^2,
- vibration speed being more than 20 mm/s,
- vibration displacement being more than 100 microns.

The condition of fans and oil pumps depends on the cooling system design. It can be estimated on a basis of the following criteria:

- indicator of defects in a blower fan is a vibration velocity of the bearings being higher than 7.1 mm/s;
- damaged condition of an oil pump is related to a vibration velocity being higher than 4.5 mm/s.

The pressing quality of the windings and magnetic conductor can be determined by examining the vibration spectrum on the surface of the transformer tank. Measurements are carried out in two modes, idle and loaded. It is assumed that in the idle mode the vibrations are caused by magnetostriction in the magnetic conductor, while in the loaded mode the additional effect of electromagnetic forces in the windings is taken into account. A decrease in the compression force of magnetic conductor sheets below a nominal value leads to an emergence of frequencies of 300…500…700 Hz. A weakening of winding pressing leads to a decrease in the 200 Hz frequency component.

The quality of winding pressing can also be estimated by measuring the natural frequencies of winding vibrations under mechanical shock. The basic method is

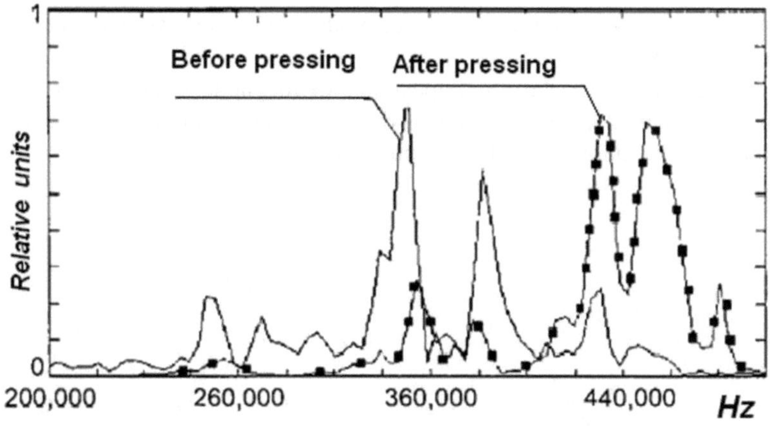

Fig. 6.1 EMF spectrum induced in the winding of a TTs-630000/500 transformer under pulsed mechanical action on the transformer winding at different degrees of winding pressing

registration of EMF induced on the windings at the bushings of a bus-bar transformer during a pulsed mechanical action. This process has the form of damped oscillations. The spectrum of EMF induced under pulsed mechanical action on transformer windings at different degrees of winding pressing is different and is shown in Fig. 6.1.

The disadvantages of this method are as follows:

- higher requirements for mounting a vibration sensor;
- dependence of vibration parameters on a large number of factors, combined with the problem of isolating a vibration signal due to malfunction, which requires the use of methods of correlation and regression analysis.

Despite these disadvantages, the vibration control of the pressing degree in transformer windings is developing in many countries with developed energy and electrical engineering.

The technical support of vibration diagnostics is high-precision means of measuring vibration and digital signal processing, the capabilities of which are constantly growing, and the cost is decreasing.

Based on the results of vibration diagnostics, signal forms and vibration spectra are compiled. Comparison of the waveforms, but already with the reference one, can be carried out using information spectral technology based on narrow-band spectral analysis of signals. When using this type of signal analysis, diagnostic information is contained in the ratio of the amplitudes and initial phases of the main component and each of its frequency multiple components [7].

Their analysis shows that the appearance of magnetic saturation of the active core is accompanied by a distortion of the shape and an increase in the vibration components at the harmonics of the supply voltage. Forms and vibration spectra of the transformer core operating in normal mode and during overload are shown in Figs. 6.2 and 6.3.

Their analysis shows that the appearance of magnetic saturation of the active core is accompanied by a distortion of the shape and an increase in the vibration components at the harmonics of the supply voltage.

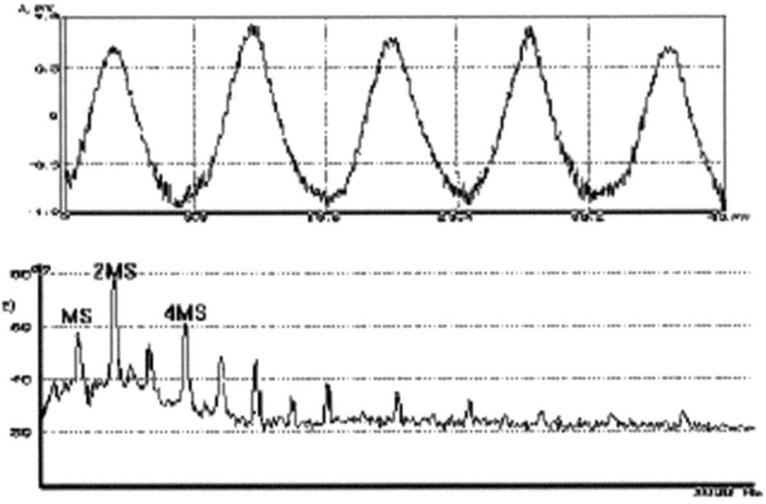

Fig. 6.2 Forms and vibration spectra of the core of a transformer operating in normal mode

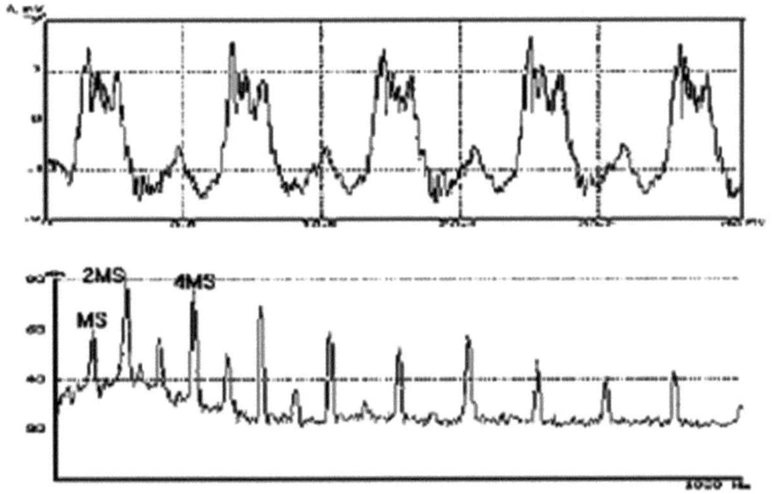

Fig. 6.3 Forms and vibration spectra of the transformer core during overload, accompanied by magnetic saturation of the core

6.3 Thermal Methods and Devices for Control and Diagnostics

Thermal control methods (TCM) are based on the measurement, evaluation and analysis of the temperature of controlled objects. The main condition application of diagnostics with the help of TCM is the presence of heat flows in the diagnosed object. Temperature is the most universal reflection of the state of any electrical equipment. Temperature reactions in different operating modes, due to their versatility, occur at all stages of operation of electrical equipment. In almost any non-normal operating mode of the equipment, a change in temperature is the very first indicator indicating a malfunction.

Infrared diagnostics (IRD) is the most promising and an effective direction for the development of diagnostics of electrical equipment. It has a number of advantages over traditional testing methods, namely:

1. reliability, objectivity and accuracy of the information received;
2. personnel safety during equipment inspection;
3. no need to turn off the equipment;
4. no need to prepare the workplace;
5. a large amount of work performed per unit of time;
6. the possibility of identifying defects in the early stages of development;
7. the ability to diagnose most types of electrical equipment;
8. energy systems;
9. low labor costs for the production of measurements of a piece of equipment.

The development of almost all types of defects causes a change in the temperature of defective elements and, as a consequence, a change in the intensity of infrared radiation (IRR), which can be recorded by thermal imaging devices.

In the power system, TCM can be used to diagnose the following equipment:

1. power and instrument transformers and their high-voltage bushings;
2. switching equipment: circuit breaker, disconnectors;
3. dischargers and surge arresters;
4. busbar distribution devices;
5. insulators;
6. contact connections;
7. generators (frontal parts and active steel);
8. overhead transmission lines (OTL) and their structural elements (for example, OTL supports), etc.

TCM for high-voltage equipment as one of the modern methods of research and control was introduced in 1998, although in many power systems was used much earlier.

Among ***the devices*** used for the implementation of TCM, a special place is occupied by a thermal imaging measuring device (***thermal imager***). This is an optical-electronic device designed for non-contact (remote) observation, measurement and

a b c

Fig. 6.4 Appearance of the thermal imager: **a** professional thermal imager; **b** stationary thermal imager for systems of continuous leg control and monitoring; **c** the simplest compact portable thermal imager

registration of the space–time distribution of the radiation temperature of objects in the field of view of the device, by generating a time sequence of thermograms and determining the surface temperature of an object using known emissivity and shooting parameters (ambient temperature, atmospheric transmission, observation distance, etc.). A thermal imager is a kind of television camera that shoots objects in infrared radiation, which allows you to get a real-time picture of the distribution of heat (temperature difference) on the surface.

Thermal imagers come in various modifications, but the principle of operation and their designs are approximately the same, as shown in Fig. 6.4a, b, c.

The range of measured temperatures, depending on the brand and type of thermal imager, can be from −40 °C to +2000 °C. The operation of all thermal imagers is based on fixing the "object/background" temperature difference and converting the received information into an image (*thermogram*) visible to the eye. A thermogram is a multi-element two-dimensional image, each element of which is assigned a color/ or gradation of one color/gradation of brightness screen, determined in accordance with the conditional temperature scale. That is, the temperature fields of objects are considered in the form of a color image, where color gradations correspond to temperature gradations. All colors on the thermograms are rather conditional and do not correspond to the real colors. The choice of the color palette of the thermogram depends on the range of use temperatures. Changing the color palette is used to increase contrast and effectiveness of visual perception (informativeness) thermograms. The number and types of palettes depend on the manufacturer of the thermal imager.

Also used for non-contact temperature measurement ***pyrometers***, the operating principle of which is based on measuring the thermal radiation power of the object being measured, predominantly in the infrared range. On Fig. 6.5 shows the appearance of two different pyrometers.

The range of measured temperatures, depending on the brand and type of pyrometer, can be from −100 °C to +3000 °C.

Fig. 6.5 Appearance of the pyrometers

The fundamental difference between pyrometers and thermal imagers is that pyrometers measure the temperature at a specific point (up to 1 cm), and thermal imagers analyze the entire object, showing the entire difference and temperature fluctuations at any point.

When analyzing the results of IRD, it is necessary to take into account the structures of the diagnosed equipment, methods, conditions and duration operation efficiency, manufacturing technology and a number of other factors.

In Table 6.3 from [8] shows the main types of electrical equipment on the substations and types of defects detected using IRD.

6.4 Magnetic Structuroscopy

Magnetic methods of non-destructive testing are mainly used to control products made of ferromagnetic materials, i.e. products that, under the influence of an external (magnetizing) magnetic field, change their magnetic characteristics: magnetic permeability, magnitude and direction of the magnetic flux, due to field disturbances caused by defects. Ferromagnetic materials include: iron, steel, cast iron, nickel, cobalt and some alloys (aluminum with nickel, etc.)—materials from which most of the high-voltage equipment of electric power systems is made [9, 10].

Magnetic non-destructive testing is based on the detection by various methods of stray magnetic fields arising above defects, or on the determination and evaluation of the magnetic properties of the test object. Their list and purpose are presented in Table 6.4 [11].

The magnetic particle method is based on the detection of magnetic fields scattering arising over defects in a part during its magnetization, using a ferromagnetic powder or magnetic suspension as an indicators. This method, among other methods of magnetic control found the greatest application. Approximately 80% of all subjects control of parts made of ferromagnetic materials is checked exactly by this method.

Table 6.3 Types of defects in high-voltage equipment detected by IRD

High voltage equipment	Detectable faults
Generators	Interleaf stator faults Deterioration of winding rations Assessment of the thermal state of the brush apparatus Malfunction of stator cooling systems Checking the elements of the excitation system
Transformers	Locations of stray magnetic fields Formation of stagnant zones in transformer tanks due to sludge formation, swelling or displacement of winding insulation, oil system failure Input defects Evaluation of the efficiency of cooling systems
Switching equipment	Overheating of contacts of current-carrying tires, working and arcing chambers The state of tank insulation Defects in inputs, dividing transformers Cracks in post insulators
Oil-filled transformers	Overheating of external and internal contact connections Deterioration of the internal insulation of the windings
Valve dischargers and surge arresters	Violation of the tightness of the elements Breakage of shunt resistances Wrong set of elements
Condensers	Breakdown of element sections
Linear high-frequency barriers	Overheating of contact connections
Complete switchgear for indoor installation (CSGII) and complete switchgear for external installation (CSGEI), conductors	Overheating of contact connections of switches, disconnectors, current transformers, cables, busbars, etc.
Cable facilities of power plants	Overheating of power cables, assessment of fire hazard of cables

High sensitivity, versatility, relatively low labor intensity of control and simplicity—all this provided it with a wide some application in industry in general and in electrical engineering in particular.

The main disadvantage of this method is the complexity of its automation.

The induction method involves the use of a receiving coil inductance moved relative to a magnetized part or another magnetized controlled object. The coil is induced EMF, the value of which depends on the speed relative displacement of the coil and the characteristics of the magnetic fields of defects. A method of magnetic flaw detection, in which the measurement of distortion magnetic field arising in places of defects in products made of ferromagnetic filament materials is carried out

Table 6.4 Types of magnetic methods

No.	Control methods	Basic control parameters	Purpose of control
1	Magnetic particle	Powder drawing over a stray magnetic field	Detection of defects such as metal discontinuity (surface, subsurface and internal defects)
2	Magnetographic	The magnetization of a tape over a stray magnetic field	
3	Induction	Instantaneous electromotive force (EMF) induced in a moving coil	
4	Ferroprobe	EMF, proportional to the gradient or strength of the measured magnetic field	Thickness measurement of iron sheets, magnetized coatings, hardened layers including superficial one
5	Ponderomotive	Force of attraction of a permanent magnet or an electromagnet to a controlled object	
6	Hall effect method	Hall EMF arising in a semiconductor with current under the influence of a transverse magnetic field	
7	Magnetoresistor	The electrical resistance of a semiconductor with current, changing under the influence of an external magnetic field	
8	Magnetic memory method	Stress concentration zones	–

by ***ferroprobes***. Measuring device and indication of magnetic fields (mostly constant or slowly flowing) and their gradients.

The Hall effect method is based on the detection of magnetic fields by Hall's converters. The essence of the Hall effect lies in the occurrence transverse potential difference (EMF Hall) in a rectangular semiconductor plate as a result of curvature of the path of the flowing through this plate of electric current under the influence of a magnetic current perpendicular to this current. The Hall effect method is used to detection of defects, measurement of the thickness of coatings, control of the structure and mechanical properties of ferromagnets, registration of magnetic fields.

The ponderomotive method is based on the measurement of the breakaway force a permanent magnet or an electromagnet core from a controlled object. In other words, this method is based on the ponderomotive interaction of the measured magnetic field and the magnetic field of the frame with the current of an electromagnet or a permanent magnet.

The magnetoresistor method is based on the detection of magnetic fields by magnetoresistive transducers, which are galvanomagnetic elements, the principle of operation of which is based on magnetoresistive Gaussian effect. This effect is associated with a change in the longitudinal resistance of a current-carrying conductor under the influence of a magnetic field. In this case, the electrical resistance increases due to the curvature of the trajectory of charge carriers under the influence of a magnetic field. Quantitatively, this effect manifests itself in different ways and depends on the material of the galvanomagnetic element and its shape. For conductive materials, this effect is not typical. It mainly manifests itself in some semiconductors with high mobility of current carriers.

Magnetic particle flaw detection is based on the detection local magnetic stray fields arising above the defect with the help of ferromagnetic particles playing the role of an indicator. A stray magnetic field arises above a defect due to the fact that in a magnetized part, magnetic field lines, encountering a defect on their way, go around it as an obstacle with low magnetic permeability, as a result of which the magnetic field is distorted, individual magnetic field lines are displaced by the defect to the surface, exit details and enter it back. The stray magnetic field in the defect zone is the greater, the larger the defect and the closer it is to the surface of the part.

Thus, magnetic methods of non-destructive testing can be apply to all electrical equipment consisting of ferromagnetic materials.

6.5 Acoustic Control Methods

Acoustic control methods (ACM) are used to control products, in the material of which radio waves do not attenuate much: dielectrics (glass-fibers, plastics, and ceramics), semiconductors, magnetodielectrics (ferrites), thin-walled metal materials.

Lack of non-destructive testing by radio wave method—low resolution devices, which are based on lies this method, due to the small depth of penetration of radio waves.

ACM are divided into two large groups: *active* and *passive* methods. Active methods are based on the emission and reception of elastic waves, passive methods are based only on the reception of waves by a source which the object of control itself serves, for example, the formation of cracks accompanied by the appearance of acoustic vibrations detected by acoustic—emission method. Active methods are divided into methods of reflection, transmission, combined (using both reflection and transmission), natural oscillations.

Reflection methods based on pulse reflection analysis elastic waves from inhomogeneities or boundaries of the control object, ***methods passing*** based on the influence of the parameters of the control object on the characteristics of the waves that have passed through it. ***Combined methods*** use the influence of the parameters of the test object both on the reflection and on the passage of elastic waves. In ***natural oscillation methods***, the properties of the control object are judged by the parameters of its free or forced oscillations (their frequencies and the magnitude of losses).

Thus, according to the nature of the interaction of elastic vibrations with controlled material, acoustic methods are divided into the following main methods:

1. transmitted radiation (shadow, mirror-shadow);
2. reflected radiation (echo-pulse);

3. resonant;
4. impedance;
5. free vibrations;
6. acoustic emission.

By the nature of registration of the primary informative parameter, acoustic methods are divided into *amplitude, frequency, spectral.*

ACM solve the following control and measuring tasks:

1. The transmitted radiation method reveals deep defects of the type discontinuity, delamination, non-riveting, non-soldering;
2. The method of reflected radiation detects defects such as discontinuity, determines their coordinates, sizes, orientation by sounding the product and receiving the echo signal reflected from the defect;
3. The resonance method is mainly used to measure the thickness of a product (sometimes it is used to detect a zone of corrosion damage, non-soldering, delamination of thin sheets of metals);
4. Acoustic emission method detects and registers only developing or capable of development under the influence of mechanical crack load (qualifies defects not by size, but by the degree of their danger during operation). The method has a high sensitivity to the growth of defects—detects an increase in the crack by (1 … 10) microns, than measurements are usually carried out under operating conditions in the presence of mechanical and electrical noise;
5. The impedance method is designed to control adhesive, welded and solder joints having a thin skin, glued or soldered to the elements that provide structural rigidity. Defects in adhesive and solder joints are detected only on the side of the input of elastic vibrations;
6. The method of free vibrations is used to detect deep defects.

Acoustic signals in equipment caused by electrical discharges, can be detected even against the background of interference: vibration knock, noise oil pumps and fans, etc. An electrical discharge is initiated at the site of damage and then sound vibrations generated by it are heard over the site of damage.

Acoustic methods are applied not only to large-sized equipment (for example, transformers), but also to such equipment as cables. When diagnosing cable lines, an electric discharge is initiated at the point of damage and then the sound vibrations generated by it are heard over the place of damage. This method is used to detect all types of damage with the condition that at the place of damage can be created electrical discharge. For the occurrence of a stable spark discharge it is necessary that the value of the contact resistance at the fault site exceeds 40 ohms. The audibility of sound from the earth's surface depends on the depth of the cable, the density of the soil, the type of cable damage and the power of the discharge pulse. The listening depth ranges from 1 to 5 m. The use of this method on openly laid cables, cables in channels, tunnels is not recommended, since due to the good propagation of sound along the metal sheath of the cable, a large error in determining the location of damage can be made.

As an acoustic sensor, piezo sensors or electromagnetic systems that converts mechanical vibrations of the ground into electrical signals entering the input of the audio frequency amplifier are used. Above the damage site, the signal is greatest.

When using ***reflection methods***, information is obtained from the reflection of acoustic waves in the control object (CO). The reflection method is implemented in several ways.

The echo method is based on the registration of echo signals from defects—discontinuities. It is similar to radar and sonar. Other methods reflections are used to search for defects that are poorly detected by the echo method, as well as to study the parameters of defects.

The echomirror method is based on the analysis of acoustic impulses, specularly reflected from the bottom surface of the CO and the defect. Variant of this method designed to detect vertical defects is called *tandem method*.

The delta method is based on the use of wave diffraction by a defect.

Part of the transverse wave incident on the defect from the emitter is scattered in all directions at the edges of the defect, and partially turns into longitudinal wave. Some of these waves are received by a longitudinal wave receiver located above the defect, and some are reflected from the bottom surface and also arrive at the receiver. Variants of this method suggest the possibility of moving the receiver along the surface, changing the types of emitted and received waves.

The diffraction-time method is based on the reception of waves, scattered at the ends of the defect. In this case, both longitudinal and transverse waves can be emitted and received.

Acoustic microscopy differs from echo method by increasing by one or two orders of magnitude of the ultrasound frequency, using sharp focusing and automatic or mechanized scanning of small objects size. As a result, it is possible to fix small changes acoustic properties in CO. The method allows reaching a resolution of hundredths of a millimeter.

Coherent methods differ from other reflection methods in that as an information parameter, in addition to amplitude and time, the pulse path also uses the phase of the signal. Thereby the resolving power of reflection methods increases by an order of magnitude and it becomes possible to observe images of defects that are close to real ones.

Passing Methods

These methods, often referred to as *shadow methods* in Russia, are based on the observation of changes in the parameters of an acoustic signal that has passed through the CO. At the initial stage of development of the method, continuous radiation was used, and the sign of a defect was a decrease in the amplitude of the through signal caused by the sound shadow formed by the defect. Therefore, the term “shadow” adequately reflected the content of the method. However, in the future, the areas of application of the methods under consideration have expanded.

Methods began to be used to determine the physical and mechanical properties of materials, when controlled parameters are not related to discontinuities that form a sound shadow. Thus, the shadow method can be considered as a special case of the more general notion of *"passing method"*. When monitoring by passing methods, the emitting and receiving transducers are located on opposite sides of the CO or the controlled area. In some methods of passage, the transducers are placed on one side of the CO at a certain distance from each other. Information is obtained by measuring the parameters of the end-to-end signal transmitted from the emitter to the receiver.

Amplitude passing method (or amplitude shadow method) is based on registering a decrease in the amplitude of the through signal under the influence of a defect that impedes the passage of the signal and creates sound shadow.

The temporal passing method (temporal shadow method) is based on the measurement of the pulse delay caused by defect rounding. In this method, the information parameter is the time of arrival of the end-to-end signal. The method is effective in testing materials with high ultrasonic scattering, such as concrete, etc.

The multiple shadow method is similar to the amplitude transmission method (shadow), but the presence of a defect is judged by the amplitude of the through signal (shadow pulse), which repeatedly (usually twice) passed between the parallel surfaces of the product. The method is more sensitive than the shadow or mirror-shadow method, since the waves pass through the defective zone several times, but it is less noise-resistant.

The variations of the passing method discussed above use to detect defects such as discontinuity.

Photoacoustic microscopy. When implementing this method acoustic vibrations are generated due to the thermoelastic effect when lighting CO with a modulated light flux (for example, pulsed laser) focused on the surface of the CO. Energy light flux, being absorbed by the material, generates a thermal wave, parameters of which depend on the thermophysical characteristics of the CO. Thermal wave leads to the appearance of thermoelastic oscillations, which registered, for example, by a piezoelectric detector.

The velocimetric method is based on registering changes in speed elastic waves in the defect zone. For example, if in a thin product a flexural wave propagates, then the appearance of stratification causes a decrease in its phase and group velocities. This phenomenon is fixed by the phase shift of the transmitted wave or the delay in the arrival of the pulse.

The term *ultrasonic tomography* is often used in relation to various defect imaging systems. Meanwhile, initially it was used for ultrasound systems in which they tried to implement the approach, repeating x-ray tomography, i.e. through sounding CO in different directions, highlighting the features of the CO obtained with rays in different directions.

Laser detection method. Known methods of visual representations of acoustic fields in transparent liquids and solid media based on the diffraction of light by elastic waves.

The thermoacoustic control method (also called *ultrasonic local thermography*) consists in introducing powerful low-frequency (~20 kHz) ultrasonic vibrations. At the defect, they are converted into heat. The greater the effect of a defect on the elastic properties of a material, the greater the elastic hysteresis and the greater the heat release. The temperature rise is recorded by a thermal imager.

Combined Methods

These methods contain features of both reflection methods and methods passing.

The mirror-and-shadow method is based on measuring the amplitude of the bottom signal. According to the execution technique (an echo signal is fixed)—this is a reflection method, and in terms of physical essence (the attenuation by a defect is measured passed CO twice) it is close to the shadow method, so it is not classified as passing methods, but as combined methods.

The echo shadow method is based on the analysis of both transmitted and reflected waves.

The reverberation-through (acoustic–ultrasonic) method combines the features of the multiple shadow method and the ultrasonic reverberation method.

On CO of small thickness at some distance from each other install direct emitting and receiving transducers. The emitted pulses of longitudinal waves after multiple reflections from the walls of the CO reach the receiver. The presence of inhomogeneities in the CO changes the conditions for the passage of pulses. Defects are registered by changing the amplitude and spectrum of the received signals. The method is used to control products made of polymer composite materials (PCM) and joints in multilayer structures.

Free Oscillation Methods

These methods are based on excitation in the CO of forced or free oscillations and measuring their parameters: natural (own) frequencies and losses. Free vibrations excite by short-term impact on CO (for example, mechanical shock), after which it fluctuates in the absence of external influences. Forced oscillations are created by the action of an external force with a smoothly variable frequency (sometimes long pulses with a variable carrier frequency are used). Resonant frequencies are recorded by increasing the amplitude of oscillations when the natural frequencies of the CO coincide with the frequencies of the disturbing force. Under the influence of the excitation system, in some cases, the natural frequencies of the CO change slightly, so the resonant frequencies differ somewhat from the natural frequencies. The oscillation parameters are measured without stopping the action of the exciting force.

There are *integral* and *local methods*. In *integral* methods analyze the natural frequencies of CO as a whole, in *local*—its individual sections. Informative parameters are the values frequencies, spectra of natural and forced vibrations, as well as characterizing the losses, the quality factor and the logarithmic damping decrement. Integral methods of free and forced vibrations provide for the excitation of vibrations in the entire product or in a significant part of it. The methods are used to control the physical and mechanical properties of products made of ceramics, metal castings and some others materials. These methods do not require scanning and are highly productive, but do not provide information about the location and nature of defects.

The local method of free vibrations is based on excitation free vibrations in a small section of CO. The method is used to control layered structures by changing the frequency spectrum in the part of the product excited by impact; for measuring thicknesses (especially small ones) of pipes and other COs by exposure to a short-term acoustic pulse. Local method of forced oscillations (ultrasonic resonance method) is based on the excitation of oscillations, the frequency of which is smoothly changed.

To excite and receive ultrasonic vibrations, combined or separate converters are used. When the excitations frequencies coincide with own frequencies, CO (loaded transceiver converter) resonances arise in the system. A change in thickness will cause a shift in resonant frequencies; the appearance of defects will cause the disappearance of resonances.

The acoustic-topographic method has features of both integral and local methods. It is based on arousal in CO intensive bending vibrations of continuously changing frequency and registration distribution of amplitudes of elastic vibrations on the surface of the CO with the help of a finely dispersed powder applied to the surface. A smaller amount of powder settles on the defective area, which is explained by an increase in the amplitude of its oscillations as a result of resonance phenomena. The method is used to control joints in multilayer structures: bimetallic sheets, honeycomb panels, etc.

Impedance Methods

These methods are based on the analysis of the change in mechanical impedance or the input acoustic impedance of the CO surface area, with which converter interacts. Within the group, methods are divided according to types of waves excited in the CO and by the nature of the interaction converter with CO. The method is used to control joint defects in multilayer structures. It is also used to measure hardness and other physical and mechanical properties of materials.

The ultrasonic method defectoscopy deserves special consideration.

Ultrasonic defectoscopy is applied not only to massive large-sized equipment (for example, transformers), but also to extended objects (for example, cable products).

The main types of equipment for ultrasonic flaw detection:

1. An oscilloscope that allows you to register an oscillogram of a signal and its spectrum;
2. An ultrasonic probe that uses heterodyning signal and listening to the converted signal spectrum through the head-phones (Fig. 6.6);
3. Ultrasonic moderator that allows you to record, slow down and hear the ultrasonic signal.

6.6 Acoustic Emission Diagnostics

Acoustic emission *(AE)* is a physical phenomenon that underpins effective non-destructive testing and material evaluation technology and involves the detection of elastic waves generated by the sudden deformation of a stressed material. These

Fig. 6.6 Probe

waves propagate from the source to the sensor(s), where they are converted into electrical signals. AE instruments measure these signals and display data, on the basis of which the operator evaluates the state and behavior of the structure under tension. Traditional methods of non-destructive testing (ultrasonic, radiation, eddy current) detect geometric inhomogeneities by exposing the structure under study to energy in one form or another.

AE provides a different approach: detecting microscopic motion rather than geometric inhomogeneities. Crack growth, inclusion fracture, and liquid or gas leakage are examples of the hundreds of processes that produce acoustic emissions that can be detected and effectively investigated using this technology. A growing defect produces its own signal, which travels meters, and sometimes tens of meters, until it reaches the sensors. Not only can a defect be detected remotely; it is often possible to find its location by processing the difference in arrival times of waves to different sensors [12].

AE-control has the following advantages:

1. The method ensures the detection and registration of only developing defects, which makes it possible to classify defects not by size, but by their degree of danger;
2. Under production conditions, the AE method makes it possible to detect a crack increment by tenths of a millimeter;
3. The property of the integrity of the method provides control of the entire object using one or more AE transducers, fixedly installed on the surface of the object at a time;
4. The position and orientation of the defect do not affect the detectability;
5. The AE method has fewer restrictions related to properties and structure of structural materials than other methods of non-destructive monitoring control;
6. Allows you to control areas inaccessible to other methods (thermal and waterproofing, design features);

7. The AE method makes it possible to prevent catastrophic destruction of structures during testing and operation by estimating the rate of development of defects;
8. Provides the ability to locate leaks.

6.7 Radiation Diagnostics

When implementing this method, X-ray, gamma radiation, neutron fluxes and other penetrating radiation are used. Passing through the thickness of the product, they are weakened in different ways in defective and defect-free sections and carry information about the internal structure of the substance and the presence of defects inside the product. Radiation control methods are used to control welded and brazed seams, castings, rolled products, etc. They belong to one of the types of non-destructive testing.

When using destructive testing methods, selective control (for example, by cut samples) of a series of products of the same type is carried out and its qualities are statistically evaluated without establishing the quality of each specific product. At the same time, some products are subject to high quality requirements, which necessitate complete control. Such control is provided by non-destructive methods, which are mainly amenable to automation and mechanization.

The quality of a product is determined by the totality of its properties, which determine its suitability to satisfy certain needs in accordance with its purpose. This is a capacious and extensive concept, which is influenced by a variety of technological and design-operational factors. For an objective analysis of product quality and management, they involve not only a set of non-destructive testing methods, but also destructive tests and various checks, and control at various stages of product manufacturing. For critical products designed with a minimum margin of safety and operated in harsh conditions, 100% non-destructive testing is used.

Radiation non-destructive testing is based on obtaining flaw detection information about an object using ionizing radiation, the passage of which through a substance is accompanied by ionization of atoms and molecules of the medium. The test results are determined by the nature and properties of the ionizing radiation used, the physical and technical characteristics of the controlled object, the type and properties of the detector, the control technology and the qualifications of the flaw inspectors.

There are *direct* and *indirect ionizing radiation. Direct ionizing radiation* is formed by charged particles (electrons, protons, α-particles, etc.) that have sufficient kinetic energy to ionize the environment upon collision. *Indirect ionizing radiation* consists of photons, neutrons, or other uncharged particles that can produce direct ionizing radiation and/or cause nuclear transformations. X-ray films, semiconductor gas-discharge and scintillation counters, ionization chambers, etc. are used as detectors in radiation methods.

Radiation methods of flaw detection control are designed to detect macroscopic discontinuities in the material (controlled defects) that occur during the manufacture of products (cracks, porosity, shells, etc.), to determine the internal geometry of parts and assemblies (variation in wall thickness and deviations of the shape of internal contours from those specified according to drawing in parts with closed cavities, improper assembly of units, gaps, loose fittings in joints, etc.). Radiation methods are also used to detect defects that appeared during operation: cracks, corrosion of the inner surface, etc.

Depending on the method of obtaining primary information one should distinguish between such methods as radiographic, radioscopic, radiometric control, as well as the method of registration of secondary electrons. The essence of these methods is as follows.

The radiographic method is based on converting a radiation image of a controlled object into a radiographic image or recording this image on a memory device with subsequent conversion into a light image. A radiographic image is a distribution of blackening density (or color) on X-ray film and photographic film, light reflectance on a xerographic image, etc., corresponding to the radiation image of the objet under control. Depending on the type of detector used, a distinction is made between radiography itself—registration of the shadow projection of an object onto an x-ray film—and *electroradiography.* If color photographic material is used as a detector, i.e., the gradations of the radiation image are reproduced as color gradations, then one speaks of *color radiography*.

Radioscopic means a method of radiation monitoring, based on radiation image transformation of the controlled object into a light image on the output screen of the radiation-optical converter, and the resulting image is analyzed in the control process. When used as a radiation-optical converter of a fluorescent screen or in a closed-circuit television system of a color monitor, a distinction is made between *fluoroscopy* and *color radioscopy*. X-ray machines are mainly used as radiation sources, less often accelerators and radioactive sources.

The radiometric method is based on the measurement of one or more parameters of ionizing radiation after its interaction with a CO. Depending on the type of ionizing radiation detectors used, *scintillation* and *radiation monitoring* are distinguished. As radiation sources, radioactive sources and accelerators are mainly used, and X-ray machines are also used in thickness measurement systems.

There is also a *method of secondary electrons*, when a flow of high-energy secondary electrons formed in as a result of the interaction of penetrating radiation with controlled object.

By the nature of the interaction of physical fields with CO distinguish methods of transmitted radiation, scattered radiation, activation analysis, characteristic radiation, field emission. Transmitted radiation methods are practically all classical methods of x-ray and gamma flaw detection, as well as thickness measurement, when various detectors register the radiation that has passed through the controlled object, i.e., useful information about controlled parameter carries, in particular, the degree of attenuation radiation intensity.

The activation analysis method is based on the analysis of ionizing radiation, the source of which is induced radioactivity controlled object, which arose because of the impact on it of primary ionizing radiation. The induced activity in the analyzed sample is created by neutrons, photons or charged particles. According to the measurement of induced activity, the content of elements in various substances is determined.

Characteristic radiation methods include methods X-ray radiometric (adsorption and fluorescent) analysis. In its essence, this method is close to the classical X-ray spectral method and is based on the excitation of atoms of the elements being determined by the primary radiation from the radionuclide and subsequent registration of the characteristic radiation of excited atoms.

Question and Tasks

1. What are the quality indicators of transformer oil and the types of defects detected with their help.
2. Decipher the abbreviation CADG and explain the essence of this technique.
3. Explain the scientific and technical foundations of vibration diagnostics and the areas of its most effective application.
4. Name the main methods and instruments of control with the general name "Thermal methods of control and diagnostics".
5. What equipment of electric power systems can be diagnosed by thermal methods?
6. Name the methods of control, which are united by the general name "Magnetic methods".
7. Explain the nature of panderomotor forces.
8. Explain the nature of the Hall effect.
9. Name 2 groups of radio wave control methods and indicate their main advantages and disadvantages.

References

1. Rogers RR (1978) IEEE and IEC codes to interpret incipient faults in transformers using gas in oil analysis. IEEE Trans Electr Insul 13(5):348–354
2. IEEE Std C57.104-1991/IEEE Guide for the interpretation of gases generated in oil-immersed transformers
3. Oommen TV, Ronnau RA, Girgis RS (1998) New mechanism of moderate hydrogen gas generation in oil-filled transformers (United States). CIGRE Paris, paper 12-206
4. Zaengl W, Kuffel E, Kuffel J (2000) High voltage engineering. Fundamentals. Butterworth–Heinemann, Boston, p 539
5. Lyutikova M, Korobeinikov S, Konovalov A (2021) New insulating fluids and diagnostic techniques for paper-oil insulated equipment. Power Technol Eng 55:136–142
6. Lyutikova MN, Korobeinikov SM, Konovalov AA, Kozlov VK, Garifullin MSh (2019) Control of the antioxidant additive (ionol) content in liquid insulation of high-voltage equipment in electric grid companies using modern instrumental methods. Power Technol Eng 53(1):118–125
7. Ushakov VYa, Mytnikov AV, Lavrinovich VA, Lavrinovich AV (2022) Transformer condition control—advanced and traditional technologies. Springer Verlag, Cham, p 150
8. Darian LA, Arakelyan VG (1998) Problem and achievement a physico-chemical diagnostics of oil-filled electrotechnical equipment. CWIEME, Berlin

9. Moflmann A, Pahlavanpour B (1999) New guidelines for interpretation of dissolved gas analysis in oil-filled transformers. ELECTRA, No. 186, October
10. Condition monitoring and maintenance of electrical machines (1989) Proc. Coll SC11 CIGRE and CEA, Montreal, QC
11. Khalyasmaa AI et al (2015) Diagnostics of electrical equipment of power stations and substations: textbook. Ural Publishing House, Ekaterinburg, p 64 (in Russian).
12. Wadhwa CL (2007) High voltage engineering, 2nd edn. New Age International Publishers, New Delhi, p 312

Chapter 7
Traditional Electrical Diagnostic Methods

Abstract As in Chap. 6, the titles of paragraphs in this chapter reflect a large arsenal of methods and tools used to diagnose high-voltage equipment, only in this case—electrical. Seven of them are discussed in this chapter, the eighth most modern (defectography method) is discussed in more detail in Chap. 8.

The sensitivity of many electrical methods is higher than that of non-electric ones, so they are more widely used at present. For example, they allow one to detect the largest number of PD characteristics and use them to estimate the condition of transformers and some other high voltage devices. Most electrical methods do not need the object under analysis to be supplied by electrical voltage being much higher than the rated operating values, so they are gentle on the insulation of electrical equipment. However, many of these diagnostic methods require the measuring instruments to be in contact with the object being diagnosed, which does not contribute to simplicity and convenience. Due to the high sensitivity of the methods, it is necessary to take a series of measures and use special equipment for debugging induced interference.

7.1 Insulation Condition Monitoring by PD Registration

Methods of PD registration are divided into two groups: non-electrical and electrical methods.

Non-electrical methods include the optical and acoustic methods.

The optical method is historically the first one and is based on a registration of PD glow. The method allows one to control PDs at electrode edges and in transparent materials. PD control inside opaque structures is impossible, which puts significant limits on its use in the diagnosis of high-voltage transformers, whose insulation is not transparent optically.

The acoustic method makes it possible to register PDs inside opaque objects. The apparent simplicity of the method does not compensate for its massive problems in determining the location of PD occurrence in high-voltage transformers. To detect such occurrences, supersensitive microphones are utilized, which detect sound waves located in the frequency range above the audibility threshold. Research on the improvement of the method is in progress.

V. Y. Ushakov et al., *High-Voltage Equipment of Power Systems*, Power Systems,
https://doi.org/10.1007/978-3-031-38252-9_7

Due to the high sensitivity of the electrical methods, it is necessary to take a series of measures and use special equipment for debugging induced interference.

The electrical methods are divided into three main types:

1. *Indirect methods of PD registration*
2. *Electromagnetic or remote (PD registration with antennas)*
3. *Registration of high-frequency (HF) electromagnetic radiation generated by PDs.*

(1) *Indirect methods of PD registration*

This method includes those that allow one to determine dielectric losses by measuring the tangent of dielectric loss angle in insulation (tgδ), or by measuring Volt–Coulomb characteristics. These methods provide an idea about the voltage of PD occurrence judging by a sharp increase in tgδ. Since the application of these methods involves a summation of different types of losses, it is difficult to single out the direct losses due to PDs. In addition, these methods have low sensitivity.

A practical implementation of this idea is the *control of an ionization curve when measuring tgδ*. The method is based on the fact that PD occurrence at an insulation defect causes a decrease in its resistance, and thereby an increase in active current (and hence tgδ); the reactive current is almost unchanged. The voltage dependence of tgδ is determined using a high-voltage Schering bridge. The voltage at which ionization starts in an insulating structure corresponds to the onset of growth in tgδ, and the inflection point of the curve is regarded as the onset voltage of ionization, Fig. 7.1.

The drawback of the method is that the recorded increase in tgδ is related not to the onset of discharges, but rather to an intense process of PDs repeated numerous times during the cycle, which covers a significant part of foreign inclusions in the dielectric volume. For this reason, the method can be classified as obsolete.

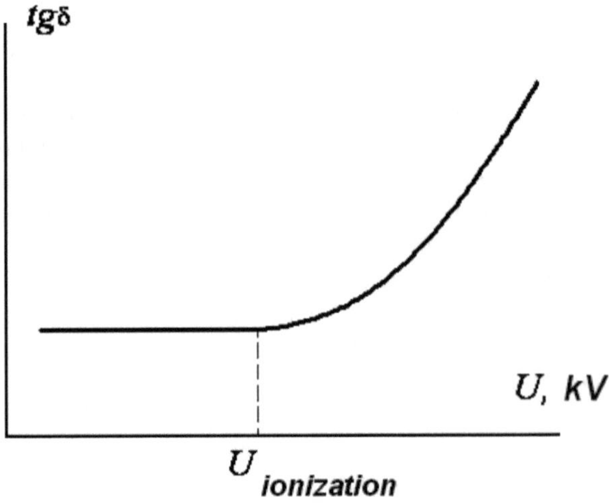

Fig 7.1 Curve for determining the onset of an ionization process

(2) *Electromagnetic or remote method (PD registration with antennas)*

This method is remote and allows one to detect PDs using a directional receiving microwave antenna device. The use of this equipment does not depend on the voltage class, which is an advantage. The disadvantage of this method is the inability to obtain any quantitative information about the processes and phenomena accompanying PDs, as well as the effect caused in the readings of a receiving device by electromagnetic radiation from external sources [1–3].

Research is currently in progress to improve the adaptation of the method to operational requirements. Even now the method allows one to carry out lasting continuous monitoring of equipment accompanied by information transfer to an operator with the aid of digital technologies.

The most active research on PDs in isolation began in the late 1950s and early 1960s. At that time, the destructive action of PDs was first encountered in a large group of ultra-high voltage equipment in the late 1950s during the operation of 400 kV transformers, and later on, 500 kV ones.

Since the 1980s, the strategy for diagnosing equipment in Europe and America has gradually changed: there has been a shift from the concept of routine testing to the concept of testing determined by estimating the technical condition of equipment. It is known from practice that positive results of testing equipment by high-voltage, which has been currently regulated in Russia, do not guarantee protracted trouble-free operation of the tested equipment. Besides, during the insulation test the condition of equipment significantly deteriorates due to the supply of voltages being 4–6 times higher than the nominal value. The diagnostic methods for PD registration, at the same time, allow one to give the most accurate estimation of the residual life of equipment, while not affecting the isolation in any significant way. This is ensured by applying lower voltages to an inspected object, in some cases being close or equal to the nominal value. In parallel, the task of creating a system of automated monitoring of equipment parameters under operating voltage has been undertaken. The largest success in this direction has been achieved by High Voltage Partial Discharge ltd (HVPD) that producing devices for on-line PD monitoring in the insulation of electrical equipment. To date, this is almost the only method that permits a detection of incipient local defects in the process of insulation degradation. A wide application of the method is constrained by the complexity of measurement techniques, the high costs of equipment, and the high requirements to qualifications of the personnel operating the equipment.

(3) *Registration of electromagnetic PD-generated HF radiation*

It has become the most widespread technique, since it allows for reliable measurements of basic PD characteristics and provides high sensitivity.

The method of registering high-frequency PDs components, implemented using partial-discharge indicators (PDI), is also widely used.

The efforts of researchers and engineers are aimed at creating instruments and techniques that should make it possible to register the weakest PDs and provide measurements of energy dissipated by single discharges. This requirement has now been met by using PDIs connected directly to a discharge circuit. Such indicators consist of a receiving circuit, an amplifier and a measuring device. This is based on the measurement of an apparent charge:

$$\Delta Q_0 = \Delta U C_0, \tag{7.1}$$

where C_0 is the insulation capacitance.

Ripples of voltage ΔU are measured and fed through an amplifier to an oscilloscope. The moment the pulses appear on the screen determines the voltage of ionization occurrence, while the amplitude and frequency of the ripples determine the PD intensity. There are several options for this scheme. Two of them are shown in Figs. 7.2 and 7.4.

(a) An active resistance scheme is shown in Fig. 7.2.

Resistance R_1 is connected in series with a tested object and a voltage drop on the object is recorded by a PDI. The resulting oscillogram is interpreted as to the presence or absence of PD. The disadvantage of this method is its low noise immunity.

On the oscilloscope, one can observe the picture shown in Fig. 7.3.

(b) A scheme with inductance and capacitance is shown in Fig. 7.4.

The PDI is connected to the tested object through the dividing capacitor C_{div}, which serves as a blocking filter for currents of the operating frequency. At the emergence of PDs in the object (C_x), chaotic voltage fluctuations in the object excite continuous

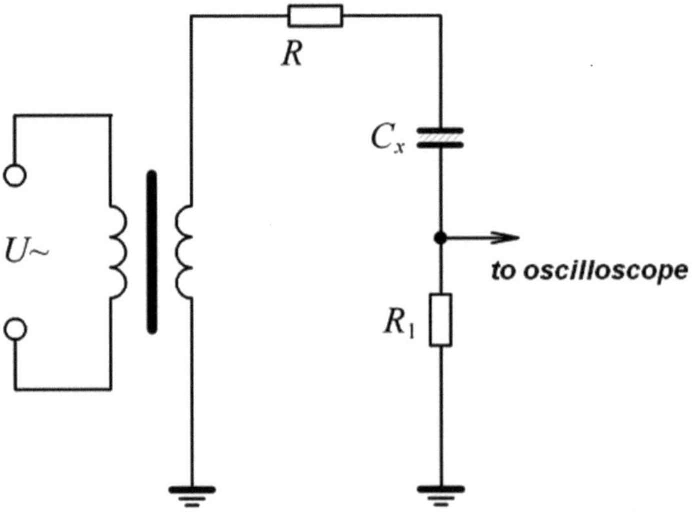

Fig. 7.2 Active resistance scheme for PD detection: R is a protective resistance; C_x is a tested object; R_1 is an active resistance

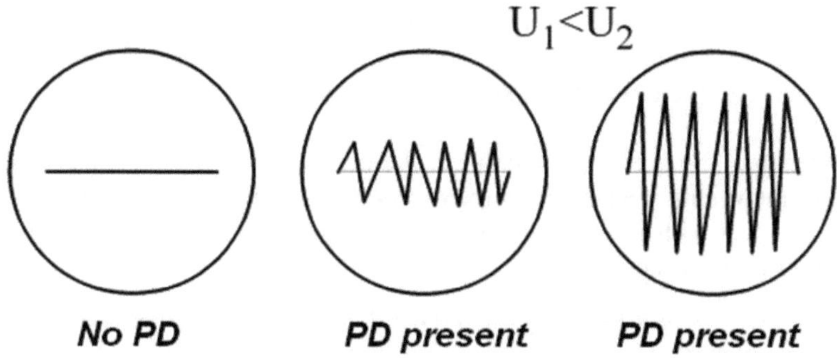

Fig. 7.3 Oscillograms for different PD intensities

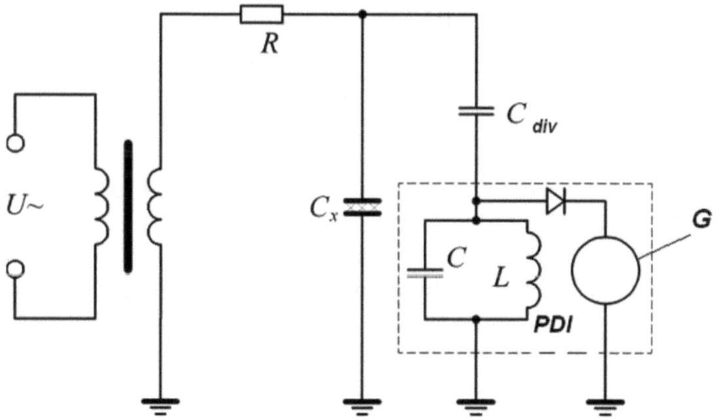

Fig. 7.4 PD detection scheme with an oscillatory circuit and a galvanometer: *R* is a protective resistance; C_x is a test object; C_{div} is a dividing capacitance; *L–C* is the oscillatory circuit; *G* is the galvanometer; *PDI* is a partial discharge indicator

periodic oscillations in the PDI with a frequency corresponding to the oscillation period of the circuit:

$$T = 2\pi\sqrt{LC}. \tag{7.2}$$

The order of the PDI tuning frequency is usually chosen as several tens of kilohertz. The apparent intensity of ionization is determined using the value of ΔU given by formula (7.1).

In conclusion, it should be noted that the use of the PDs method for preventive tests is very promising, since it allows for continuous monitoring under operating voltage. Unfortunately, it also has some drawbacks that demand it to be used in combination with some other methods:

1. The presence of a large amount of interference that makes it problematic to decipher the results obtained (the source of interference may be the appearance of a corona on the wires, the sparking of the collectors of electrical machines, etc.);
2. Rather than the presence of failures, the method detects the presence of PDs, while there may well exist dangerous defects without any PDs involvement: cracks filled with water or some other conductive liquid, carburized pores in which PDs have stopped.

Technical implementations of the concept of insulation diagnostics by PD registering are extremely diverse.

Measurement of PDs electrical characteristics and determination of PDs origin location are successfully implemented by many high-voltage laboratories at large energy companies in the USA, Australia, Japan, Germany, China, Canada, Brazil, as well as transnational energy companies EDF and Enel, and others, where this control method is included in the mandatory insulation monitoring programs. Preference is given to automated systems and devices operating in real time. The principal contribution to these developments is due to such large companies as ABB, Siemens, Mitsubishi, Tettex, General Electric, Megger, Omicron, Haefely. To monitor the insulation condition of a power transformer of the highest voltage class (1000 kV), Mitsubishi has developed a highly sensitive PDs monitoring system.

Tettex Instruments AG produces the multichannel PD analyzer Digital Partial Discharge Detector DDX9101, a digital meter with computer data processing. The software of this device allows displaying PDs both as Lissajous figures (with reference to the phase of the operating voltage) and as three-dimensional images showing the amplitude-frequency characteristics of discharges (Fig. 7.5).

Acoustic methods, despite their inherent disadvantages, have found extensive use in PD detection and location during measurement on operating equipment due to their lower sensitivity to extraneous electrical signals (interference). An acoustic assessment of PD intensity is recommended as part of diagnosing the condition of power transformers. Particularly effective are acoustic PD sensors placed inside the transformer tank. Such sensors are used to control of particularly critical transformers and also in type tests.

Research on acoustic waves arising from discharges in oil has been conducted by a number of scientific institutions in Japan. Techniques for locating the PD site in a transformer have also been developed. In particular, a 1000 kV high-voltage power transformer produced by Mitsubishi is equipped with acoustic sensors in the tank.

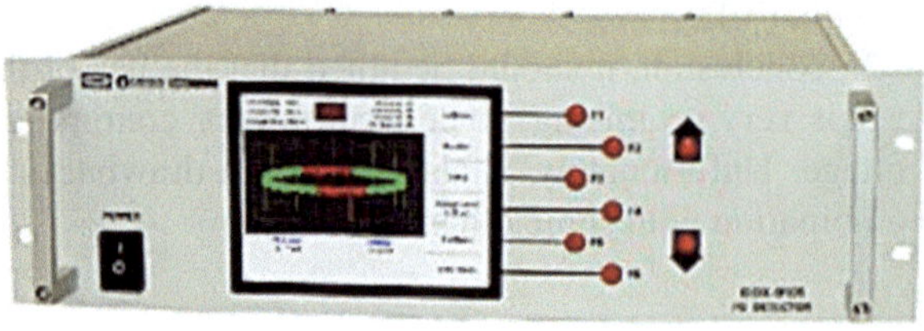

Fig. 7.5 Digital Partial Discharge Detector (DDX9101)

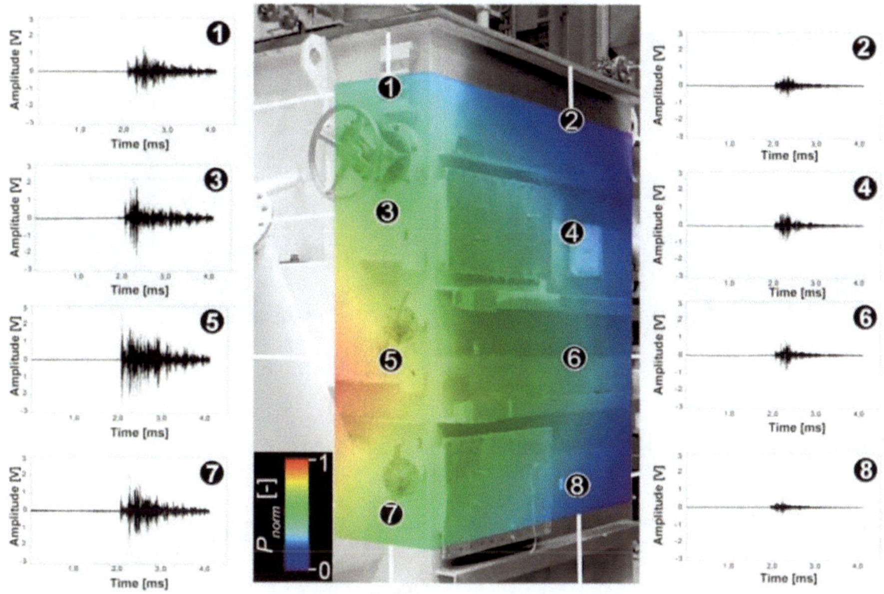

Fig. 7.6 The result of partial discharge source location with advanced auscultatory technique and example of AE waveforms recorded in 8 locations on the transformer tank [1]

At the Polytechnic Institute in Poznan (Poland), the mechanism of sound wave propagation from a PD source has been studied in detail. An innovative technique for measuring and locating the PD site in a transformer has been developed using an acoustic method [1]. The result of PD registration with the site localization is shown in Fig. 7.6.

Acoustic PD sensors, both inside and outside the tank, have been part of the EPRI control system for testing a 525/345 kV, 300 MVA transformer at the Ramapo substation in the ConEd electric power system.

PD sensors (electrical and acoustic) are included in automated systems for a continuous monitoring the condition of large power transformers, developed by Westinghouse, Siemens, ABB, see [2].

To date, several models of devices and system architectures have been developed and used for detecting and measuring PD parameters by using their electromagnetic radiation. For example, Fig. 7.7 shows a directional antenna device with an HVPD Longshot operating unit and an HFCT sensor. The system allows detecting PDs by oscillations of electromagnetic waves in the frequency range from 300 MHz to 3 GHz. Its disadvantage is the impossibility of controlling the intensity of the discharges.

TEV (Transient Earth Voltage) sensors, which are capacitively coupled devices, are effective for measuring local PDs in insulation, it is shown in Fig. 7.8. For PD measurements, a TEV sensor must be fixed to a grounded metal surface.

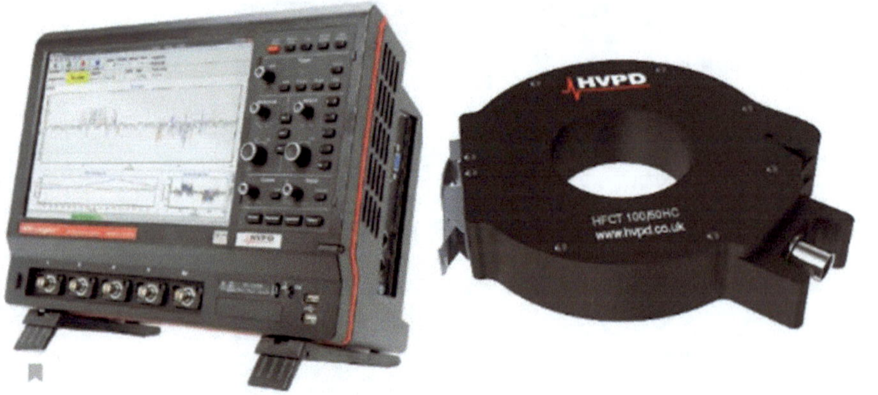

Fig. 7.7 HVPD antenna device by Longshot

Fig. 7.8 HVPD Longshot TEV sensor

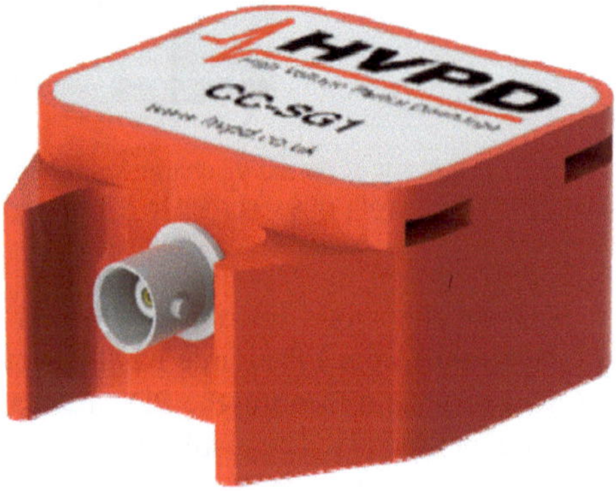

Among the promising methods of PD detection, a technique based on a registration of ultrasonic PD vibrations stands out, using sensor detectors and data analysis through wireless communication in ON-LINE mode. A technology based on the acoustic method is being developed in China [3]. The concept of the method implementation is shown in Fig. 7.9.

High-performance sensor detectors located at a certain angle in relation to the equipment and to each other capture the slightest ultrasonic waves and transmit the data to a cloud server, where they are analyzed and decoded in detail.

An effective system for analyzing PD pulses in the time and frequency ranges has been proposed in Switzerland.

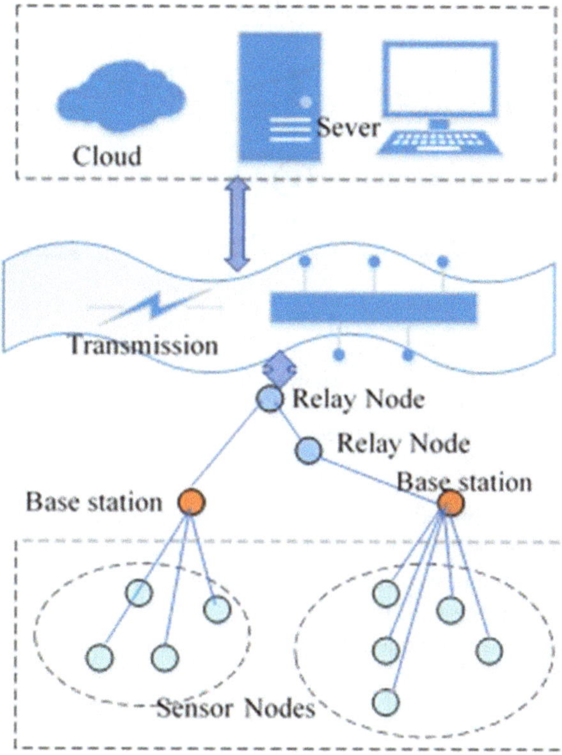

Fig. 7.9 PD registration based on highly sensitive detectors and wireless communications

7.2 Measurement of Idling Losses

The losses in a power transformer are known to consist of so-called copper and steel losses. Copper losses are associated with a flow of load current through winding conductors, having a certain electrical resistance. Losses in the steel of a core are due to eddy currents, being magnetizing currents arising in a magnetic conductor.

According to the Rules for Electrical Installations (REI), measurements are made for transformers with a capacity of 1000 kVA and higher, at a voltage on the LV winding equal to that specified in the factory test report (passport), but not higher than 380 V. Idling losses of three-phase transformers are measured at a single-phase excitation according to the manufacturer scheme. When commissioning a transformer, the ratio of losses on the different phases of three-phase transformers should not differ from the factory data by more than 5%, and for single-phase transformers, the difference of the measured of losses from the original ones should not exceed 10%.

An open-circuit test for single-phase transformers is made at a voltage of 380 (220) V or less, supplied from the LV side at a frequency of 50 Hz in accordance with Figs. 7.10a, b. The supplied voltage must not exceed the rated voltage [4–6].

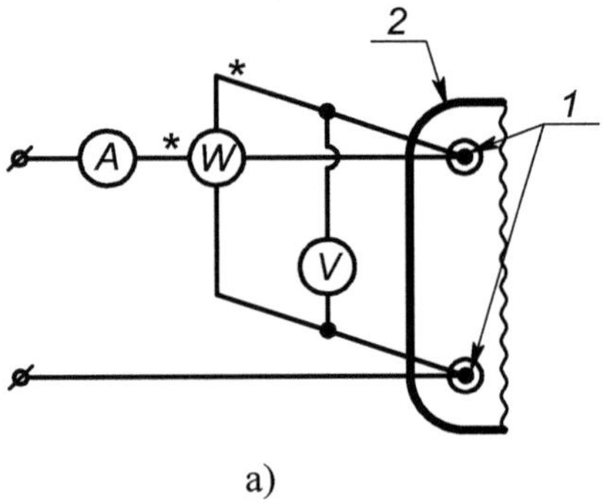

Fig. 7.10 Scheme for measuring the losses of a single-phase transformer XX: **a** measuring the losses of a tested transformer. **b** measuring the losses of measuring instruments: A—ammeter, V—voltmeter, W—wattmeter

The applied voltage, the current, and the power P consumed by the tested transformer and the measuring devices are measured in accordance with Fig. 7.10a. Then the power consumed by the measuring devices $\sum P_{MES.DEV.}$ is measured in accordance with Figs. 7.10a, b.

Losses in the transformer (P_0) are calculated using the formula

$$P_0 = P - \sum P_{MES.DEV.}. \tag{7.3}$$

Note. In the circuits of Fig. 7.10, it is allowed to use current-measuring instrument transformers.

Experiment of a three-phase transformer at low voltage is carried out in as the following three single-phase experiments made in accordance with experiment for single-phase transformers.

The first experiment: the windings of phase A are short-circuited; phases B and C of the transformer are excited and the losses are measured [7].

The second experiment: the windings of phase B are short-circuited; phases A and C of the transformer are excited and the losses are measured.

The third experiment: the windings of phase C are short-circuited, phases A and B of the transformer are excited and the losses are measured.

Short-circuiting of any phase is made at the corresponding terminals of any of the transformer windings (high, medium or low voltage).

Any of the transformer windings can be used to short-circuit one phase. The connection diagrams of transformer windings are shown in Fig. 7.11.

A measurement is usually made by applying voltage to two of the phases and short-circuiting the third phase of the low voltage (LV) winding, which results in more excitation of the magnetic system. The lower voltage makes it easier to measure voltage, power and current, therefore it is common to conduct experiment from the side of the LV winding [8].

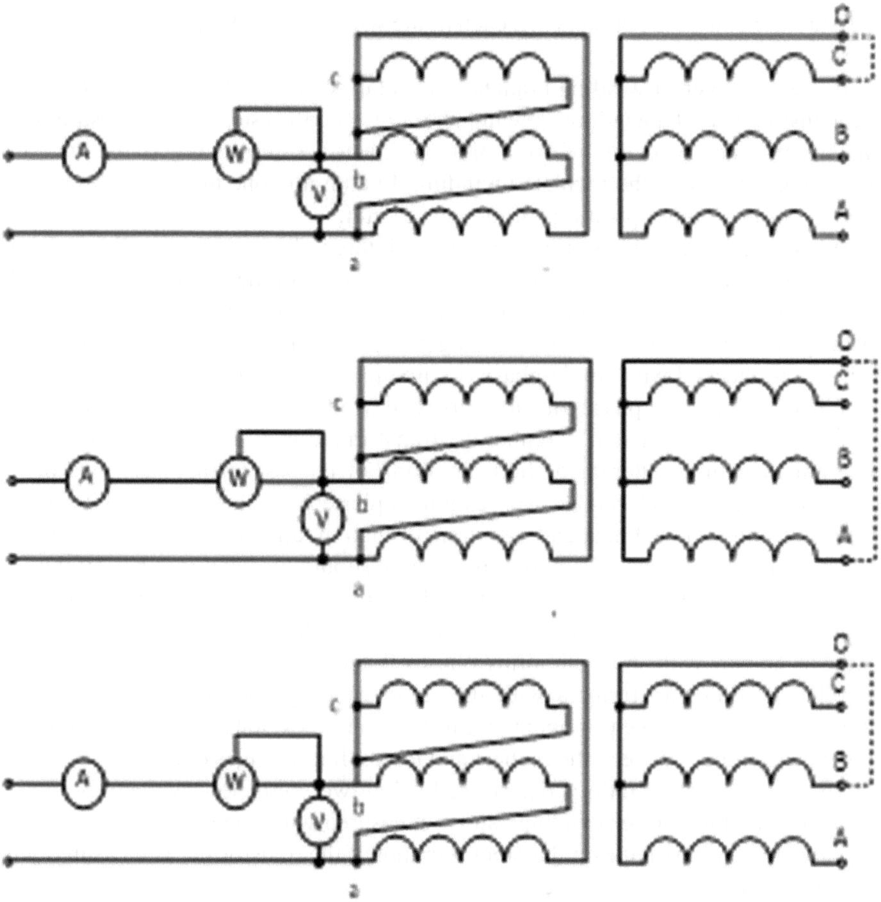

Fig. 7.11 Connection diagrams for the windings of a three-phase transformer

The disadvantage of this method is the limited accuracy and reliability of identifying the winding defects.

7.3 Measurement of the Transformation Ratio

The transformation ratio (TR) of power transformers is tested in order to confirm the compliance of the actual TR with the nominal one. TR is determined on all the winding branches for all the phases; on those winding branches that are inaccessible for switching on an assembled transformer this is determined before the complete assembly of a transformer. When testing three-winding transformers and transformers with split windings, TR may be examined for two pairs of windings, and measurement on all the branches of each of the windings is sufficient to be conducted once.

If a tap changer of winding branches has a tap selector, which is used to reverse the adjusting part of a winding or switch the coarse-control steps, then measurements can be carried out with one position of a tap selector corresponding to the lowest of the voltage values on the regulated winding. In addition, one more measurement is carried out in all other positions of the tap selector.

When testing three-phase transformers with three-phase excitation, line voltages are measured, corresponding to the same line terminals of the tested windings. If it is possible to measure phase voltages, it is allowed to determine the transformation ratio by the phase voltages of the corresponding phases. The transformation ratio for phase voltages is checked with single-phase or three-phase excitation of the transformer.

When testing three-phase transformers with windings connected according to the "star-delta" and "delta-star" circuits, the transformation ratio for phase voltages is determined with an alternated short-circuiting of phases. In this case, one of the phases connected in a "triangle" (for example, phase A) is closed, then, with single-phase excitation of the linear ends, the transformation ratio of the remaining free pair of phases is determined, which, in this method, should be equal to 2 K_{ph} (if the HV winding is connected as a "star") and 0.5 K_{ph} (if the LV winding is connected as a "star"), where K_{ph} is the phase transformation ratio. Measurements are carried out in the same way with short-circuiting phases B and C.

When testing transformers with the same winding connection schemes, it is allowed to carry out measurements with three-phase excitation, if it is established that the difference between the highest and lowest line voltages does not exceed 2%.

The transformation ratio is determined using two voltmeters. Voltage is applied to one of the transformer windings and is measured with one of the voltmeters. At the same time, the voltage on the other winding of the transformer is measured with the other voltmeter.

It is allowed to use measuring voltage transformers, as well as external resistors additional to voltmeters. The accuracy class of voltage transformers and additional resistors must be at least 0.2. The supplied voltage must not exceed the rated voltage of the transformer, but must be no less than 1% of the rated voltage. It is allowed to supply a voltage of less than 1% of the rated value if a voltage transformer is required when supplying a voltage exceeding 1% of the rated value. A voltmeter on the side of supplied voltage is allowed to be connected to the supply wires if this has no significant effect on the measurement accuracy. When measuring the transformation ratio, the resistance of the measuring circuit wires should be no more than 0.001 of the internal resistance of the voltmeter.

Conclusions on the consistency or inconsistency of measurement results are made on the basis of analysis for the measured value of transformation ratio.

The main disadvantage of the method is a limited number of identifiable defects, having a high degree of development.

7.4 Insulation Resistance Monitoring for Transformer Windings

The insulation resistance measurement method is the simplest and most accessible one. It is based on the characteristics of a change in the electric current passing through the insulation after applying a constant voltage to it. The insulation of transformer windings is a non-uniform dielectric. In the general case, an electrical equivalent circuit of insulation can be represented in the form of three branches (Fig. 7.12): the geometric capacitance of the dielectric C_1, the geometric capacitance of the dielectric layers C charged through the resistance of the dielectric layers R, and the insulation resistance of the dielectric R_1.

When a constant voltage is applied to the terminals of a circuit, the current will consist of an arithmetic sum of three components:

1. Capacitive current I_r, due to the so-called geometric capacitance C_r; the current I_r almost instantly drops to 0, since the capacitance C_r is connected to a source without resistance and does not affect the results of measuring the resistances R_{15} and R_{60};
2. Absorption current I_{abs} flowing along the R_{abs}–C_{abs} branch; this current reflects the process of charging a number of dielectric layers through the resistance of the previous layer. When the insulation is moistened, the resistance R_{abs} decreases, and the capacitance of C_{abs} increases, and therefore, for a more humidified insulation, the current I_{abs} has a greater value and drops faster to 0. In dry insulation, the resistance R_{abs} is high, the charge of the capacitor C_{abs} flows slowly, so the initial value of the current I_{abs} is small, and the current takes a long time to drop;
3. Through conduction current I_{thr}, flowing through a resistance R_{thr}, depending on the characteristics of both external pollution of the insulation and on the paths of through leakage; this current is established almost instantly and does not change over time.

The total insulation resistance, measured by a megohmmeter, is inversely proportional to the sum of the indicated current components; at the beginning of measurement it has the lowest value, and then, as the current falls, I_{abs} increases, reaching a

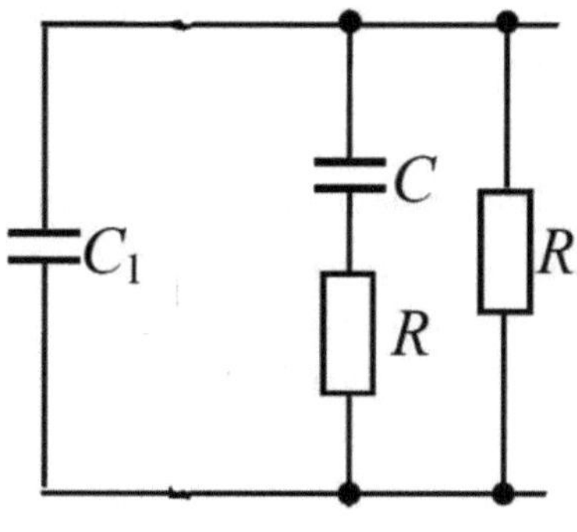

Fig. 7.12
Isolation-equivalent circuit

steady-state value determined by the current I_{thr}. In order to have comparable results, the insulation resistance is measured 60 s after the voltage is applied, although in some cases the current I_{abs} has not completely dropped by this time.

Measurement of the insulation resistance of transformer windings is carried out in accordance with GOST 3484.3-88. The insulation resistance value indicates the average condition of insulation and decreases as this condition worsens, mainly due to moisture and pollution. To estimate the condition of the transformer insulation, the insulation resistance of all windings is measured. When measuring, all the winding leads of the same voltage are connected together. The rest of the windings and the transformer tank must be grounded.

When assessing the condition of insulation, simultaneously with a measurement of the resistance R_{60} the absorption coefficient is measured. The absorption coefficient is the ratio of insulation resistance measured 60 s after the voltage is applied to the resistance measured after 15 s; its values do not depend on the geometric dimensions of insulation and characterize only the intensity of the drop in the absorption current. With the removal of moisture from the insulation, the absorption coefficient increases, and with moisture it decreases.

The value of absorption coefficient $K_{abs} = R_{60}/R_{15}$ must be no less than 1.3 at temperatures of 10 to 30 °C. For well-dried insulation, the values of absorption coefficient normally range from 1.3 to 2.0. During commissioning and preventive tests, insulation resistance is measured in accordance with schemes used by the manufacturer, and additionally also in the insulation zones. During preventive tests, measurements are only allowed in the isolation zones.

The insulation temperature when measuring the insulation resistance of windings must be no less than

10 °C—for transformers with voltage up to 150 kV inclusive;
20 °C—for transformers with a voltage of 220–750 kV.

The insulation resistance of the windings of transformers up to 35 kV inclusive should be no lower than the values presented in Table 7.1.

Being combined with other control methods, this method can be useful for detecting contamination and moisture in insulation at relatively early stages. Measurements are usually carried out with a 2.5 kV megohmmeter using a clamp on the "shield" for measurement by zones. The insulation resistance is highly dependent on the winding temperature and the stability of the megohmmeter voltage. Thus, the difficulty of determining the temperature with reliability (by its resistance to direct

Table 7.1 Insulation resistance of transformer windings

Winding temperature, t °C	10	20	30	40	50
Insulation resistance 35 kV, Mohm	450	300	200	130	90
Insulation resistance 110 kV, Mohm	900	600	400	260	180
tgδ winding insulation,%	1.8	2.5	3.5	5.0	7.0

current) reduces the value of insulation resistance as a parameter for estimating the winding condition [7].

The disadvantages of this method include restrictions on the temperature range at which the insulation resistance is measured, while the detection of damage occurs at the stage of insulation aging, which is close to critical. This method allows one to determine only gross defects in equipment: through burnout, strong moisture or contamination of insulation.

7.5 Measurement of Winding DC-Resistance

The measurement of winding resistance to direct current is made to examine the condition of electrical contact connections and the integrity of the electrical circuit of transformer windings. Defects identified during this measurement are caused by the following reasons:

- breakage of one or more of parallel wires in the taps;
- violation of soldering;
- poor-quality contact of connecting the winding taps to the bushings;
- poor-quality contact in the off-circuit tap-changers and on-load voltage regulation devices (OLVR) of the transformer;
- incorrect installation of an off-circuit tap-changer.

In installation conditions, resistance is usually measured by the ampermeter and voltmeter method. Figures 7.13a, b show two schematic diagrams of connecting instruments for measurements. The diagram in Fig. 7.13a is used when measuring small resistance values: starting from fractions of Ohm to several Ohms, and the circuit in Fig. 7.13b is used when measuring large values of resistance. A correct choice of measurement circuit eliminates significant errors due to voltage drops in the devices, which are usually not taken into account when calculating the resistance value.

Basically, in practice, the circuit shown in Fig. 7.13a is used. In this scheme, the current and voltage circuits are separated. They are made with separate wires in order to exclude from the measured resistance the resistance of the wires of the current circuits and the transition resistances at the points of connection of the voltage circuits to the inputs of the transformer. The voltage measuring circuit must be connected directly to the current-carrying pins of the bushings of the tested winding. Typically, resistance is measured at voltages up to 24 V and currents up to 10 A. In this case, the excess current should be no more than 20% of the rated winding current.

For single-phase transformers, the resulting values should not differ by more than 2% of the values specified in the passport, at the same temperature and with the same control taps.

For three-phase transformers, the resistances measured on the same branches of different phases should not differ from each other by more than 2%, unless there are special instructions in the passport.

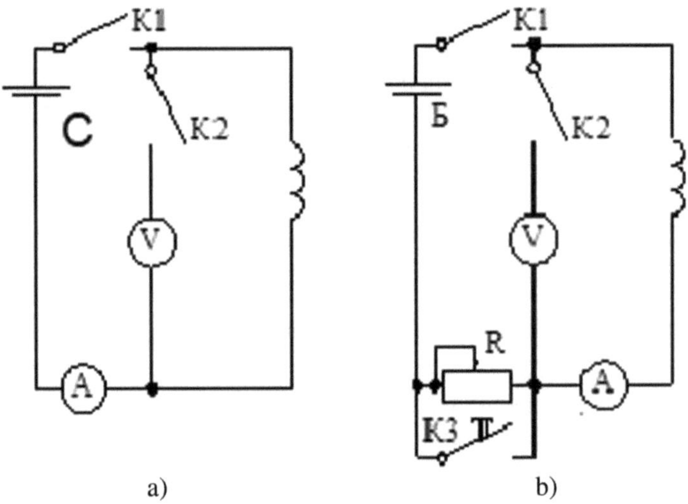

Fig. 7.13 Winding DC resistance measurement circuits

The resulting values of the winding resistance to direct current lead to the temperature specified in the transformer passport. They are determined using the formula

$$R_x = \frac{R_0(235 + t_x)}{235 + t_0}, \tag{7.4}$$

where: R_x is the resistance value at the temperature t_x specified in the passport, Ohm; R_0 is the resistance value at the measurement temperature t_0, Ohm; t_0 is the measurement temperature, °C; t_x is the temperature $^\circ C$ specified in the passport.

The temperature of the upper layers of oil is taken as the temperature of an oil transformer which has not yet been turned on and heated, provided that the resistance is measured no earlier than 30 min later after oil filling for transformers up to 1000 kVA inclusive, and no earlier than 60 min for transformers of higher power.

The considered circuits for measuring the resistance of windings and the method itself have the following disadvantages:

- measurements must be made at temperatures no lower than + 10 °C;
- all devices used must be of high accuracy class;
- only highly developed defects are registered.

7.6 Short-Circuit Resistance Monitoring

This control method, due to its relative simplicity, is widely used in power systems, both when testing transformers for short-circuit resistance, and in operation [8].

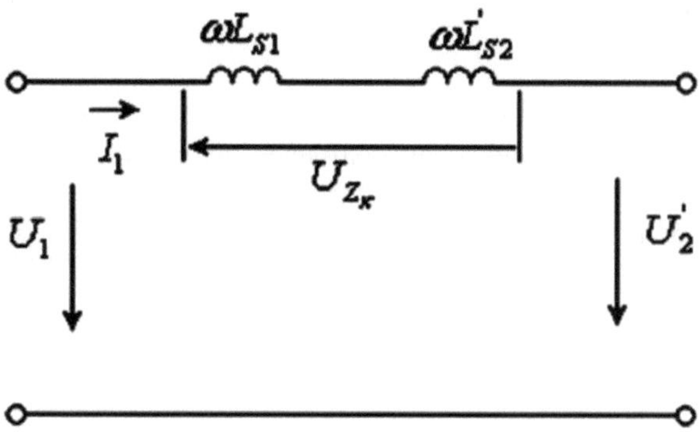

Fig. 7.14 Equivalent circuit of a transformer

The essence of the method is exemplified in Fig. 7.14. It consists in measuring the voltage drop across the resistances in each phase of a transformer and the current flowing through them, and in determining the resistance value from the result obtained.

Let us denote by R_1 and L_{S1} the active resistance and leakage inductance of the primary winding of the transformer; R_2 and L_{S2} are the values of active resistance and scattering inductance of its secondary winding reduced to the primary winding; R_0 and L_0 are the active resistance and inductance of the primary winding, which determine the no-load current of the transformer.

Under these assumptions, the electrical circuit of a transformer can be presented in a simplified form, Fig. 7.14.

By measuring the voltage U_{Z_K} and current I_1, one can calculate the modulus of Z_K, since the leakage inductance of a winding is directly related to its geometric dimensions; changes in the latter are unambiguously reflected in changes in the values of $L_{S1} + L_{S2}$. Therefore, monitoring changes in leakage inductances is a direct method for measuring winding deformation. It should be noted that the leakage inductance of transformer windings is small, and it is necessary to measure its change with high accuracy. Therefore, a measurement should be organized according to the first harmonic, taking measures to minimize the level of higher harmonics, first of all, the third harmonic, which occurs when the magnetization reversal of a transformer core is due to the nonlinearity of the magnetization curve of a steel core. This creates additional inconveniences in control. The main disadvantage of the method is that it is highly sensitive only to those types of deformations of transformer windings that lead to a change in the size (volume) of the scattering channel. Therefore, having a relatively good sensitivity to the loss of radial stability of windings, due to the fact that this increases the volume of the main scattering channel, the Z_K method turns out to be insensitive to other types of damage, such as lodging of turns, slipping of winding turns, and unwinding.

7.7 High Voltage Test

This type of test is carried out at the manufacturing plant to control the quality of products, and is the final stage of preventive insulation tests in order to determine the electrical strength margin and to detect defects that were not revealed by previous tests. Taking into account the natural aging of the insulation, the test voltage under operating conditions is assumed to be 10–15% lower than the voltage during acceptance tests at the factory [9].

Alternating voltage is used in testing various types of insulating structures and makes it possible to detect PD, sliding (surface) discharges and corona. The most important indicator of the dielectric properties of the original (fresh) and operating transformer oil is its dielectric strength at alternating voltage. It is characterized by a breakdown voltage U_{br} in a standard test cell (special vessel with normalized electrode sizes) or corresponding electric field strength E_{br}. Since this characteristic changes with moistening and contamination of the oil, it can serve as a diagnostic characteristic. When the temperature decreases, excess water is released in the form of an emulsion, which causes a decrease in U_{br}, especially in the presence of impurities. In accordance with state and departmental regulations, the dielectric strength of oil is determined in a special vessel when a power frequency voltage is applied.

When testing objects with a large electrical capacitance that require high-power transformers for testing at alternating voltage, *direct* or *rectified* voltage is used.

Impulse voltage is used to test the turn insulation of transformers and rotating machines in order to check the resistance of longitudinal insulation to gradient overvoltages. This type of test is especially important for the insulation structures of overhead transmission lines and outdoor substations that are subject to lightning and switching overvoltages (see Chap. 11).

Question and Tasks

1. What is PDs? Name the causes and places of their occurrence ("ignition").
2. Name the main methods of registration and measurement of PDs characteristics and their purpose.
3. Name the methods and purposes of measuring no-load (idle) losses in power transformers.
4. Name the 3 components of the conduction current when testing equipment with high direct voltage.
5. What transformer defects are detected when measuring the resistance of DC windings?
6. Explain the methodology for testing a transformer for short circuit resistance.
7. Describe high voltage tests: (a) goals, (b) the magnitude and types of voltage used, (c) the difference in voltage values used in factory and operational tests.

References

1. Sikorski W, Siodla K, Moranda H, Ziomek W (2012) Location of partial discharge sources in power transformers based on advanced auscultatory technique. IEEE Trans Dielectr Electr Insul 19(6):1948–1956
2. Carlson A, Fuhr J, Schemel G, Wegscheider F (2010) Testing of power transformers and shunt reactors, 2nd edn. ABB Ltd. Transformers, Zurich, Switzerland, ISBN 3-00-010400-3
3. Jia Y, He P, Huo L (2020) Wireless sensor network monitoring algorithm for partial discharge in smart grid. Electr Power Syst Res 189:1–12. https://doi.org/10.1016/j.epsr.2020.106592
4. Webster JG (2004) Electrical measurements, signal processing and displays. CRC Press. LLC
5. Ushakov VY, Mytnikov AV, Lavrinovich VA, Lavrinovich AV (2022) Transformer condition control—advanced and traditional technologies. Springer Verlag, Cham, p 150
6. Ushakov VY (2001) High voltage engineering. Textbook. TPU Press, p 253
7. Wadhwa CL (2007) High voltage engineering, 2nd edn. New Age International Publishers, p 312
8. Fink DG, Beaty HW (2006) Standard handbook for electrical engineers, 15th edn. McGraw-Hill, p 2000
9. Hauschild W, Lemke E (2019) High-voltage test and measuring techniques, 2nd edn. Springer, Cham

Chapter 8
Diagnostics of High-Voltage Equipment by Defectography Methods

Abstract This chapter is devoted to monitoring the condition of transformers, rotating machines and cables by the pulse method. The historical milestones in the development of the impulse method are described. The stages of modification of the technology of impulse diagnostics are given. The origin of frequency analysis as a development of impulse diagnostics is considered. New approaches to impulse diagnostics are explained. The possibilities of this method are illustrated by examples from the practice of diagnosing power transformers and rotating machines.

A significant part of the chapter is devoted to new diagnostic methods, which are still poorly covered in the literature. For this reason, the material is presented in sufficient detail.

8.1 Diagnostics of the Transformer Windings Condition by Probing Pulses

8.1.1 Physical Basis and Development of Pulsed Diagnostics

Low-voltage pulsed diagnostics (LVPD) was proposed in 1966 at the Institute of Electrical Engineering (Warsaw, Poland) and was instantly recognized by the world's leading centers for the development and implementation of new control technologies for transformer condition [1]. The measurement scheme originally used to implement this technique is shown in Fig. 8.1.

A standard lightning impulse with parameters 1.2/50 μs was used as a probe pulse. It was suggested that changes in the curves being compared indicate mechanical displacements in the windings or short-circuited turns.

By the nature of changes in the curves, it is possible to draw some conclusions regarding the type of damages and their gradual amplification during next short circuits, as well as regarding the site of their occurrence in the winding. Changes in the voltage curves are caused by changes in capacitance and inductance (mainly longitudinal) at the site in the winding where short-circuit or mechanical movement of the turns have occurred. These changes are reflected in the shape of the curves.

V. Y. Ushakov et al., *High-Voltage Equipment of Power Systems*, Power Systems,
https://doi.org/10.1007/978-3-031-38252-9_8

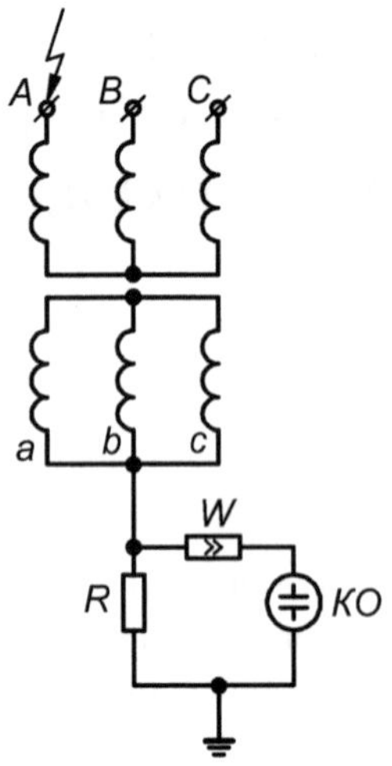

Fig. 8.1 Measurement scheme first used to implement the low-voltage pulsed method. R is a measuring resistance; W is a broadband amplifier; KO is a cathode oscilloscope

Comparison of the current and previous measurements allows one to make a well-grounded conclusion about the winding condition. Main features of the method are following:

- Fault detection methods based on measuring the effects caused by changes in the transversal magnetic scattering flux can be quite sensitive, but they are also very laborious.
- Measurements of the higher harmonic content in the idling current do not allow detecting small displacements of windings, and, at the same time, require a supply of voltage that is stable in both the magnitude and the sinusoidal shape. The measurement results allow one to identify the phase that has been damaged, but do not indicate the site of damage in this phase.
- The method of low-voltage pulses does not have the mentioned drawbacks and is easy to use. Measurements can be carried out after each SC, and the occurrence and growth of displacements in the windings can be observed. This permits a dynamic strength test to be terminated even before the transformer has been destroyed. When using this method, there is no need to open a transformer. The sensitivity of the method is very high. It allows detecting mechanical movements of even individual turns, and, with some experience, gaining valuable information about the size and type of damage.
- The LVP (low-voltage pulsed) method can also be used in transformer operation. Making records of voltage oscillograms before putting a transformer into operation and then during periodic control examinations and comparing them with each other provide valuable data on the mechanical condition of the windings.

The drawbacks of this method include the fact that high reproducibility of measurement results is reachable only when ensuring complete identity of measurements, the time between which can span years (scheme and procedure of measurements, cables and connectors used, their relative position during tests, etc.). In addition, normograms obtained on a new or known-good transformer are required. Interpretation of measurement results calls for highly qualified operating personnel. As shown by further practice, the use of a lightning impulse as a probe is ineffective. The inefficiency is explained by the relatively low content of high-frequency components in the integral spectrum of a lightning impulse. Due to this physical limitation at the initial stage of development, the LVP method did not succeed in detecting incipient faults. As further research has demonstrated, it is an expansion of the high-frequency range of a probing pulse that has a decisive effect on the sensitivity of a diagnostic procedure. Therefore, a gradual shift of the pulse shape from lightning to rectangular was a natural stage in the development of the LVP method and was brilliantly embodied in nanosecond probing.

8.1.2 Development of Pulsed Diagnostics Technology

Substantial experience of using this method was accumulated in the USSR, where its development and application began in 1973 at the Laboratory of transformer diagnostics of the All-Union Electrotechnical Institute, presently, the All-Russian Electrotechnical Institute. The method was improved and given the name of **pulsed defectography (PDG)**. Improvement was carried out in the direction of standardizing the probe pulse, namely, its shape and duration, as well as in the development of a system for recording response signals based on electronic circuits. Attention was paid to improving the accuracy of diagnostic results, in particular, the parameter "pair correlation ratio" was suggested as a criterion for comparing the normogram and the defectogram.

The essence of the method is as follows. A voltage pulse with an amplitude of 100…500 V is applied to one of the terminals of a transformer, which does not pose a danger to the insulation of the transformer windings. Then, the response signals of the windings of the other phases are recorded and analyzed. The response signals obtained during preventive tests are compared with the signals obtained when probing the windings of a known-good transformer. Their differences are due to the fact that even with insignificant mechanical movements in the windings, the capacitance of individual elements (inter-turn and inter-coil capacitance, capacitance for a tank or magnetic conductor) can change significantly, and with significant deformations, the inductance of deformed elements can also change. This leads to a change in the natural frequencies of oscillations in individual circuits, which is reflected in the oscillograms of pulsed currents and voltages. Monitoring the mechanical condition of windings using the LVP method has a higher sensitivity than measuring the short-circuit resistance. Measuring device method was modified during whole period of pulsed method time and row of "Impulse" facilities was successfully developed. The measurement scheme is shown in Fig. 8.2. The method is based on sequential impulse defectography. First, normograms are recorded from a known healthy transformer at a manufacturing plant or at a power plant before it has been put into operation. They are subsequently compared with defectograms, being oscillograms obtained during subsequent measurements after the transformer has been put into operation. The condition of transformer windings is determined by comparing normograms and defectograms between themselves according to a certain method. Deviations of the defectogram from the normogram indicate the presence of electrical damage or mechanical displacement of the windings. If there are no normograms available, then the analysis of winding condition is carried out by comparing the oscillograms of different phases of the transformer [2].

The high efficiency of the LVP method in detecting residual deformations of power transformer windings is provided by a high sensitivity of the method to changes in geometrical dimensions. Even small mechanical movements in the windings significantly change the capacitance of their individual elements (turn-to-turn, inter-coil), and significant deformations lead to changes in the inductance of the deformed elements. All of the above changes the natural frequencies of oscillations, which is recorded in oscillograms of current and voltage pulses. Figure 8.3 shows a diagram of the pulse defectography of a transformer, and Fig. 8.4 shows an equivalent circuit of one of its windings.

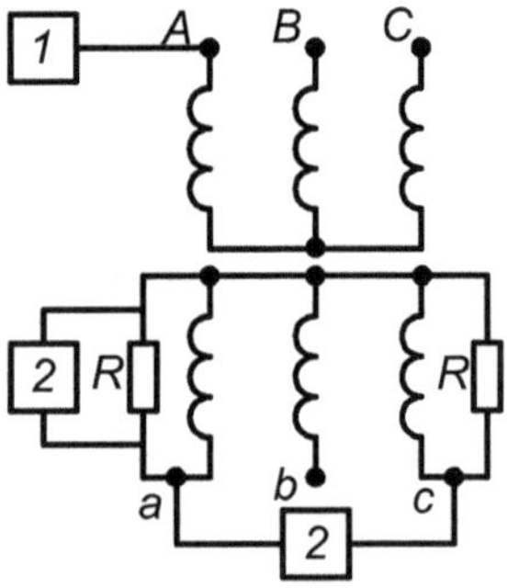

Fig. 8.2 Measurement scheme when implementing the LVP method: 1—is a pulse generator; 2—is an oscilloscope; R—is a measuring resistor

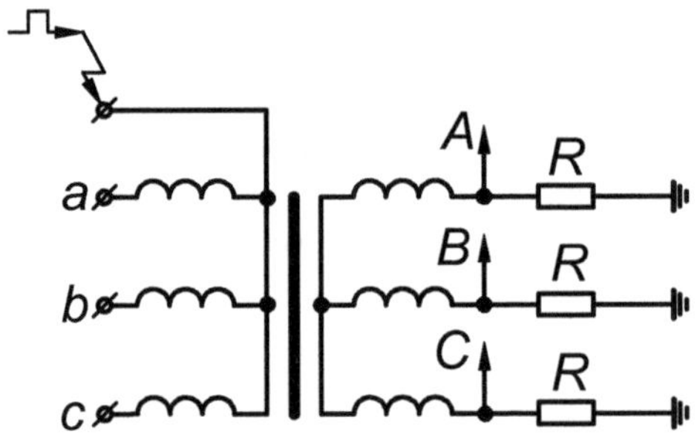

Fig. 8.3 Scheme of pulse defectography of a transformer

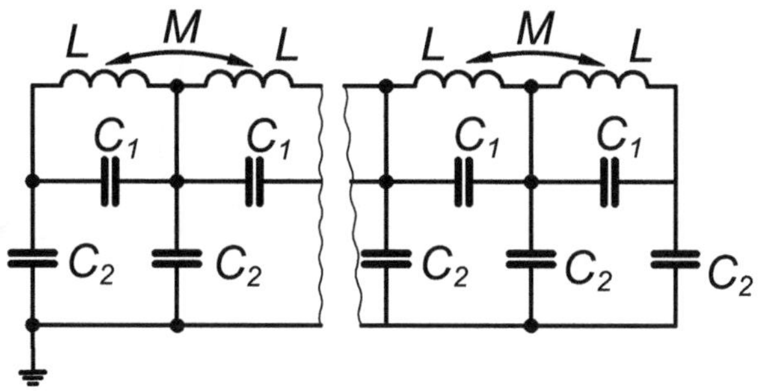

Fig. 8.4 Equivalent circuit of a transformer winding

Despite the relatively high (in comparison with other techniques) sensitivity of the LVP method and its long-term improvement (in the period from 1966 to 1988), its error remained unacceptably large in a number of cases of practical application. This was due to inaccuracy and subjectivity in the analysis of defectography results. Despite the demonstrated technological advantages of pulsed defectography and its recognition as the most accurate and perfect technique (as of the end of the 1980s), it did not continue and stopped developing in the early 1990s for a number of technical, organizational and economic reasons.

8.1.3 Frequency Analysis as a Development of Pulsed Diagnostics

The method was proposed and implemented in Canada in 1978 by a research laboratory at Ontario-Hydro. ***The FRA (Frequency Response Analysis)*** method is based on comparing a currently obtained frequency response of a transformer winding with characteristics previously taken using the same transformer, or a transformer of the same type, before it has been put into operation [3]. In practice, the most common measurement is that of reactive conductivity in the frequency range up to 2 MHz. The measurement procedure using the frequency analysis method involves recording the applied pulse and the corresponding response using high-precision analog-to-digital converters (ADC). The results are then converted to the frequency domain using the Fast Fourier Transform (FFT) algorithm. The transfer function calculated in this way will depend only on the parameters of the test object and will not depend on the applied signal and the measurement scheme. Changes in the object can be clearly identified and separated from the manifestations of external factors, which greatly facilitates the analysis of the results.

In parallel to this method, another one was developed: instead of a pulsed action, a sinusoidal voltage was applied to a winding input, which varied in frequency over a wide range, and the response voltage was recorded from the other inputs. Based on the results, an amplitude-frequency characteristic was built, being the response of the windings to the impact. Both methods are widely used. Figure 8.5 shows a diagram

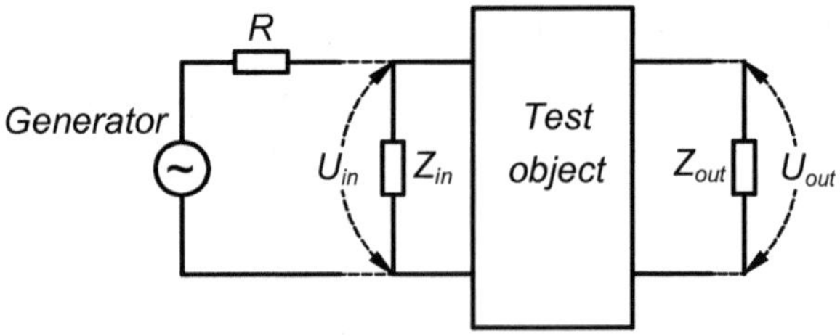

Fig. 8.5 Schematic diagram of frequency analysis measurements

of taking the amplitude-frequency characteristics of an inspected transformer using the frequency analysis method.

The response of windings is analyzed by the following manifestations:

- difference between the response signals of individual phases of a transformer;
- difference between response signals of transformers of similar design;
- alteration in the amplitude-frequency characteristics obtained on one transformer, but at different times.

A two-channel analog-to-digital converter records a signal through channel No. 1, which is fed to a winding input, and through channel No. 2, the response of the winding to the effect of the signal (probe pulse). After that, the transfer function is calculated, which is the ratio of the spectra of the input and output signals. The degree of difference between the transfer functions, calculated before and after exposure of the transformer to electrodynamic short-circuit forces or other mechanical influences, is carried out using different analysis tools.

Figure 8.6 shows an example of frequency response and phase-frequency response (PFR) of the higher voltage windings of a 25 MVA transformer with a measurement range of $0 \div 2$ MHz, recorded at different times.

The type of frequency response and phase frequency response, shown in Fig. 8.7, indicates a change in the mechanical condition of transformer windings. When analyzing the results of measurements of the frequency spectrum, the influence of various parameters is taken into account, such as: the grounding of cables, especially manifested at high frequencies, the presence or absence of oil in the transformer during its examination.

Figure 8.8 shows the frequency response of a 50 MVA transformer in the frequency range from 0 to 1 MHz with and without oil in the transformer tank. It can be seen that there is a shift in the characteristic over the entire range due to a change in the values of the corresponding values of capacitance.

Figure 8.8 shows the amplitude-frequency characteristics of a transformer operating in a "generator-transformer" unit. The transformer windings are star-connected. In order to observe a defect, experience is required, since even when using highly sensitive and very high-quality measuring equipment (for example that of the Swiss company *Haefely Test AG*) there is a high probability of missing a defect. This is one of the reasons for the lack of sensitivity of this diagnostic method. It should be noted that the frequency response shown in Fig. 8.8 exemplifies the initial stage of a winding defect development. As shown by the subsequent opening of the transformer

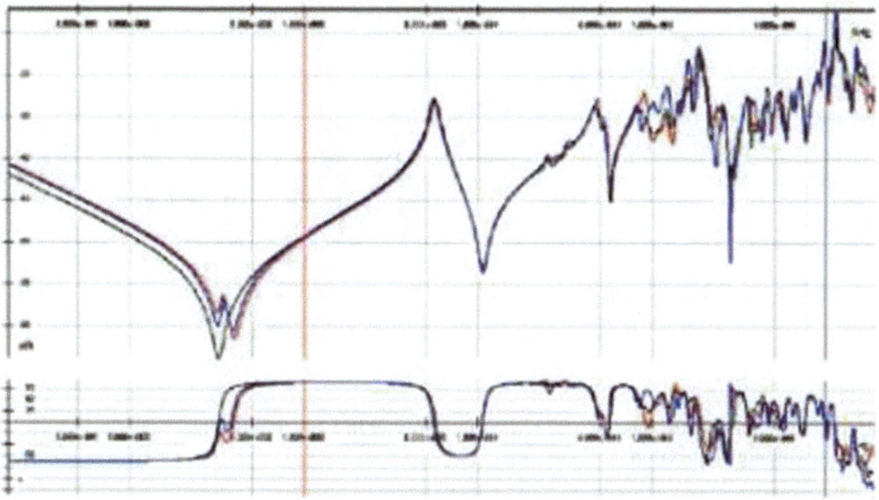

Fig. 8.6 Amplitude-frequency and phase-frequency characteristics of the higher voltage windings of a 25 MVA transformer

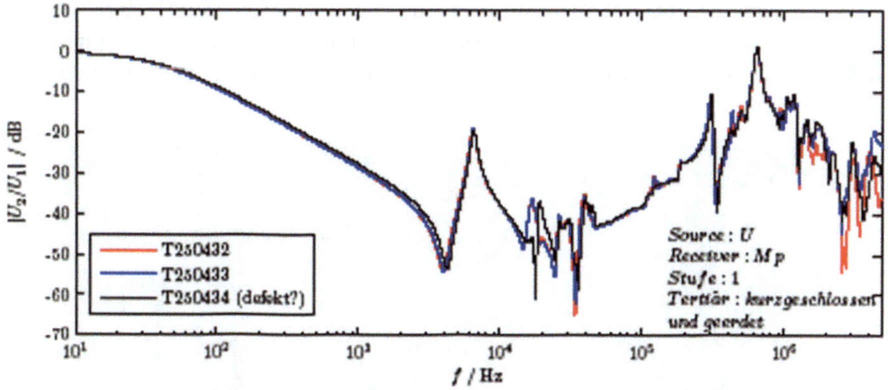

Fig. 8.7 Frequency characteristics of a single-phase 110/10 kV, 40 MVA transformer; curve T 250434 corresponds to a defective winding

and a thorough visual inspection of the winding, the defect was due to poor-quality pressing. The damage had possibly occurred during transportation of the transformer. Subsequent disassembly of the transformer fully confirmed the presence of a winding defect caused by short-circuit currents.

The nature and degree of development of an identified damage do not raise doubts that an emergency mode is inevitable if such a transformer continues to operate. Consequently, the malfunction has been detected in due time: at the initial phase. Obviously, at the first exposure to short-circuit currents, the transformer would fail. The characteristic "overshoot" / "burst" at the bottom right, observed in Fig. 8.8,

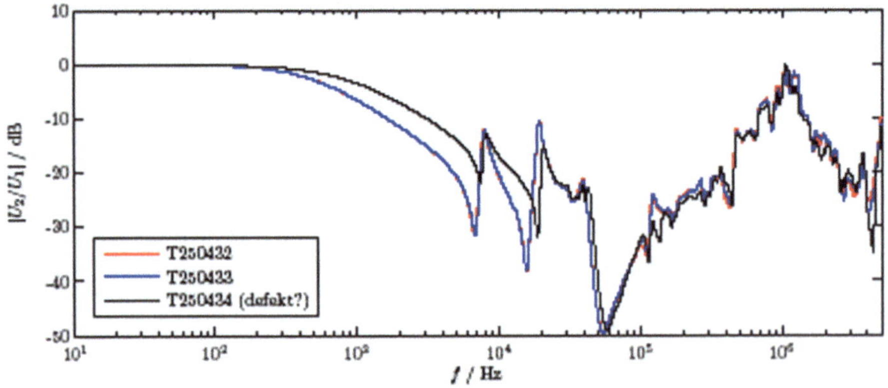

Fig. 8.8 Frequency response of a 273 MVA transformer in a generator-transformer unit with a star type of winding connection

indicates the presence of a defect in the winding. The result of analysis according to the given frequency characteristics are fully confirmed by opening the transformer, which is recorded in the photograph shown in Fig. 8.9.

When carrying out such measurements, it is necessary to carefully and correctly ground when connecting the measuring system in order to avoid interference and distortions that penetrate the measuring circuits. During measurements, the

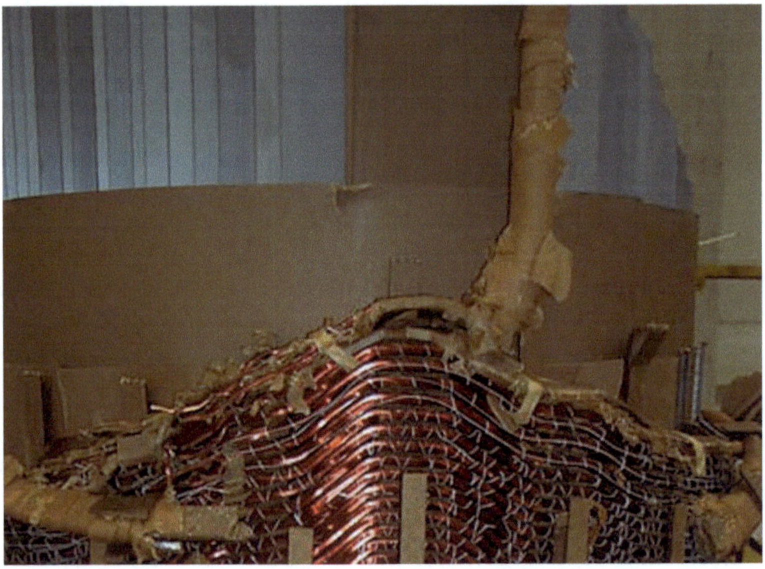

Fig. 8.9 Photograph of a defective winding corresponding to the frequency characteristics shown in Fig. 8.8

grounding of the transformer housing and the recording equipment should be carried out at the same point. Measurements taken at different times with grounding at different points distort the frequency response, which leads to an incorrect assessment of the winding condition. The cost of neglecting such, at first glance, secondary factors can lead to an incorrect diagnosis and, as a consequence, subsequent unjustified expenses for the revision of the transformer. The popularity of the *FRA* method is growing, and a large number of studies are devoted to it in the world's leading high-voltage laboratories. *FRA* technology is discussed at CIGRE colloquia and IEC meetings on the formation of test and measurement standards. Despite the absolute advantages of the *FRA* method, the percentage of errors remains quite high. In a number of cases, in the presence of a defect, the diagnosis was "healthy", i.e., the defect was not recognized. At the same time, there were situations in which a transformer with serviceable windings was rejected and taken out for repair.

To create a new technology for monitoring the condition of active parts of power transformers that should fully meet the requirements of the electric power industry of the twenty-first century, the efforts of developers should be focused on increasing the sensitivity, accuracy and reliability of the method for detecting winding defects at an early stage of their development. As will be shown below, monitoring the condition of active parts can be performed without using normograms, as well as without removing the operating voltage of a high-voltage network and disconnecting.

8.1.4 Probing Transformer Windings with Nanosecond Pulses

This part of the chapter presents the technologies for transformer windings condition controlled by means of nanosecond pulses. It is shown that short probing impulse diagnostics technology has potential advantage for application in electrical energy systems. Below different winding fails detection by means of nanosecond probing impulse technology is demonstrated.

Radial displacement of winding turns is one of the types of mechanical defects. To simulate a diagnostic procedure for identifying this type of defect, the following experiment was performed. The radial displacement was simulated on the HV winding by introducing a dielectric insert between the winding body and the turns. The turn displacement was 3 mm. Coils 12–15 were displaced. The external view of the HV winding with a simulated defect is shown in Fig. 8.10. The probe pulse generator was connected to the input of the HV winding. The response signal was recorded at the LV winding. The oscillograms of response signals in the absence and presence of a defect are shown in Fig. 8.11.

The first part of the response signal is practically identical with the normogram. The difference in the waveforms appears in the second part of the curve. A change in the waveform at a certain section indicates a change in the winding. Similar deviations of the response signal from the normogram are observed when the turns of

Fig. 8.10 Simulated defect "radial displacement of winding turns"

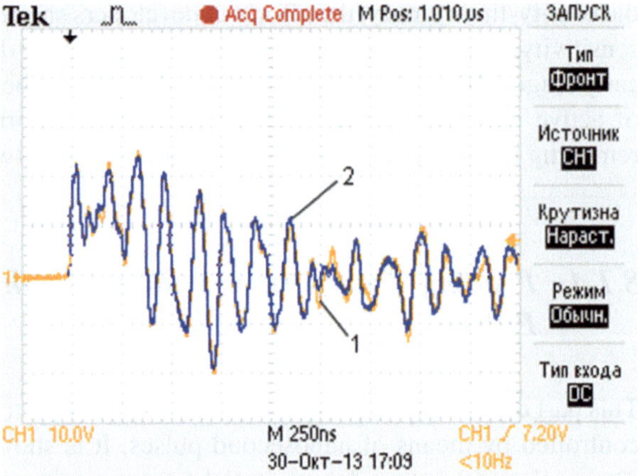

Fig. 8.11 Oscillograms of response signals: 1—no defect, 2—buckling of turns according to Fig. 8.10

the HV winding are displaced in the axial direction. Figure 8.12 shows the combined oscillograms of a defect-free winding (curve 1) and a winding with a shift of several turns in the lower part (curve 2).

The oscillograms shown in Figs. 8.11 and 8.12 demonstrate that the deviations of a normogram from response signals at a radial displacement of turns or at their axial displacement are but slightly different, so that for an objective assessment of the difference between oscillograms a special program for digital data processing has been developed.

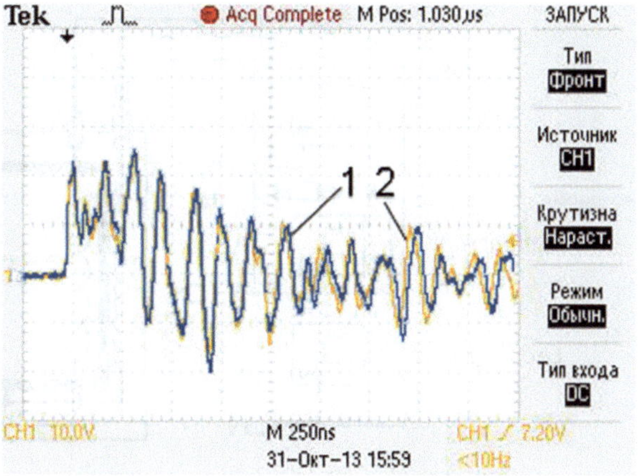

Fig. 8.12 Oscillograms of response signals: 1—no defect, 2—displacement of several turns in the axial direction in the lower part of the HV winding

8.1.4.1 Diagnostic Complex Implementing the PDs Technology in Nanosecond Mode

To show the level of sensitivity, reliability and efficiency in the use of a diagnostic complex under industrial conditions, research was conducted on serviceable and defective transformers of the same brand manufactured at the same enterprise and operated in practically the same conditions by using two methods: the one developed by the authors of this monograph and a method being the most similar to ours in analyzing the frequency characteristics. Three-phase oil-filled power transformers with a capacity of 160 kVA, 10 kV were located at a specialized repair shop of an energy enterprise in one of the northern energy systems of the Russian Federation. The scheme of monitoring the condition of transformer windings, including a generator of probing pulses of the nanosecond range, oscilloscopes to control the parameters of probe pulses and response signals, as well as the location of the phases of the LV and HV windings, is shown in Fig. 8.13.

Below it is presented the results of comparing the two response signals for different winding conditions [4].

8.1.4.2 Choice of Efficiency Criterion

One of the main efficiency indicators of a condition monitoring method is sensitivity. The sensitivity, in turn, can be estimated using the concept of the average integral of voltage pulse. The latter was determined by averaging the integral of voltage pulse for 27 oscillograms of a defect-free transformer. The average integral of voltage pulse on the HV winding was 8.75×10^{-5} (Vb). The deviation from the average of an integral of voltage pulse did not exceed 2% from pulse to pulse. This may be considered as an error (noise) in determining an integral of voltage pulse on the HV winding. The average integral of voltage pulse on the LV winding when processing

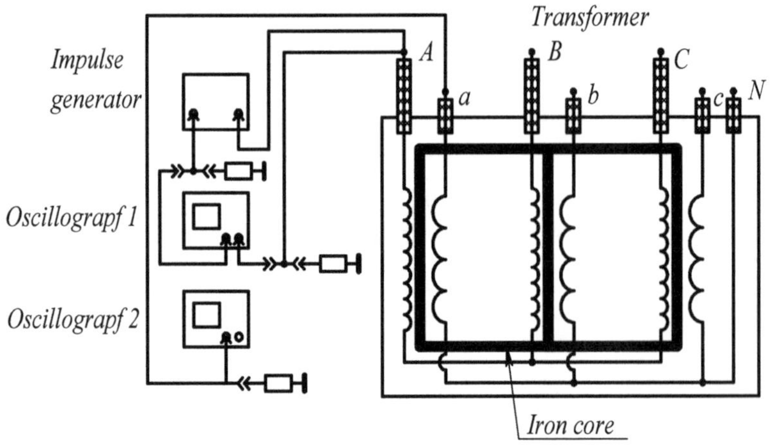

Fig. 8.13 Scheme of the experiment on monitoring the state of transformer by the nanosecond pulse method

27 oscillograms was 7.52×10^{-5} (Vb), and the deviation from the average was 1.6%. The deviation of an integral of voltage pulse from the average for a response pulse over three oscillograms ranged within 3–4%. Thus, it can be assumed that during a probing the error (noise) lies within 3–4% of an integral of voltage pulse. If the integrals of voltage pulse for the normogram and defectogram differ by more than 5% (excess over the noise), then this deviation can be considered as an indication of the presence of a defect condition in the transformer winding.

8.1.4.3 Effectiveness Comparison for the FRA and PDG Technologies

To compare the sensitivity of the proposed method (PDG) and the FRA method, a defect of the "turn-to-turn short circuit" type was arranged in the HV winding of phase ***C***. The appearance of a defective winding is shown in Fig. 8.14.

Before pulsed probing, the winding condition was tested by resistance measurement. The resistance before short-circuiting three turns was, respectively, ***A***-***C***—0.1046 Ω, ***B***-***C***—0.1042 Ω, and after short-circuiting, ***A***-***C***—0.1046 Ω, ***B***-***C***—0.1043 Ω (measured with a PTF-1 milliohmmeter). Thus, the resistance measurement did not detect the presence of three short-circuited turns.

Diagnosis by nanosecond pulses is described in sufficient detail below.

Figure 8.15 shows, as an example, the result of comparing the response signals to a nanosecond pulse: a blue curve in the absence of defects, a red curve in the case of a turn-to-turn short circuit. In all the cases, the neutral terminal is grounded, and the rest of the winding leads are in idle mode.

Processing oscillograms according to the developed program shows that the integral of the difference between the voltage pulses of the normogram and defectogram is 3.4×10^{-6} in the time interval from 0 to 2.2×10^{-6} s, the integral of voltage pulse of

Fig. 8.14 Defect of the "turn-to-turn short circuit" type on the HV winding of phase ***C***

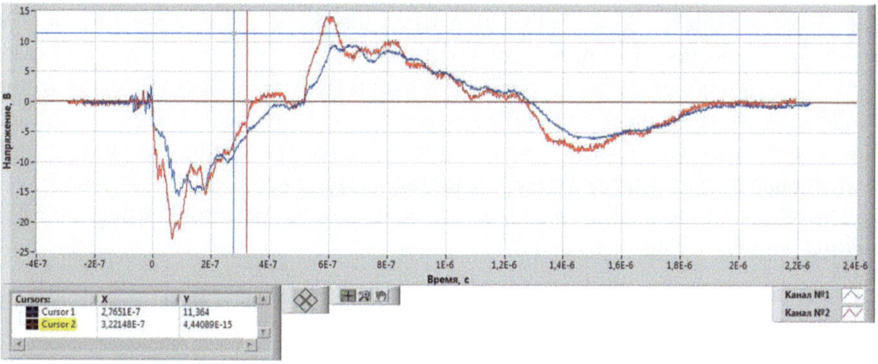

Fig. 8.15 Oscillograms of probe pulses applied to the winding of phase ***A*** and the response signal recorded in phase ***A***

the normogram in the same interval equals 1.01×10^{-5}. The integral of the difference between the voltage pulses of the normogram and defectogram, as a percentage of the integral of transformer response voltage pulse, equals 34%. This result shows that the use of the proposed oscillograms processing technique enables one to register a significant difference between the normogram and defectogram when 3 turns of the HV winding are short-circuited.

Table 8.1 shows the results of processing oscillograms according to the above method for various combinations of the probe pulse and response action.

Resistance measurement of the HV windings using an ohmmeter did not reveal any difference between a defect-free winding and a winding with three short-circuited turns in a total number of turns on the HV winding being 1312. Using the method of nanosecond pulses, a difference is recorded in the range from 35 to 55%, which depends on a combination of the winding acted upon by the probing pulse and the winding on which the response pulse is recorded. One can see that the highest

Table 8.1 Difference between normograms and defectograms for various combinations of applying the probe pulse and response

Winding acted upon by probe pulses	Winding of response pulse recording	Integral of difference between response voltage pulses of a defect-free winding and a winding with 3 short-circuited turns of phase ***C***, Wb	Integral of voltage pulse difference, in % of the integral of response pulse for a defect-free winding
A	***B***	8.7×10^{-7}	45
A	***C***	8.9×10^{-7}	47
B	***C***	9.3×10^{-7}	47
C	***A***	8.7×10^{-7}	44
C	***C***	3.5×10^{-6}	55

sensitivity is achieved in a combination when the probe pulse applied to the HV winding and the response pulse received from the LV winding are recorded from a pair of windings of the same phase. For example, if the defect is located in the winding of phase ***C***, then the largest integral of the difference of response voltage pulses is obtained when a probe pulse is applied to the HV winding of phase ***C***, and the response is recorded from the LV winding of phase ***C***.

The results of the measurements described above, but performed using the ***FRA method***, shown in Figs. 8.16 and 8.17.

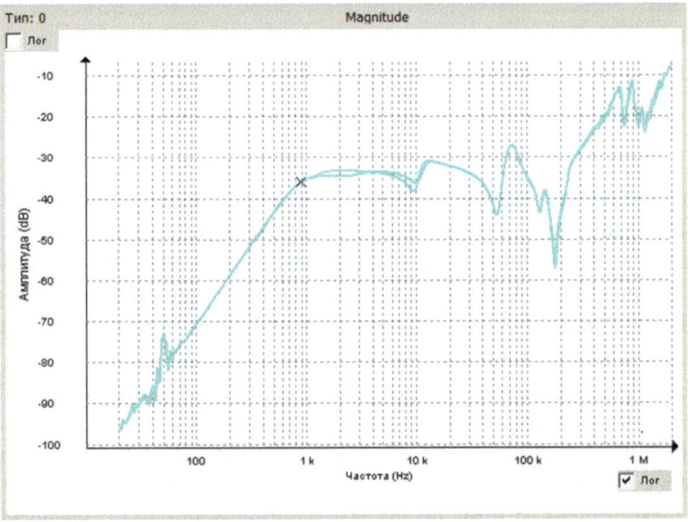

Fig. 8.16 A probe pulse is applied to the winding of phase ***A***, the response signal is recorded on phase ***a***

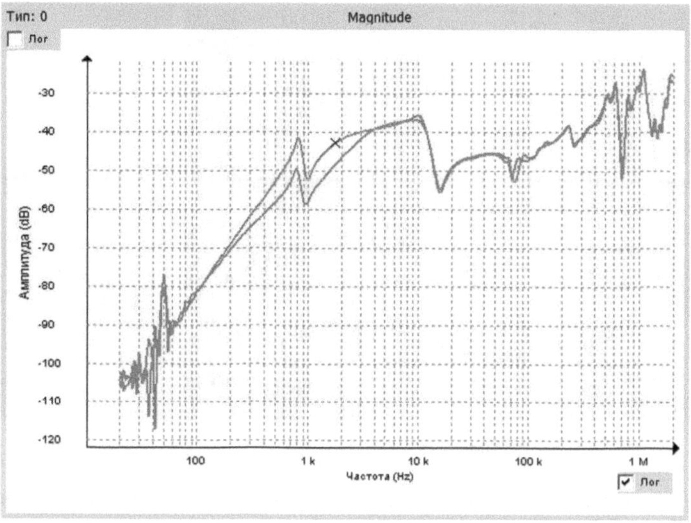

Fig. 8.17 A probe pulse is applied to the winding of phase ***A***; the response signal is recorded in phase ***C***

The results show that the largest sensitivity is achieved in the combination when a probe voltage and a response are recorded on a pair of windings of the same phase, i.e., the same as in the case of using nanosecond pulses. Differences in the amplitude-frequency response are observed in the frequency range from 200 Hz to 300 kHz. The FRA method and the nanosecond pulse method are sensitive to short circuits of several turns in the HV winding and allow them to be detected.

The procedure ***for diagnosing the "axial displacement of turns" defect in the HV winding of phase A by means of nanosecond pulses*** is described below.

To assess the diagnostic sensitivity in the presence of the "axial displacement of turns" type of defects, a defect was created on the winding of phase ***A*** covering 14 turns. The appearance of the HV winding affected by such a defect is shown in Figs. 8.18a, b.

Below we present the results of comparing the signals of response to the action of a nanosecond pulse in the absence of defects (the blue curve) and the signals in the case of axial displacement of turns (the red curve), with various combinations of the probe pulse location and the response registration location. In all cases, the neutral terminal is grounded; the rest of the winding leads are in idle mode. Figure 8.19 shows combined response oscillograms obtained from the high voltage winding.

Table 8.2 shows the results of processing oscillograms similar to those shown in Table 8.1, but for the case of "axial displacement of turns" in the HV winding of phase ***A***. Signal processing was carried out according to the above method: the difference of the integrals of response pulses for a defect-free winding and a winding with displaced turns (Fig. 8.22) was compared with the integral of a response pulse for a probe pulse.

(a) (b)

Fig. 8.18 HV winding of phase ***A*** with an “axial displacement of turns” defect: frontal (**a**) and profile (**b**) views of the coil

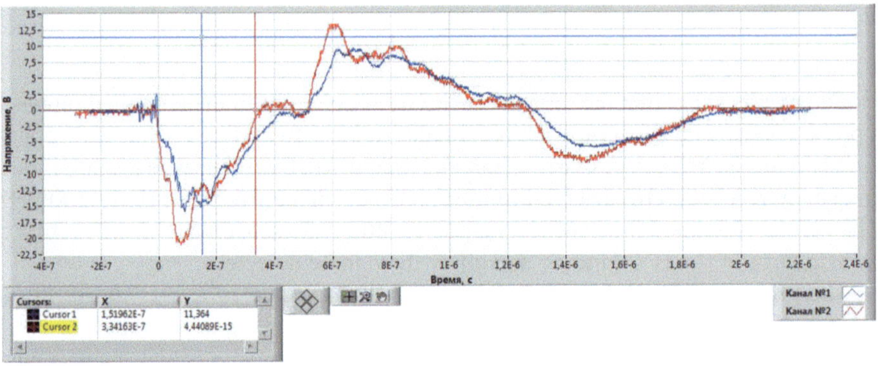

Fig. 8.19 A probe pulse is applied to the winding of phase ***A***, the response signal is recorded on phase ***a***

Table 8.2 Difference between normograms and defectograms for various combinations of the impact of a probing pulse and a response for an axial displacement of 14 turns in the HV winding of phase ***A***

Winding acted upon by probe pulses	Winding of response pulse recording	Integral of difference between response voltage pulses of a defect-free winding and a winding with 14 axially displaced turns of phase ***A***, Wb	Integral of voltage pulse difference, in % of the integral of response pulse for a defect-free winding
A	***A***	3.4×10^{-6}	38
A	***C***	9.3×10^{-7}	48
C	***A***	9.5×10^{-7}	48

One can see that the difference between the amplitude-frequency characteristics is difficult to grasp. Similar patterns are observed with other combinations of the schemes. Based on the results of studies, it should be recognized that probing with nanosecond pulses and using the FRA method react with approximately the same sensitivity to turn-to-turn circuits, but the former method is more sensitive to the displacement of turns in the high-voltage winding of the transformer.

8.1.5 Technology of Single-Stage Pulsed Defectography

The principle of "classical" defectography is based on the fact that a rectangular probe pulse of relatively low voltage is fed from a special generator to a winding input, and responses are recorded from the inputs of the other windings.

The principle of single-stage technology involves the supply of only a probe pulse and its registration as a response pulse, with a subsequent spectral analysis. Figure 8.20 shows a diagram for monitoring the condition of windings.

In order to quantify the difference between signals, it is necessary to determine the number of signal harmonics being investigated. As mentioned above, the minimum number of components of a Fourier series into which it is necessary to decompose the initial pulse is 50. In this regard, the frequency response was decomposed at a frequency from 0 to 50 MHz. The object of study was a three-phase oil-filled power transformer with a capacity of 160 kVA and a rated voltage of 10 kV. The first unit of spectral comparison of probing pulse signals applied to a healthy HV winding

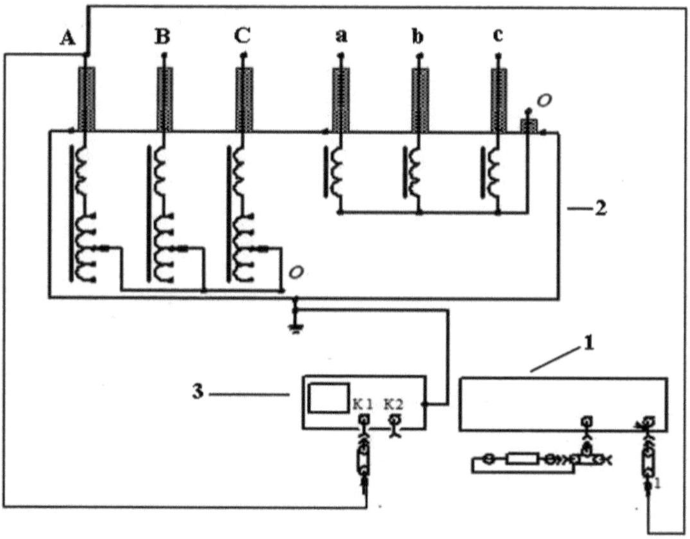

Fig. 8.20 Scheme of single-stage technology implementation in experiments: 1—generator of probing pulses, 2—high voltage 10/0.4 kV transformer, 3—Tektronix TDS 1012 oscilloscope

of phase ***A*** was performed. Figure 8.21 depict the AFC of the signals for which it was found by experiment that the average level for the spectral ratio ***SR*** has an error within 1–1.2. This may be due to different measurement losses.

Consequently, if the AFC of a defectograms has a maximum amplitude value exceeding the margin of error, then such a deviation can be regarded as a sign of damage present in the windings of a transformer under study.

To illustrate the possibilities of the new diagnostic method, ***the control of the displacement (lodging) of the turns in the HV winding for the case of a two-failure stages*** is described below.

Spectral analysis was performed for a normogram and defect No. 1. Defect No. 1 is a mechanical displacement of ***4 turns*** at the bottom of a winding. It is called the lodging of conductors and is implemented on the HV winding of three-phase transformer. Figure 8.22 shows the oscillograms of pulses and Fig. 8.23 the spectra

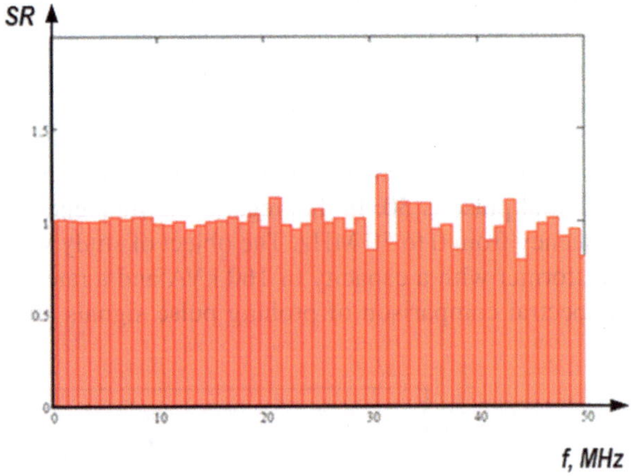

Fig. 8.21 AFC ratio for normograms of a healthy high voltage winding

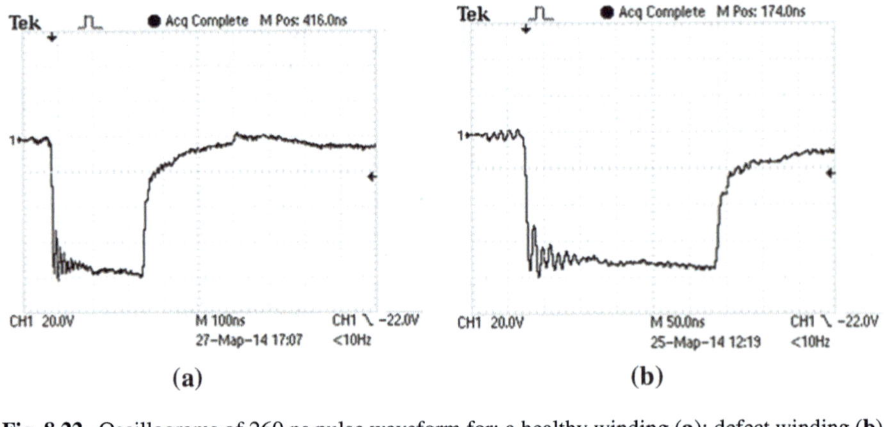

Fig. 8.22 Oscillograms of 260 ns pulse waveform for: a healthy winding (**a**); defect winding (**b**)

ratio of a defect to the healthy winding (SR—Spectrum Ratio) of a transformer at the corresponding probe pulse durations of 260 ns.

Control spectral analysis has been made for a normogram and defect No. 2, being a mechanical displacement of 8 turns at the bottom of a winding ("Lodging of conductors") of the high voltage winding. Waveforms look the same like fordefect No. 1, but spectrum allow detecting the fail condition growth (Fig. 8.24).

During the spectral analysis of signals obtained using the single-stage pulsed defectography of a transformer under investigation, and due to the obtained experimental results the following has been established:

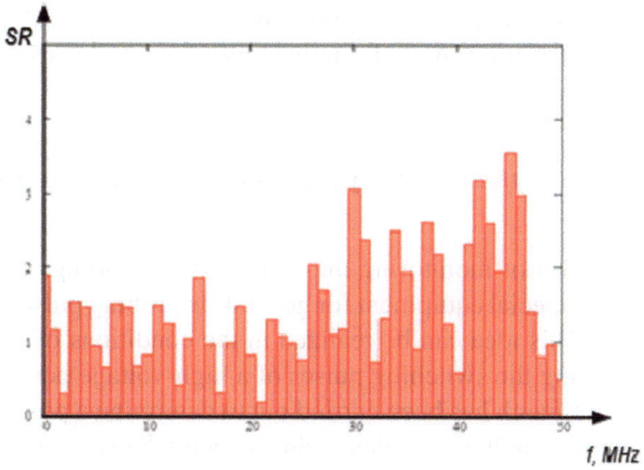

Fig. 8.23 Frequency response ratio of a spectra 260 ns defect No. 1/healthy winding

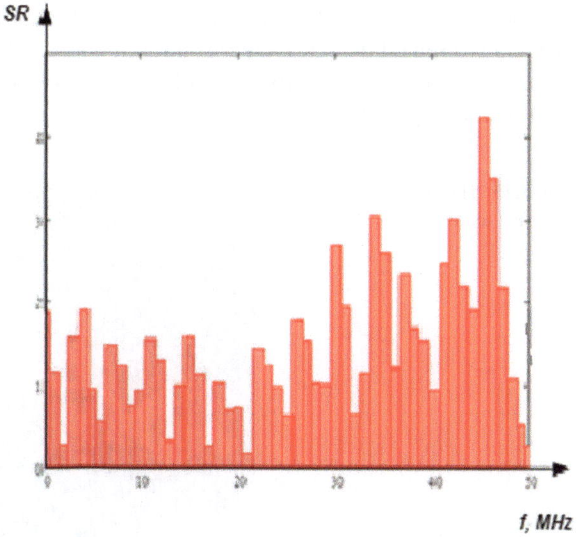

Fig. 8.24 Frequency response ratio of a 260 ns defect No. 2/healthy winding

- in principle, single-stage defectography is possible without registration of response signals from adjacent windings;
- This technology is not inferior in efficiency to the classical LVP method, and, due to its simplicity, surpasses the classical method in reducing the duration of a diagnostic process and the likelihood of making an error during diagnostics in industrial environment, as well as in the potential for monitoring the condition under operating voltage.

The technology of “single-stage defectography” allows detecting failures in a simpler and more convenient way under industrial conditions. It is possible to implement this method under operating voltage without disconnecting a transformer from a high-voltage network and unbussing by means of supplying a probe pulse to the winding under study through a voltage control sensor of the input.

8.1.6 ON-LINE Monitoring of Winding Condition

The idea of monitoring the condition of windings in the ON-LINE mode, without using special equipment for generating probing pulses seems to be the most attractive one. This idea can be realized in the conditions of an actual electric power system by using the switching pulses of a high-voltage power grid as probing ones.

Research has been carried out to determine a possibility of monitoring the condition of windings and identifying defects when a transformer is connected to an electric power grid. In this case, a switching pulse arriving at the winding as a result of regular switching plays the role of a probing one. The tested three-phase transformer has two phases in good order, i.e., they are fault-free, whereas the third phase has an artificially created defect. Defect condition was created in the HV windings of of a three-phase oil-filled transformer with a voltage of 10 kV and a capacity of 66 kVA, as shown in Fig. 8.25.

Fig. 8.25 HV winding damage: axial shift

The probe pulse and the response of the windings were recorded using a Tektronix TDS 2012 digital two-channel oscilloscope. Switching operation "turn on" from the ~ 220 V mains on the LV side—a response was recorded on the HV side of the same phase using a capacitive voltage divider. The control of the windings is carried out according to the "LV-HV" scheme. Oscillograms of diagnostic results are shown below. The response of the intact pairs, **в-*B*** and ***c*–*C***, differ significantly from the situation for the pair ***a-A***. The response signals were recorded on a "5 μs/division" sweep to observe the initial stage of the transient process (Fig. 8.26), and on a "25 μs/division" sweep, to observe the entire transient process in the winding (Fig. 8.27).

A visual analysis of the obtained oscillograms, without the use of specialized software, makes it possible to ascertain the fact that the responses on the damaged phase differ when the responses on the two undamaged phases are identical and the switching pulse is equal when switched on.

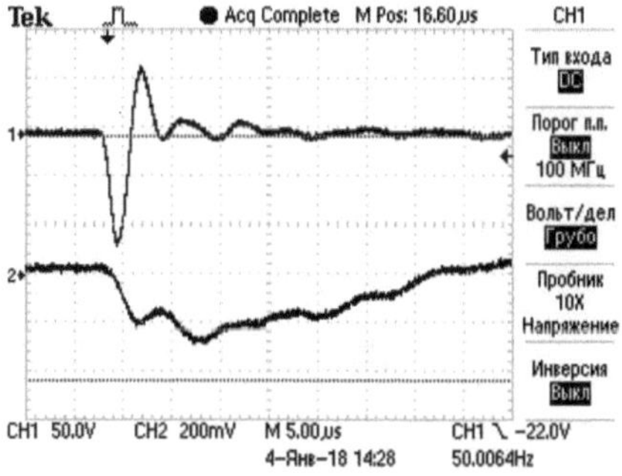

a-A

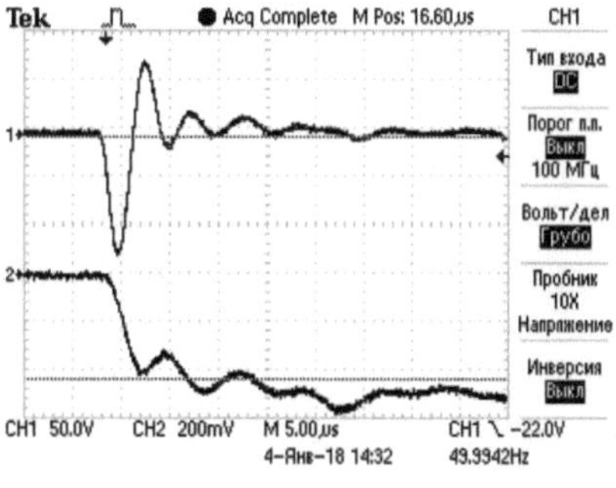

в-B

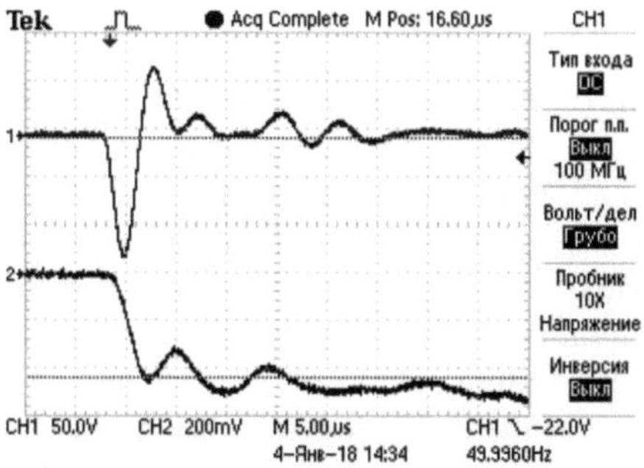

c-C

Fig. 8.26 Waveforms of commutation pulse (beam 1) and response (beam 2) for different phases. Oscilloscope sweep is "5 μs/division"

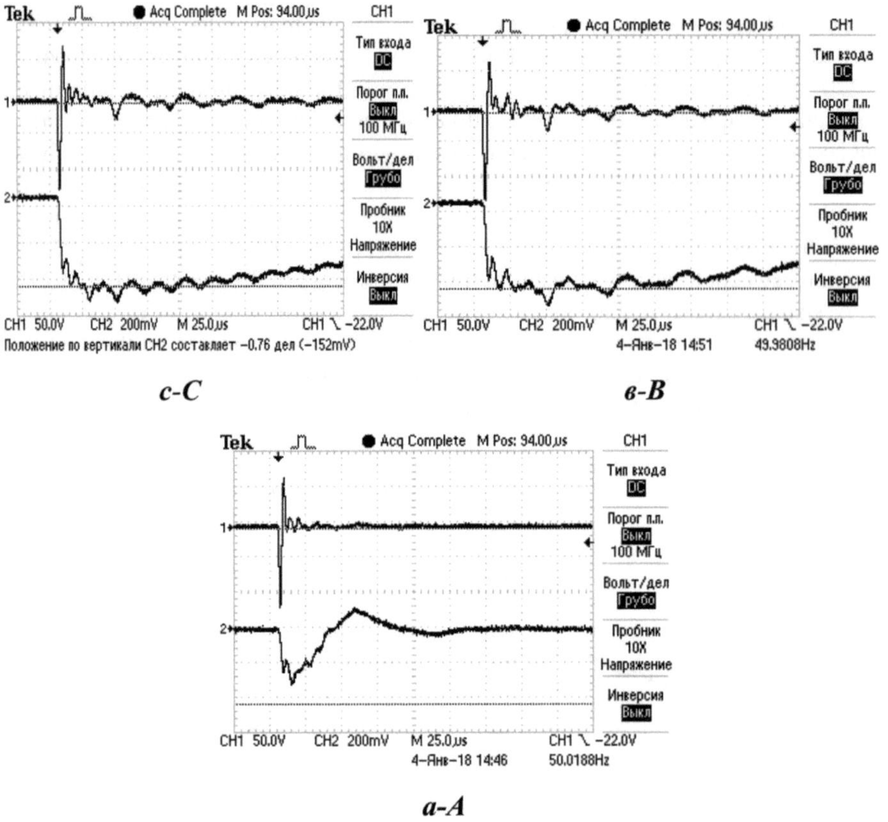

Fig. 8.27 Waveforms of commutation pulse (beam 1) and response (beam 2) for different phases. Oscilloscope sweep is "25 µs/division"

Switching operation "turn on" from the ~ 220 V mains on the HV side. The load was simulated by a rheostat with a maximum resistance of 40 ohms. Observation of a response oscillogram according to the "HV-LN" scheme (switching operation "turn on" from the ~ 220 V mains on the HV side). Oscillograms for various values of resistive loads of the secondary transformer winding are shown in Fig. 8.28 shows similar oscillograms of responses at different values of capacitive loads of the secondary transformer winding.

The widespread use of the technology for monitoring the condition of windings under operating voltage will make it possible to bring the diagnostics of transformers to a new technological level. One of the ways to solve the problem of condition monitoring without removing the voltage is to use the switching impulses of electric power grid as probing ones.

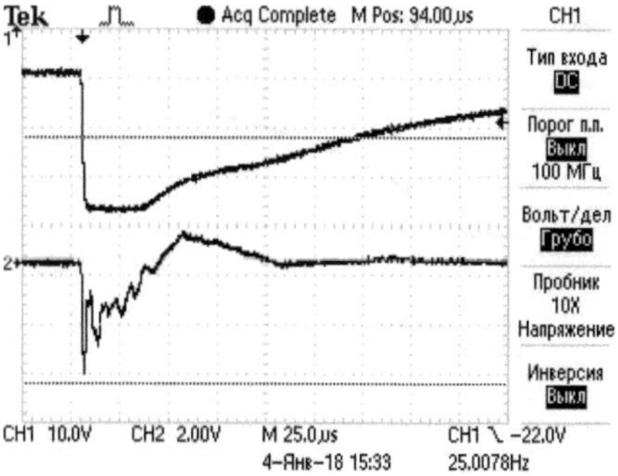

A-a; XX,

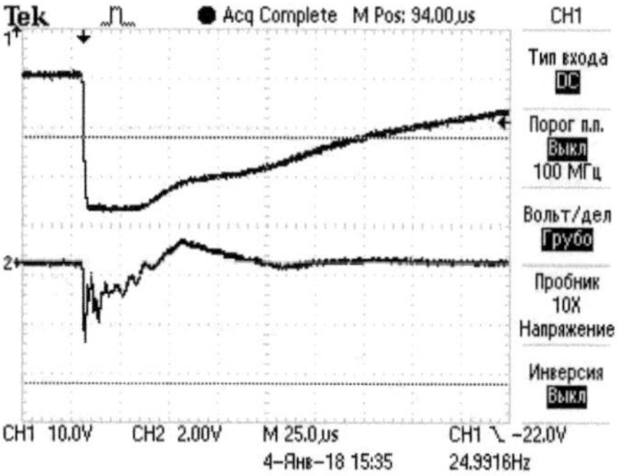

A-a; 100% load

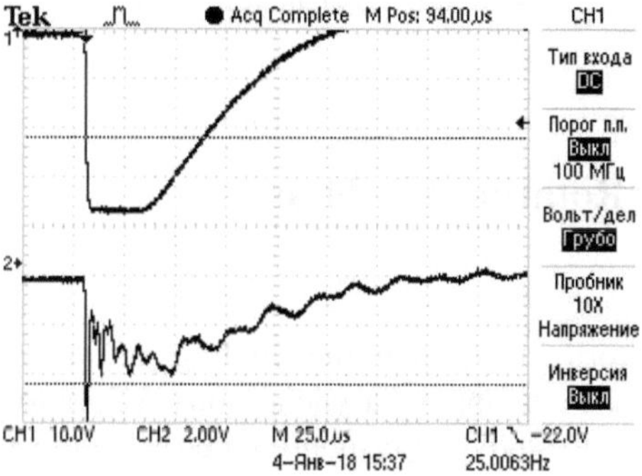

B-в; 100% load

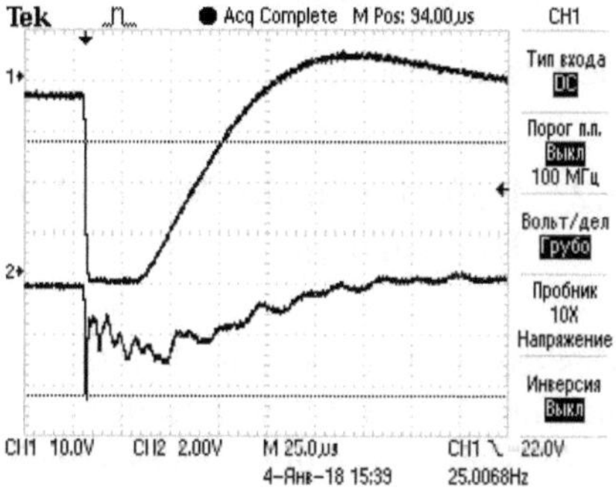

B-в; 50% load

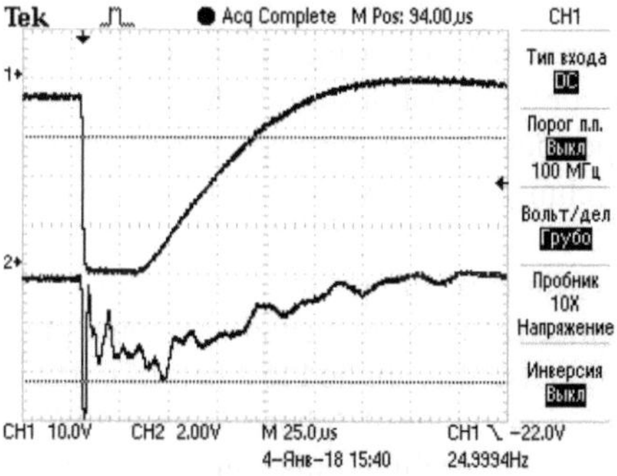

C-c; 100% load

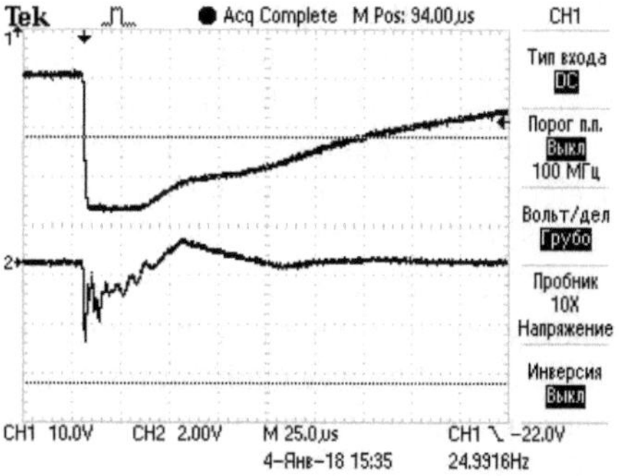

A-a; 50% load

Fig. 8.28 Waveforms of commutation pulse (beam 1) and response (beam 2) for different phases and different resistive and capacitive loads of low voltage winding loads

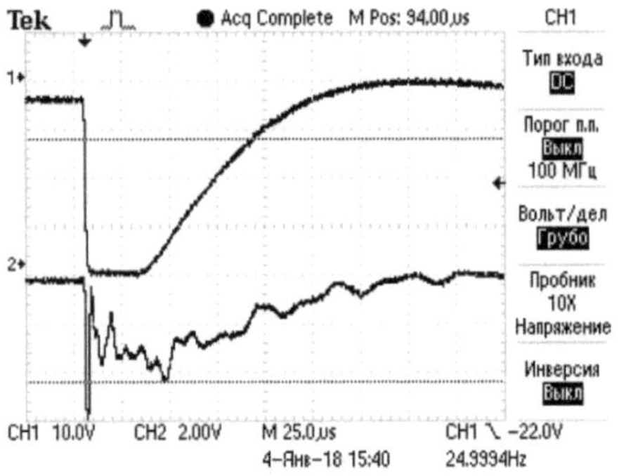

C-a; load – capacitance 0.05 μF;

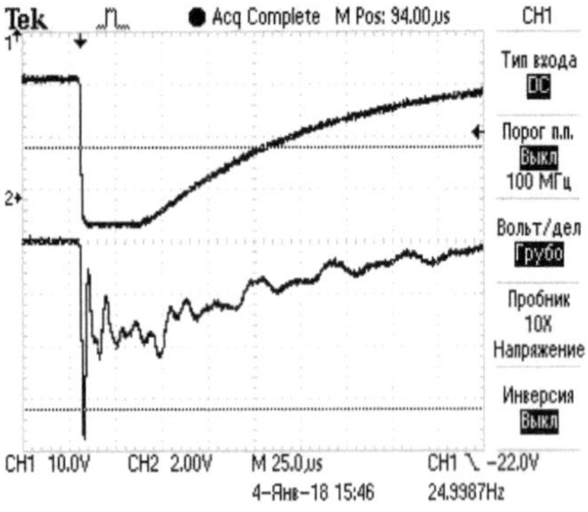

C-a; load – capacitance 0.15 μF

Fig. 8.28 (continued)

8.2 Defectography Method for Rotating Machines Diagnostics

This section describes a method for monitoring the state of the windings of rotating machines by pulse flaw detection using the example of asynchronous AC motors [5, 6]. The essence of the measurements is to study the response recorded between different phases as a response to a nanosecond rectangular pulse. Registration of signals, both probing pulses and response pulses, was carried out using Tektronix TDS 2012 oscilloscopes.

A comprehensive condition control was carried out for two electric motors—a new serviceable electric motor and an electric motor taken out of service due to defective windings. The engines are identical in type, design and purpose. Characteristics of electric motors, on which the created technique was worked out, are given in Table 8.3.

On Fig. 8.29, and as an example, oscillograms of the pulse at the input of the winding are given when a probing pulse is applied to this winding and a response pulse for the case of a serviceable motor. A probing pulse was applied to a pair of contacts "phase C—neutral", the response was recorded on a pair of "phase A—neutral".

On Fig. 8.29, b shows similar waveforms of the probing pulse and the response pulse for a faulty electric motor. It can be seen from the above waveforms that, firstly, the shape of the pulse on the probed winding depends on the state of the winding, and secondly, the pulse-response on the other winding of the electric motor also undergoes significant shape changes with a defect in the winding to which the probing pulse is applied. Thus, the pulsed sounding technique is quite sensitive for diagnosing the state of electric motor windings.

In the event that the neutral of the motor winding is not available for any reason, then this technique can be used by applying a probe pulse to the beginning of the motor winding and the housing ("ground") of the motor. As confirmation of the

Table 8.3 Parameters of investigated electrical machines

Parameter name	Asynchronous motor
Type	A4-400Y-4MY3
Number of phases	3
Frequency	50 Hz
Power	630 kW
Stator connection diagram	Star
Working voltage	6 kV
Current	72.5 Amps
Efficiency	95.2
Cos φ	0.88
RPM	1500
The weight	2290 kg
Technical conditions	3334-001-00213109-95

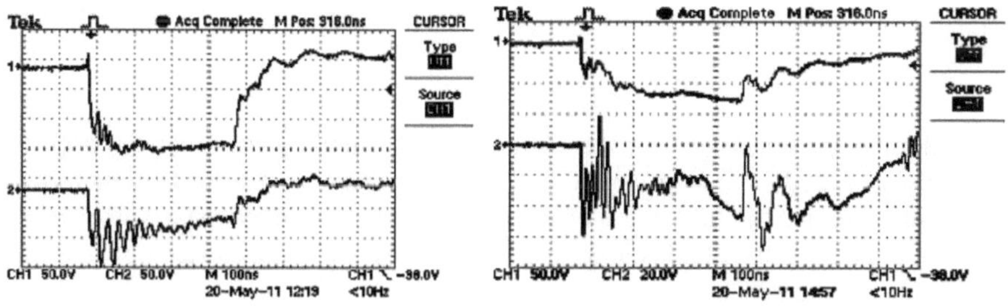

(a) (b)

Fig. 8.29 Waveforms on the winding of phase ***C*** and on the winding of phase ***A*** when a probing pulse is applied to the "C-neutral" winding: new electric motor (**a**); faulty electric motor (**b**)

efficiency of this method, Fig. 8.30, but the waveforms of the pulse between the beginning of the winding of phase ***A*** and the "ground" and the response between the beginning of the winding of phase ***B*** and the "ground" for a working electric motor are shown. On Fig. 8.30, b shows similar waveforms for an electric motor with a defective phase ***A*** winding.

Comparison of the shape of the response signals on a serviceable electric motor Fig. 8.30, but with a response signal on a knowingly defective winding Fig. 8.30b revealed the distortion of the pulse both at the input of the faulty winding, and the distortion of the pulse-response on the adjacent winding. Characteristic interrelated changes in the pulse shapes on the probed winding and responses on other windings of the electric motor make it possible to reliably determine the defective winding without using normograms for this electric motor.

This fact allows us to state that when this method is improved in relation to electric motors, it is possible to obtain a method for diagnosing high-power electric motors

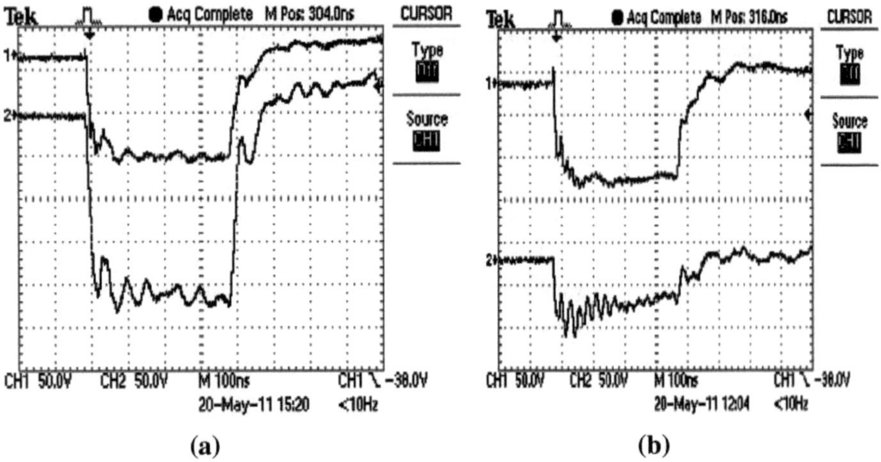

Fig. 8.30 Waveforms in diagnosing of the motor windings when a probing pulse is applied to the beginning of the phase ***A*** winding and "ground" and the response from the beginning of the phase ***B*** winding and "ground": a serviceable electric motor (**a**), an electric motor with a fault at the phase ***A*** (**b**)

that differs in its sensitivity for the better from the method for diagnosing power transformers based on comparing waveform responses with normograms, where conclusions are drawn only on the basis of an analysis of the response signal.

(It should be noted that this paragraph presents research data carried out at Tomsk Polytechnic University.)

When analyzing the measurement results, we did not apply the expansion of the response signal into a Fourier series, which is commonly used in the frequency response analysis technique. Measurements made both on a model electric motor and on real electric machines have shown that it is possible to determine defects in the windings of electric machines quite reliably without expanding the response pulses into a Fourier series. Characteristic changes in the form of responses indicate the presence of a defective state.

The results obtained in the course of this series of experiments can be formulated as the following statements:

- An improved method of low voltage pulses, based on the use of fairly short diagnostic pulses with nanosecond fronts, brought to the level of industrial technology for monitoring the state of active parts of power transformers, can be successfully applied to diagnose windings of asynchronous AC motors.
- Reducing the duration of the probing pulse to the level of hundreds of nanoseconds makes it possible to increase the sensitivity of the method and the reliability of defect detection.
- The proposed method of probing with low-voltage nanosecond pulses for the purpose of diagnostics the windings of AC motors is based on a comparison of pulse signals at the input of the probed winding and the pulse-response on other

motor windings. According to the characteristic forms and specific differences, one can conclude the presence or absence of defects.

- To interpret the measurement results, there is no need to expand the response pulse into a Fourier series, which simplifies the traditional method of amplitude-frequency characteristics.
- When performing diagnostics using nanosecond low-voltage pulses, there is no need to compare the received signals with normograms taken at the factory or after major repairs. This greatly simplifies the process of monitoring the state without reducing the accuracy and sensitivity of the method.

Question and Tasks

1. Briefly explain the essence of low-voltage pulse diagnostics.
2. When and where the LVPD method was proposed?
3. What is the drawbacks of LVPD method?
4. Give a general characteristic of the FRA technology.
5. Which two approaches are used to realize FRA?
6. Which are main features of nanosecond pulsed probing?
7. Make a comparison of nanosecond pulse probing and FRA.
8. What is a technology of single-stage pulsed defectography?
9. Describe the way of pulsed defectography winding control under operating voltage.
10. Which are main advantages of nanosecond pulse probing for rotative machines condition control?

References

1. Lekh V, Tyminskiy L (1966) Novyi metod indikaschii povrezhdenii pri ispytanii trasformatorov na elektricheskuyu prochnost' [New method of fail diagnostics at the dynamic reliability transformer test]. Elektrichestvo [Electricity] 1(1):77–81 (in Russian)
2. Ushakov VY, Mytnikov AV, Lavrinovich VA, Lavrinovich AV (2022) Transformer condition control advanced and traditional technologies. Springer Verlag, Cham, p 150
3. Measurement of frequency response (2012) IEC 60076-18
4. Lavrinovich VA, Mytnikov AV (2015) Development of pulsed method for diagnostics of transformer windings based on short probe impulse. IEEE Trans Dielectr Electr Insul 22(4):2041–2045
5. Boldea I, Nasar Syed A (2002) The induction machine handbook. CRC Press, p 968
6. Boldea I (2005) Synchronous generators. CRC Press, p 448

Chapter 9
Dielectric Spectroscopy Technology

Abstract This chapter is devoted to advanced insulation diagnostics technology based on specific response from dielectric medium. Physical basic of dielectric spectroscopy is considered in detail. Dielectric response in time and frequency domain processes are explained. Three main dielectric spectroscopy methods—recovery voltage method, polarization and depolarization currents measurement and frequency domain technology are described.

9.1 Basic Fundamental of Dielectric Spectroscopy

The main aim of this paragraph is to present and to recall the fundamentals of advanced method insulation diagnostic of high voltage equipment, first power transformers. It is based on dielectric spectroscopy. These are ***dielectric response in time—and frequency–domain*** by introducing into the basic equations. From these fundamentals, dentifing and diagnostic methods can be derived which become more and more attractive for a condition based maintenance of electric power equipment which has to be applied in future.

One of the innovations in the electric power engineering is the maintenance of high-voltage equipment based on its actual condition, evaluated by the results of monitoring or testing, and not by the duration of its operation.

The reasons for this evolution are well known. In North America, for instance, their roots can be found in restructuring and re–regulation of the Electric Utility Industry, and here in Europe, it was the adoption of the "Single Market Directive" which became the norm in February 1999, whereas it was still the exception many years before. The main issue of this regulation of the electricity market, which becomes now subdivided into independent power producers, transmission companies, system operators and distribution companies, is to encourage competition still maintaining basic public policy and service objectives. All partners within this new scheme are thus forced to cut costs in maintenance and operation without endangering steady supply of electricity to demanding customers. Costs can be reduced first of all by a transition from time–based maintenance (TBM) to condition based maintenance (CBM), if the actual conditions of the expensive high voltage (HV) components

V. Y. Ushakov et al., *High-Voltage Equipment of Power Systems*, Power Systems,
https://doi.org/10.1007/978-3-031-38252-9_9

within the electric power transmission systems are reliably known. The application of "unscheduled maintenance", the philosophy of which is based on a reactive mode of operation, will not at all reduce costs. Unscheduled maintenance means that only then repairs and maintenance will be made if the equipment, as e.g. a transformer or cable, breaks down. But this causes in general a downtime of the electricity supply so that breakdowns become much more costly than planned maintenance. A CBM based on diverse, reliable diagnostic tools should be applied today. Driving forces for the development and application of improved diagnostic methods are the steadily increasing age of expensive HV components. In many parts of the world, the majority of e.g. large power transformers have been installed in the 60s of the last century. Also, cable technology did change at this time and the first generations of PE or XLPE cables are still prone to breakdowns. All these facts are very well known. It is not the aim of this contribution to discuss the full complexity of all the different diagnostic techniques as already successfully applied to different HV components. Most of the dangerous breakdowns, however, are caused by ageing effects of the HV insulation systems as used within these components, and there is still some lack of appropriate tools to diagnose insulation materials non–destructively and reliably in the field of application. New methods have been published in the last decade and even before, for which reliable diagnostics are claimed. Many of these methods are based on changes of dielectric properties of the insulation, whose basics are often not very well known. Therefore, this contribution will predominantly introduce into such basics but will also provide some hints to more recent instruments, which are able to quantify dielectric properties in the field [1].

Experience shows that most of electric power engineers are not too familiar with the fundamentals of "dielectric response functions" and their background. The following short introduction may contribute to close this gap. Every kind of insulation material consists at an atomic level of negative and positive charges balancing each other in microscopic as well as in more macroscopic scales (if no uni–polar charge was deposited within the material before by well known charging effects). Macroscopically, some localised space charge may be present, but even then an overall charge neutrality exists. As soon as the material is exposed to an electric field (as generated by a voltage across electrodes between which the dielectric is embedded), very different kinds of dipoles become excited even within atomic scales. A local charge imbalance is thus "induced" within the neutral species (atoms or molecules) as the "centres of gravity" for the equal amount of positive and negative charges, $\pm q$, become separated by a small distance d, thus creating a dipole with a dipole moment, $p = qd$, which is related to the "local" or "microscopic" or a slow process as e.g. active in cable material electric field E acting in close vicinity of the species. Thus, the dipole moment can also be written as $p = \alpha E$, where α is the *"polarisability"* of the species or material under consideration. Note that p, d and E are vectors not marked here. As the distance d will be different for different species as well as the number of dipoles per unit volume, also their polarisability is different. Due to chemical interactions between dissimilar atoms forming molecules, many molecules will have a constant and stabledistance d between the charge centers thus forming "permanent dipole moments", which are in general not uniformly distributed within the matter as long as no external field is applied. (Note, that any kind of "permanent polarisation", as effective in *electrets* or *ferroelectrics,* is not considered here). The macroscopic effect of the polarisation–ability of individual species is finally manifested in a general relation between the ***macroscopic polarisation P*** and the number of polarised species N per unit volume of the matter. These relationships are known, but not treated here. But let us recall now the main mechanisms producing polarisation P: ***"Electronic Polarisation"*** is effective in every atom or molecule as the centre of gravity of the electrons surrounding the positive atomic cores will be displaced by the electric field E. This effect is extremely fast and thus effective up to optical frequencies.—***"Ionic (or molecular) Polarisation"*** refers to matter containing molecules forming ions, which do not separate by low electric fields or working temperatures.—***"Dipolar (or orientation) Polarisation"*** belongs to matter containing molecules with *permanent* dipole moments with orientations statistically distributed due to the action of thermal energy. Under the influence of E, the dipoles will be oriented only partially, so again, a linear dependency of P and E exists. Ionic and dipolar polarisation are still quite fast effects and may follow AC frequencies up to MHz or GHz.—***"Interfacial Polarisation"*** is predominantly effective in insulating materials composed of different dielectric materials *as e.g. oil impregnated paper/cellulose.* The mismatch of the products permittivity by electric conductivity for the different dielectrics forces movable positive and negative charges to become deposited on the interfaces of different insulating materials, thus forming dipoles. This phenomena is often *very slow* and in general active in a very low frequency range up to power frequencies.—Finally, ***hopping of charge carriers*** between localised sites of charges may occur creating also polarisation. This is in general also a slow process as e.g. active in cable material [2].

So, the dielectric polarisation is the result of a relative shift of positive and negative charges in the matter. During all of these processes, the electric field is therefore *not able to force the charges to escape from the matter*, which would cause inherent electric condition.

Dielectric Response Methods are based on fundamental interactions between well known electric quantities: Usually, HV insulation materials (also called dielectrics) are isotropic and in general homogeneous, at least at macroscopic scales. Then, the vectors of (macroscopic) polarisation P and the electric field E are of equal direction and interrelated by:

$$P = \chi \varepsilon_0 E \tag{9.1}$$

where χ is the dielectric susceptibility of the matter, a dimensionless number which is zero for ideal vacuum. Thus, the susceptibility χ accounts for all kinds of polarisation processes as effective within a dielectric, ε_0 is the permittivity of vacuum.

This provides already a hint that all polarisation processes induce also electric charges at the electrodes, between which the dielectric is sandwiched. From Eq. (9.1) it can well be deduced that the polarisation P will change or vanish if the field E is changed or set to zero.A reduction of E will thus lead to a *depolarisation* process, which will follow with some delay or retardation to the reduction of E. Dielectric properties become thus *dynamic* events which can be detected in **time**- as well as in ***frequency domain.***

In a *vacuum–insulated* electrode arrangement, the vector of *electric displacement* (or *"dielectric flux density"* or *"electrical induction") D* is exactly proportional to the electric field *E*,

$$D = \varepsilon_0 E \tag{9.2}$$

or, if the electric field is generated by a time-varying voltage,

$$D(t) = \varepsilon_0 E(t) \tag{9.3}$$

here, ε_0 is again the permittivity of vacuum. The origin of *D* and *E* is in general provided by a voltage source connected to the electrodes of any electrode arrangement under consideration. No time delay at all will exist between both magnitudes, if the time scales considered produce still "electrostatic field conditions". Note, however, that *D* represents the (positive and negative) electric charges per unit area as induced at the electrode surface, these charges are the origin—sources and sinks—of all electric field lines. For time–varying fields, *E*(t), the so called "displacement current" must be supplied by the voltage source to maintain the area charge density at the electrodes. This current is governed by *dQ/dt*, where *Q* is the total electric charge deposited on each of the electrodes. If now the vacuum is replaced by any kind of isotropic dielectric material, the electric displacement *D* of Eqs. 9.2 and 9.3 increases by its inherent (macroscopic) polarisation *P* as defined in Eq. (9.1):

$$D(t) = \varepsilon_0 E(t) + P(t) = \varepsilon_0(1 + \chi)E(t) \tag{9.4}$$

This equation is quite essential as it separates the two kinds of charge induction. As for *isotropic* materials both vectors, *P* and *D*, are still in parallel to *E*, we can further on avoid bold letters to mark vectors. However, the time dependency of *P(t)* will not any more be the same as that of *E(t),* as the different polarisation processes have different time delays with respect to the appearance of *E*. This delay is obviously caused by the time-dependent behaviour of the susceptibility $\chi = \chi(t)$. This time delay may be understood best with the following reflections: Let us assume, that a step-like constant electric field of magnitude E_0 is applied within the dielectric at any time t_0 and that this field remains constant for $t \geq t_0$. The dielectric can then be completely characterised by its time dependent susceptibility $\chi(t)$ or its specific polarisation *P(t)* as a ***response in time domain***, i.e. the formation and evolution of the different kinds of polarisation processes, which develop as well within extremely short times (as e.g. electronic polarisation) or are much slower or even very slower (as e.g. interfacial polarisation). For $t \leq t_0$, the magnitude of the susceptibility or polarisation is still zero [2–4]. Figure 9.1 illustrates this situation:

This, according to Eq. (9.1), can be expressed as

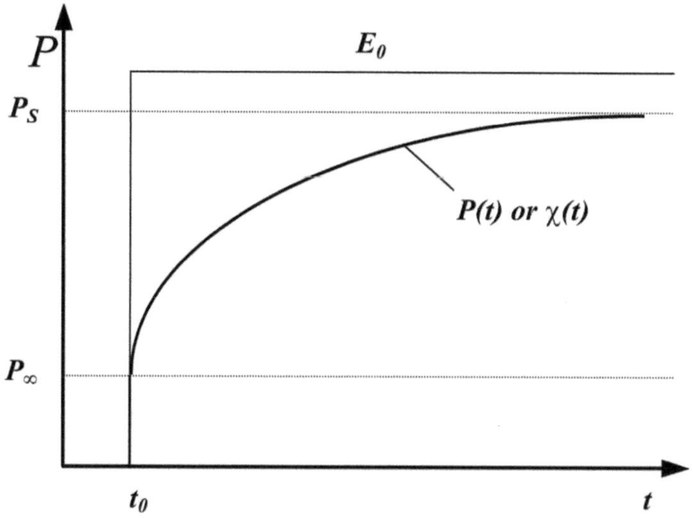

Fig. 9.1 Polarization of a dielectric exposed to a step field magnitude E_0

$$\frac{P(t)}{E_0} = \varepsilon_0 \chi(t) l(t) \tag{9.5}$$

where $\chi(t)$

as well as *P(t)* represent "*step response functions*". The factor *1(t)* is used to indicate the unit step. It is necessary to note, however, that in Fig. 9.1 the first part of these functions are simplified by an ideal step to account for the very fast polarisation processes, marked by an *"instantaneous polarisation"*, $P(t = t_o) = P\ \infty$. This step can in general not be recorded neither in time—nor in an adequate frequency domain. As all kinds of polarisation become finite and will settle at longer times, the polarisation becomes finally "static", $P(t \to \infty) = P_s$, i.e. for a selected time static. Due to Fig. 9.1, the step response of this somewhat simplified polarisation can now be written as:

$$P(t) = P_\infty + (P_s - P_\infty) g(t - t_0) \tag{9.6}$$

where *g(t)* is dimensionless, monotonically increasing function.

The Eq. (5.4) may also be written as:

$$P(t) = \varepsilon_0 [\chi_\infty + (\chi_s - \chi_\infty) g(t - t_0)] E_0 \tag{9.7}$$

As known from general circuit theory, it is now possible to compute any other time dependent polarisation *P(t)* for *any other* time dependent excitation *E(t)* of a test object, as the special solutions for the step excitation are already known. This can be done by means of *"Duhamel's Integral"* or convolution in the time domain. For the quantities as used in Eq. (9.7), the result is:

$$P(t) = \varepsilon_0 \chi_0 E(t) + \varepsilon_0 \int_{-\infty}^{t} f(t-\tau) E(\tau) d\tau \tag{9.8}$$

where *f(t)* is the so called dielectric response function.

$$f(t) = (\chi_s - \chi_\infty)^{\partial g(t)}/_{\partial t} = (\varepsilon_s - \varepsilon_\infty)^{\partial g(t)}/_{\partial t} \tag{9.9}$$

f(t) is obviously a ***monotonically decreasing*** function and inherent to the dielectric investigated.

9.2 Polarization and Depolarization Current Measurement

The polarisation *P(t)* is not an observable magnitude by itself, but it produces the main part of the ***polarisation*** (or absorption, or charging) ***current*** in a test object if the electric field, *E(t),* is ***suddenly*** applied. But up to now, we have not yet considered any inherent or "pure" dc conductivity σ_0, which represents the movement of the free charges in the dielectric and which is ***not*** involved in polarisation. As already postulated by Maxwell in 1891, this field *E(t)* generates a total current density *j(t),* which can be written as the sum of conduction, vacuum and polarisation displacement current:

$$j(t) = \sigma_0 E(t) + \frac{\partial D(t)}{\partial t} = \sigma_0(E) + \varepsilon_0 \frac{\partial E(t)}{\partial t} + \frac{\partial P(t)}{\partial t} \tag{9.10}$$

Taking into account $\varepsilon_\infty = 1 + \chi_\infty$ Eq. 9.9 can be written as:

$$j(t) = \sigma_0 E(t) + \varepsilon_0[\varepsilon_\infty \delta(t) + f(t)]E(t) \tag{9.11}$$

Equation 9.11 is thus a first basis for the measurement of the dielectric response function *f(t)*. For doing so, a step–like dc charging voltage of magnitude U_C which must be constant and free of ripple, is suddenly switched to the test object which has been totally discharged before. Then *a* ***polarisation current*** $\boldsymbol{i_{pol}(t)}$ through the test object can be recorded according to the equation:

$$i_{pol}(t) = C_0 U_C \left[\frac{\sigma_0}{\varepsilon_0} + \varepsilon_\infty \delta(t) + f(t) \right] \tag{9.12}$$

where C_0 is the geometric or vacuum capacitance of the test object and $\delta(t)$ is the delta function originated from the applied step voltage at $t = t_0$.

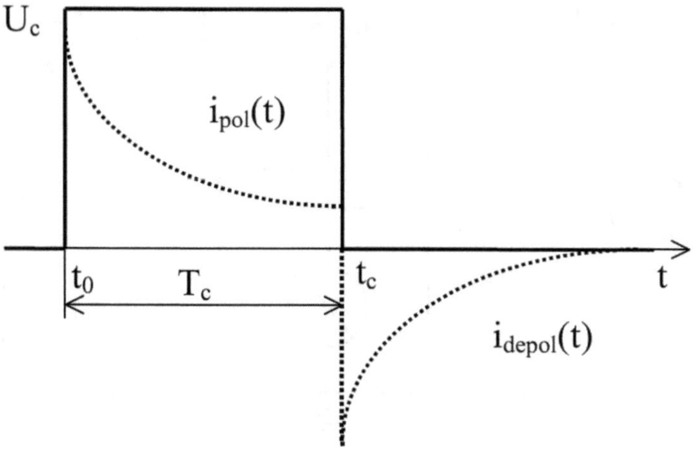

Fig. 9.2 Principle of relaxation current measurement

This polarisation current contains three parts. The first one is related to the intrinsic conductivity of the object, the last one represents all polarisation processes as activated during the time of voltage application and the middle part with the delta function can not be recorded in practice due to the large dynamic range of the current amplitudes.

A polarisation current measurement can be stopped if the current becomes either stable due to the *dc* term or very low [5]. Immediately following the polarisation, the ***depolarisation*** (or discharging) ***current*** i_{depol} can be measured by a subsequent short–circuiting of the sample, as shown on Fig. 9.2.

According to the superposition principle and neglecting the second term in Eq. 9.12 which is again a very short current pulse, we get for $t \geq (t_0 + T_C)$:

$$i_{depol}(t) = -C_0 U_C [f(t) - t(t + T_C)] \tag{9.13}$$

where T_C is the time during which the step voltage for polarisation was applied. ***This current is of opposite polarity.***

In practice, the polarisation and depolarisation or relaxation currents are measured with a "two electrode" technique as shown in Fig. 9.3.

The measured currents are then defined by the selected electrode arrangement and can be sensed at virtual earth potential. The complex insulation system of a power transformer is a typical example for such an application. Finally, an example of recently performed relaxation current measurements is shown in Fig. 9.4.

It is taken from investigations concerning the dielectric response of oil-impregnated pressboard with the ***moisture content (m.c.)*** as a parameter. Here, all measurements started 1 s after voltage application (i_{pol}) and after short circuit (i_{depol}). From the selected results it can be recognised, that only for the higher moisture content the final value of the polarisation current could be reached, though the

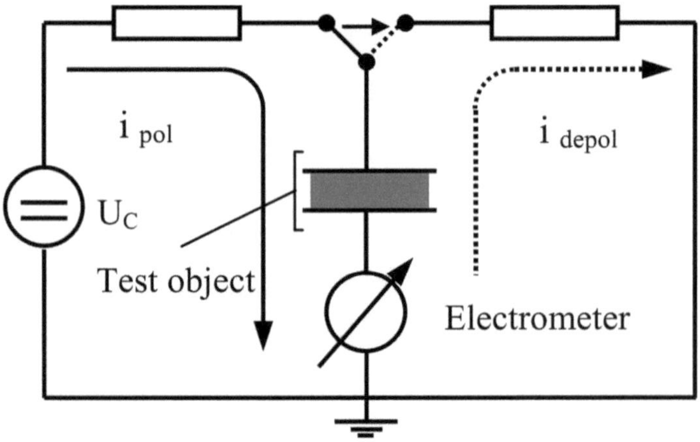

Fig. 9.3 Principle of test scheme for the PDC measuring technique

Fig. 9.4 Relaxation currents of unaged samples with different moisture contents

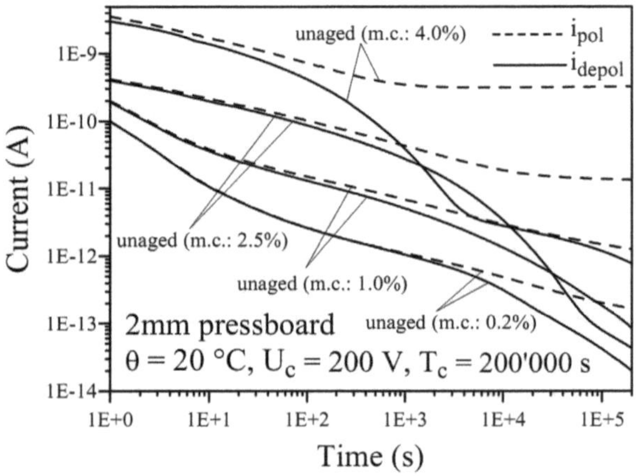

measurements lasted up to 200000s, which is about 56 h. Representation of these results in log–log scale is paramount due to the large dynamic of the process.

This is shown y means of Eqs. (9.8) and (9.11) in which an ideal step response for the total current density of an ideal dielectric response function *f(t)* considering also instantaneous polarisation processes are assumed:

$$j(t) = \sigma_0 E(t) + \varepsilon_0 \frac{dE(t)}{dt} + \varepsilon_0 \frac{d}{dt} \int_o^t f(t-\tau)E(\tau)d\tau \tag{9.14}$$

Taking into account $j(\mathrm{t}) \Rightarrow j(\mathrm{p})$; $E(\mathrm{t}) \Rightarrow E(\mathrm{p})$; $E'(\mathrm{t}) \Rightarrow p\,E(\mathrm{p})$; $f(\mathrm{t}) \Rightarrow F(\mathrm{p})$; and considering the convolution of the last term in this equation we get with p being the Laplace Operator:

$$j(p) = \sigma_0 E(p) + \varepsilon_0 p E(p) + \varepsilon_0 p F(p) E(p) \tag{9.15}$$

As p is nothing else than the ***complex frequency i*** ω, we can reduce the equation to:

$$\overline{j}(\omega) = \overline{E}(\omega)\left[\sigma_0 + i\omega\varepsilon_0\left(1 + \overline{F}(\omega)\right)\right] \tag{9.16}$$

Thus it becomes obvious, that $\overline{F}(\omega)$ is the Fourier Transform of the dielectric response function *f(t)* or the complex susceptibility $\overline{\chi}(\omega)$:

$$\overline{\chi}(\omega) = \overline{F}(\omega) = \chi'(\omega) - i\chi''(\omega) = \int_0^{\propto} f(t)\exp(-i\omega t)dt \tag{9.17}$$

Combining Eqs. (9.16) and (9.17) shows the well known relationship:

$$\overline{j}(\omega) = \left\{\sigma_0 + \varepsilon_0\omega\chi''(\omega) + i\omega\varepsilon_0\left[1 + \chi'(\omega)\right]\right\}\overline{E}(\omega) \tag{9.18}$$

The complex electric displacement $\overline{D}(\omega)$ can now be expressed by the *relative*, but *complex dielectric permittivity*, $\overline{\varepsilon}(\omega)$, with the relation:

$$\overline{D}(\omega) = \varepsilon_0\overline{\varepsilon}(\omega)\overline{E}(\omega) = \varepsilon_0\left[1 + \chi'(\omega) - i\chi''(\omega)\right]\overline{E}\omega \tag{9.19}$$

where:

$$\overline{\varepsilon}(\omega) = \varepsilon'(\omega) - i\varepsilon''(\omega) = \left(1 + \chi'(\omega)\right) - i\chi''(\omega) \tag{9.20}$$

Actual measurements of this dielectric response in frequency domain in difficult to perform if the frequency range become large. Usually such measurement is applied for a single power frequency. New approach can cover some decades in frequency range. Note, that according to Eq. (9.18) such instruments can not distinguish between the current contribution of the "pure" DC conductivity σ_0 and that of the dielectric loss $\varepsilon''(\omega)$. This means that the effective measured relative dielectric permittivity $\overline{\varepsilon}_R(\omega)$ is different from the relative permittivity $\overline{\varepsilon}(\omega)$ as defined in Eqs. (9.19) and (9.20) Then the effective relative dielectric permittivity $\overline{\varepsilon}_R(\omega)$ is defined form the following relations:

$$\overline{\varepsilon}_R(\omega) = \varepsilon'_R(\omega) - i\left[\varepsilon''_R(\omega) + {}^{\sigma_0}\!/_{\varepsilon_0\omega}\right] = 1 + \chi'(\omega) - i\left[\chi''(\omega) + {}^{\sigma_0}\!/_{\varepsilon_0\omega}\right] \tag{9.21}$$

and the "dielectric dissipation factor" ***tan***δ can be written as:

$$\tan\delta(\omega) = \frac{\varepsilon''_R(\omega) + {}^{\sigma_0}\!/_{\varepsilon_0\omega}}{\varepsilon'_R(\omega)} \tag{9.22}$$

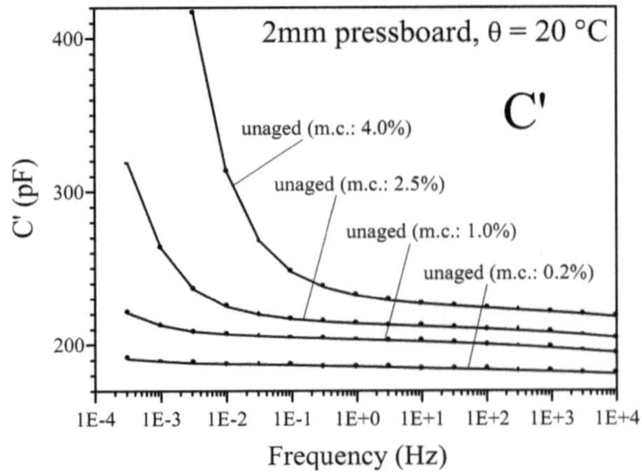

Fig. 9.5 Real part of the complex capacitance of the pressboard sample

The real part of Eq. (9.21) represents the capacitance of a test object, whereas the imaginary part represents the losses. Both quantities will depend on frequency. Often, this fact is not appreciated, if only one single frequency is applied for the measurement. As ageing effects will change these magnitudes in quite different and specific frequency ranges, new diagnostic tools have to take care for this effect. Measurements in the frequency domain need voltage sources of variable frequencies. So, these measurements become very lengthy if low frequencies are considered. At least 3 cycles of an AC voltage are in general necessary to determine of amplitude and phase shift between voltage and current. Therefore, up to 3000s can be necessary to get a single value only of C and $tan\ \delta$ for a frequency 1 MHz. Figures 9.5 and 9.6 display the frequency range for the (real part of the) capacitance $C' = \varepsilon' C_0$ and dissipation factor of the pressboard samples as used in Fig. 9.4. The measurements have been made with a special "dielectric spectrometer" for selected individual frequencies, which can be identified by the straight lines between measuring points.

The results show, that the m.c. of the pressboard affects the low and very low requencies much more than the power frequency, this for the dissipation factor as well as for the capacitance, *C';* this increase is obviously caused by the conductivity of water and thus increased interfacial polarisation inside the board.

Ageing of insulation materials can be detected by its dielectric response in any of the two domains. But also many other methods exist as based on other physical reactions, as e.g. changes in chemical, mechanical or optical behaviour. Since many years such methods are applied to power transformer diagnostics, as e.g. by oil parameter analysis including DGA or HPLC, or Carl Fischer tests for detecting moisture in oil and paper, for which a very large amount of publications does exist. This paper, however, is completely restricted to dielectric response techniques, which can be applied to any kind of equipment. The test objects are treated as "black boxes" accessible only by their terminals. ***The disadvantage of dielectric response measurements in either of the two domains is their "off–line" character, i.e. the equipment must be switched off from the power transmission voltage.***

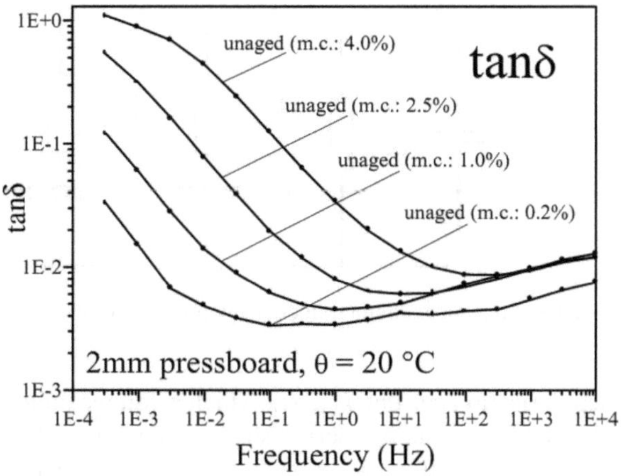

Fig. 9.6 Dissipation factor tan δ of the pressboard sample

9.3 Recovery Voltage Measurement

A single ***Return*** *or* ***Recovery Voltage (RVM)*** measurement is the oldest method to qualify dielectric properties (permittivity and losses) of "anomalous" dielectrics, defined by not only a single relaxation time. The principle is shown in Fig. 9.7.

To get complete information, similar to Fig. 9.1 a step voltage U_0 is applied across the electrodes of the completely discharged object. After a short grounding (short–circuiting) period, a recovery voltage, $U_R(t)$, can then be measured under open circuit conditions, if the input impedance of the voltmeter is very high. The source of the recovery voltage are the relaxation processes inside the dielectric material, i.e. the

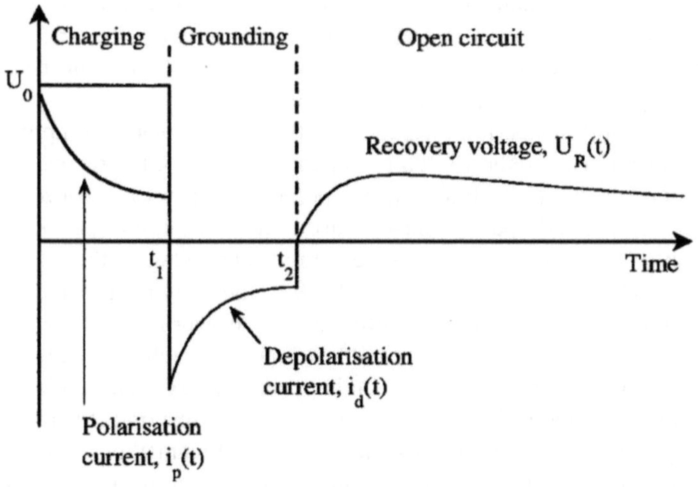

Fig. 9.7 Principle of a Return Voltage Measurement

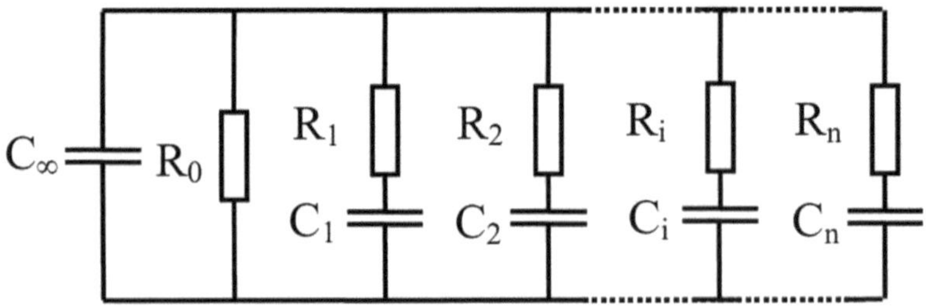

Fig. 9.8 Equivalent circuit to model any linear dielectric

depolarisation current, $i_{depol}(t)$, which a voltage builds up on the electrodes of the test object. The grounding period shall be much less than 1 s so that only the "instantaneous" polarisation processes disappear. $U_R(t)$ can be calculated numerically if the dielectric response *f(t)* of the test object has been measured and if the duration of the grounding period, (t_2—t_1), together with other boundary conditions are taken into account. The phenomena, however, can easiest be explained if we represent the dielectric by an *equivalent circuit* which is able to explain all measurements in time domain: As already shown earlier, for "slow" polarisation processes the function *f(t)* and thus the relaxation currents are monotonically decreasing, see Fig. 9.4. These currents can be simulated by a sum of exponential functions as shown in [4, 5] and elsewhere. This sum is, together with the power (high) frequency capacitance $C_\infty = \varepsilon_\infty C_0$ as determined by conventional methods and the final value of the insulation resistance R_0 (equivalent to σ_0), the origin of this well known and general equivalent circuit, see Fig. 9.8.

If this circuit is charged with a voltage source U_0 during $0 \leq t \leq t_1$, the individual polarisation currents including the constant current through R_0 will flow into the circuit, charging instantaneously C_∞ and, with some delay, the *RC*-elements according to their time constants. Depending on how long (t_1) the object is charged, the different polarisation processes as represented by the *RC* elements become either fully or only partly activated. A short grounding period from $t_1 < t \leq t_2$ will only discharge C_∞ but if this period is larger, also the slower polarisation processes start to relax. As soon as this short circuit is opened for $t > t_2$, the recovery voltage is measured at open circuit conditions, as the polarisation processes (or *RC* elements), which either not or only partly relaxed during the short circuit period, will partly discharge into C_∞ and R_0. The magnitude of the return voltage is thus always proportional to U_0. This method can be traced back to the beginning of the last century and was originally applied due to the difficulties in measuring small (depolarisation) currents. It was, however, always possible to measure voltages with electrostatic voltmeters, the input impedance of which is extremely large. The application of this method is of advantage if the "ground" terminal of the test object is not accessible. In Germany, at least one measuring system based on this method is available.

If this ancient method is nowadays again applied, it was triggered by the appearance of a similar, but specialised method called simply ***"RVM-technique"***. The method, originally proposed in [4] and somewhat later commercialized [5], became attractive for diagnosing transformer insulation as it was claimed that the moisture content in the pressboard of the complete insulation system can be quantified by analysing a so–called ***"polarisation spectrum"*** resulting from the measurement. This "spectrum" is performed by applying a series of charging voltages U_0 to the test object, followed by the short circuiting as explained before, at each step increasing the charging time $t_1 = t_c$ and the short circuiting time $(t_2—t_1) = tg$ and using a fixed ratio of $(t_c/t_g) = 2$. During these periods, nothing is measured. After t_g has elapsed, the recovery voltage is recorded and from its peak value, the amplitude U_{Rmax} is quantified and plotted as a function of t_c. This dependency is called "polarisation spectrum" as its maximum appears indeed at a t_c–value, which agrees with a single time constant *RC* of our equivalent circuit, Fig. 9.8. The method, however, is very lengthy as after each measuring cycle (the longest of which is 10 000s the insulation system must be discharged to prepare the next cycle. The method is thus fully based on *depolarisation* currents only which are, depending on the charging period, incomplete as only a part of these currents is used to build up the individual return voltages. Instead of performing such a measurement, it is possible to compute the "polarisation spectrum" and its parameters if the dielectric response function *f(t)* is known.

A quite new instrument, the ***PDC Analyser,*** as described recently in [5] and manufactured by a Swiss company, measures a complete set of polarisation and depolarisation currents and thus also the response function *f(t),* examples of which have already been shown. The default software of this instruments calculates all other quantities as e.g. insulation resistance in function of time, all kinds of "Polarisation Indexes", single Return Voltages or "polarisation spectra" as applied by the ***RVM-technique***, and complex capacitance including loss factor in the frequency domain. Some examples of application and evaluation are shown below. This is basic fundamentals of ***Principles in Frequency Domain***. The measurement of *"C–tan δ"* at power frequency (i.e. at one single value in the frequency domain) by means of bridge circuits, which are based on "standard capacitors", is well known. Since some years this quantity is sometimes measured at 0.1 Hz in combination with a low frequency HV test equipment as used for diagnostic tests of medium—voltage PE cables, the loss factors of which are sensitive to chemical treeing. A quite large frequency range is now covered by a new instrument called "*IDA 200*" manufactured by a Swedish company. This "Insulation Diagnostic Tool" measures *C-tan δ* from 0.0001 Hz—1000 Hz and covers thus also the low and very low frequency range which is most prone to ageing effects. Below, an example for the application of one of the new diagnostic tools is presented based on ***time domain (PDC) measurements***. The example is taken from very recent investigations in the **CIGRE Task Force 15.01.09.** In the course of the investigations, a test object modelling power transformer insulation systems was designed, constructed and used. The measurements as performed on this model are based on the 3 methods as explained in Chapter 5 (RVM–technique, PDC Analyser and IDA 200). The design of the test object and the goal of the investigations was already published in [5].

The model can be characterised as follows: metal tank (ca. 490 × 1040x1560 mm) with bushings to connect the inside flat electrodes between which different insulation

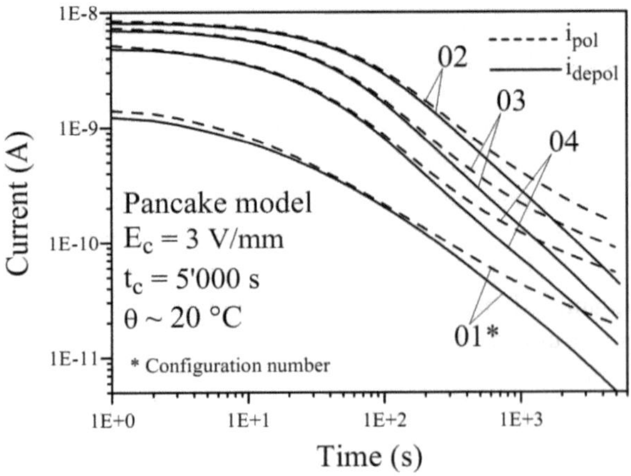

Fig. 9.9 Relaxation currents measured on four different configurations of the "pancake" model. The electrode spacing ratio is 3 V/mm for all measurements

configurations have been sandwiched. Each insulation configuration consisted from flat plates of pressboard ("Kraft Thermo 70") with or without oil gaps between the electrodes. Pressboard plates have been fixed in the middle of the whole gap between electrodes by means of pressboard spacers.

In the following figures, the curves are marked by codes indicating the insulation configurations. The effective area of the electrodes is about 1 m^2. The "instantaneous" capacitance as measured with 50 Hz ranges from about *4.8nF to 2.35nF*. Figure 9.9 shows all four sets of relaxation currents as measured on the either "pure board"—configuration 01 or the multi–layer configurations of code 02 to 04.

The currents have been recorded just *1 s* after voltage application or short–circuiting, and measurement periods have been *5 000s each.* The shapes of the currents for these 4 configurations can be subdivided into 2 groups: The multi–layer configurations 02 to 04 are characterised by their *pronounced exponential shape* at short times of less than about 200s. ***This kind of shape is typical for PDC measurements on all kinds of HV power transformers.*** They are essentially produced by the formation of the *interfacial polarisation* between oil gaps and pressboard barriers. It can well be identified from these shapes in log–log–scales that their dominating time constants increase with the thickness ratios of oil gaps to pressboard barriers. The dielectric response of the pressboard becomes more apparent at long times, at which the depolarisation currents are quite well in parallel indicating the identity of the dielectric response function of the pressboard material as used to reach different thickness (from 2 to 10 mm) of the barriers. The (nearly) homogeneous gap of configuration 01 behaves quite different, as the initial shape of the currents is not pronounced exponential.

Figure 9.10 presents the calculated *"polarisation spectra"* for all 4 configurations 01 to 04 of the model and for a charging voltage of *1 V*.

The results show that the position of the peak value in *"polarisation spectrum"* does change in dependence of geometry. For the configurations 02 to 04, this maximum is due to the "macroscopic" interfacial polarisation and its geometric

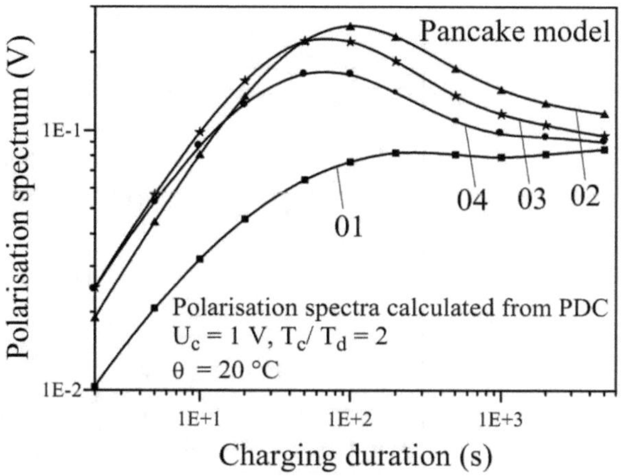

Fig. 9.10 Calculated polarisation spectra of all configurations

position at the interface between oil gaps and pressboard. This fact can also be calculated by assuming a simple Maxwell–Wagner interfacial polarisation process. In this model, 2 series–connected dielectrics with losses are assumed. Each dielectric can then be modelled by—for our example—the conductivity and relative permittivity of the oil ($\sigma_{oil}, \varepsilon_{roil}$) and of the pressboard ($\sigma_{Board}, \varepsilon_{rBoard}$). Then the time constant τ of the interfacial polarisation is given quanitatively by the following equation, where d_{oil} and d_{Board} are the thickness of oil gap and pressboard barrier.

$$\tau = \varepsilon_0 \times \frac{d_{Board}\varepsilon_{rOil} + d_{Oil}\varepsilon_{rBoard}}{d_{Board}\sigma_{Oil} + d_{Oil}\sigma_{Board}} \tag{9.23}$$

Though in Eq. (9.22) the much more complex dielectric properties of the pressboard are not taken into account and only simulated by a single tie constant, a numerical evaluation with approximate values for both components would already show quite good agreement with the results as measured. This equation can not simulate the spacers as applied to fasten the board within the gaps, but it shows already for the different ratios of oil gaps to board thickness the change of the "dominant" time constants, which decrease in magnitude if this relationship decreases. The default software of the "PDC Analyser" calculates for all measured currents (see Fig. 9.8) the specific equivalent circuits (see Fig. 9.9) so that the transition to frequency domain is easy to perform. In Fig. 9.11 the results of such calculations are displayed for the most interesting configurations 02 to 04 and compared with the measured values of capacitance *C'(ω)* and dissipation factor *tan δ(ω)* as registered by the dielectric spectrometer "IDA 200".

The comparison shows the good agreement of measured and calculated values down to very low frequencies. The maxima in the tan δ curves and the significant increase of capacitance at low frequencies confirm again the predominant influence of interfacial polarisation on the total dielectric response of multi–layer arrangements. All measurements as shown in Figs. 9.9 and 9.11 have been made with the original

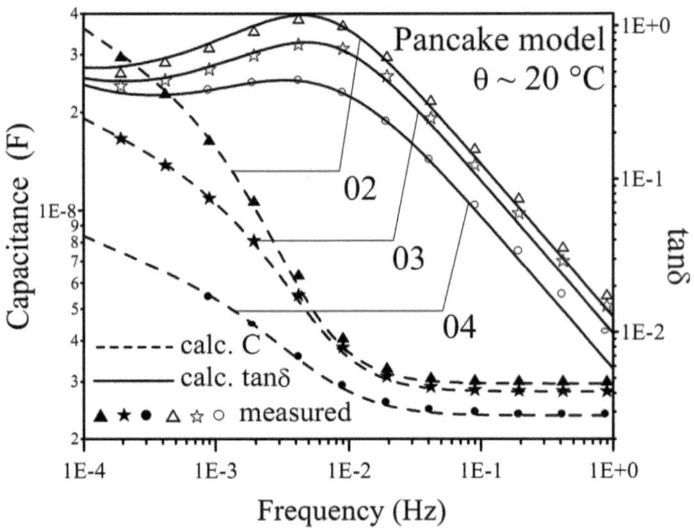

Fig. 9.11 Comparison between calculated and measured capacitance and dissipation factor tan δ

new oil of type "Nynas Nytro 10GBN" with a conductivity of about *2 pS/m (20 °C)*. As also the influence of the oil and its conductivity should have been investigated, it was decided to replace it by an oil of different origin with lower conductivity. The new type was "*Shell Diala D*" with $\sigma = 0.24$ pS/m (21 °C). For configuration "02" Fig. 9.12 shows now the measured relaxation currents before and after oil exchange.

Now, the initial amplitudes dropped by a factor of about 9 and the duration of the nearly exponential decay of the currents is heavily prolonged. This decrease of the initial current amplitude and the shift of the time constant of interfacial polarisation to long times is thus only due to the decrease of oil conductivity.

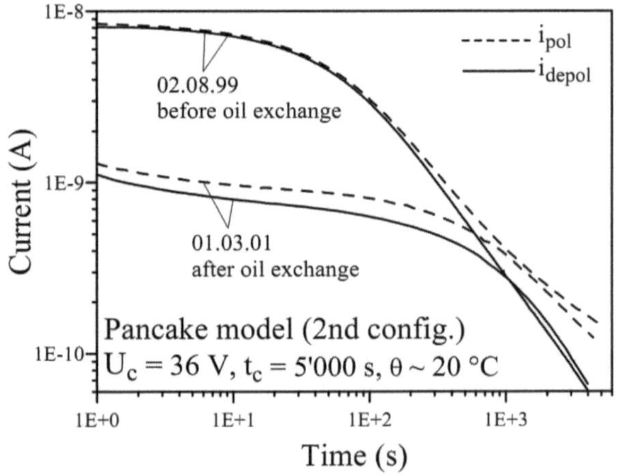

Fig. 9.12 Measured relaxation currents on configuration 02 before and after oil exchange

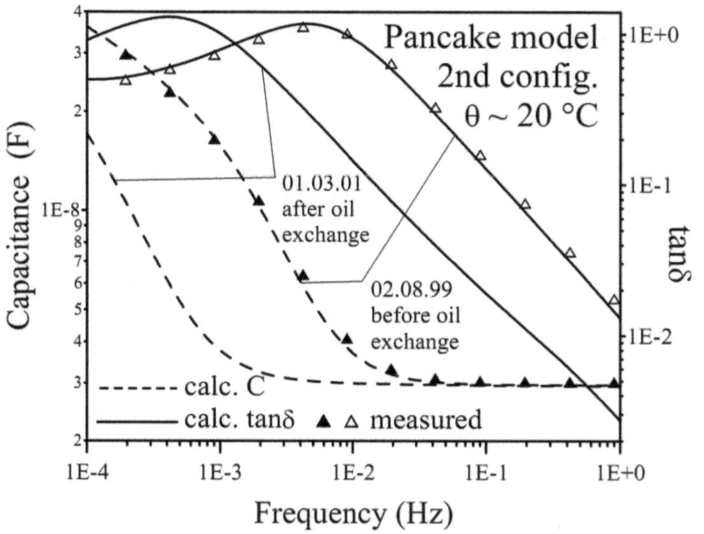

Fig. 9.13 Comparison of calculated capacitance and tan δ curves before and after oil change

The significant changes must also appear in the frequency domain. Figure 9.13 shows the calculated and partly measured dependencies of capacitance $C'(\omega)$ and dissipation factors $tan\ \delta(\omega)$.

The calculated values are determined from the relaxation currents as displayed in Fig. 9.12. In the low frequency domain, the increase of the capacitance takes now place for much lower frequencies for the low conductivity oil as the process of interfacial polarisation between oil and pressboard needs much more time. This increase in capacitance is due to the simple fact that the electric field in the oil gap will more or less vanish with time; therefore, for extremely low frequencies, the capacitance of the pressboard alone will appear.

Finally, Fig. 9.14 displays the two polarisation spectra as calculated by the *PDC* Analyser from the measured relaxation currents. For this calculation the charging voltage was set to 500 V permitting the direct comparison with measured values as performed by the *RVM* technique, marked by triangles and stars.

This figure shows again the significant influence of oil conductivity on the time position of the first maximum in the polarisation spectrum and the good agreement between calculated and measured values, although the RVM meter quantifies a somewhat different insulation system than the PDC Analyser.

These investigations confirm that the main (or first) maximum in the *"polarisation spectrum"* of the *RVM* technique for multi–layer arrangements is due to the interfacial polarisation and that its time position is dependent on the geometrical layout (thickness ratio of oil gap to pressboard barrier) and conductivity of oil. Therefore, the time position of the first maximum cannot be correlated with the moisture content of the pressboard material.

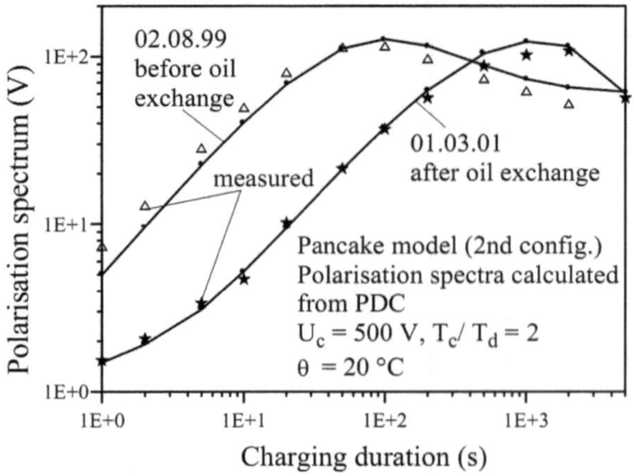

Fig. 9.14 Comparison of calculated and measured polarisation spectra of configuration 02 before and after oil change

The results as presented confirm already the possibility to distinguish between oil and impregnated pressboard as applied for insulation systems in power transformers. Much more quantitative data can be evaluated if the geometry of the insulation system including spacers is quantitatively taken into account. This can be done by a specific software for the PDC Analyser which is based on dielectric response functions $f(t)$ of otherwise investigated materials and its comparison with measured data. As an example, the actual oil conductivity of the test object can immediately be determined as the first part of the polarisation current is completely governed by this conductivity. Similar evaluations are possible if frequency domain measurements have been made.

So, principally new technologies for power transformer insulation state control have been developed. Examples of this technologies application with detailed explanations are shown below.

Measurements of resistance, the dielectric loss factor $tg\delta$ at power frequency, and calculation of the absorption coefficient R_{60}/R_{15} have been used for many years to assess the condition of paper-oil insulation. Capacitance between two windings has also been measured at 2 Hz and at 50 Hz, to derive C_2/C_{50} ratio that reveals the moisture content in transformer main insulation. Recently, usefulness of these parameters to transformer diagnostics was contested. New methods, based on dielectric polarization relaxation in a broad frequency range have been proposed. Moisture content cellulose and its thermal degradation can be assessed using these methods. In consequence, the remaining technical life of the transformer can be assessed. Polarization RVM and PDC records are taken in the time domain, whereas frequency characteristics of the insulation capacitance and dielectric loss factor $tg\delta$ are recorded in the range from 0.1 MHz to 100 Hz.

PDC method may be perceived as a modification of traditional insulation resistance measurement and calculation of $R_{60/15}$ quotient. A direct voltage is applied to the transformer insulation, and the resulting current charges capacitance of the examined insulation. This current is composed of polarization current i_{pol} that decays, and the constant current determined the insulation conductivity. Subsequently, the charged insulation is short–circuited and an opposite–polarity depolarization–current i_{dep} decays to zero. The rate of current decay depends on polarization relaxation of the paper–oil insulation, and contains several exponential components. Test scheme is shown on the Fig. 9.15.

The transformer oil conductivity determines the polarization current magnitude during first 100s. This conductivity depends on the water content in oil, its acid number, contamination and temperature. The polarization current intensity is proportional to oil conductivity that can be derived from the initial value of $\boldsymbol{i_{pol}}$. Water content in cellulose (barriers, spacers) can be derived from the $\boldsymbol{i_{pol}}$ and $\boldsymbol{i_{dep}}$ characteristic after 100s. ***Increased water content in cellulose accelerates depolarization and results in a faster decay of*** $\boldsymbol{i_{dep}}$, as well as in an increase of i_{pol} after 100s (Fig. 9.16).

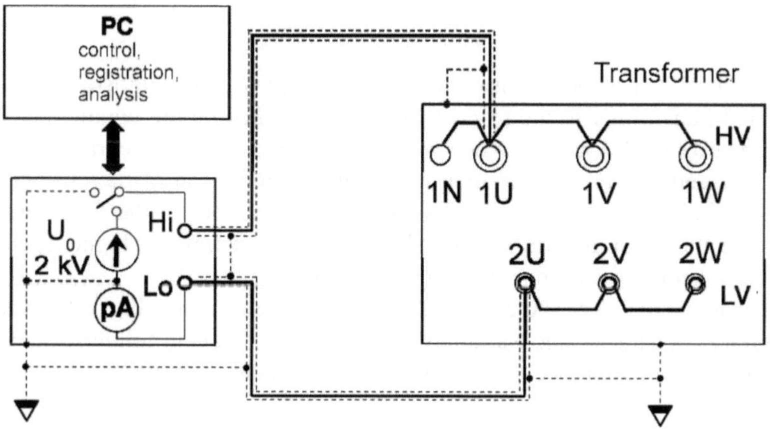

Fig. 9.15 Test scheme for PDC measurements on real transformer

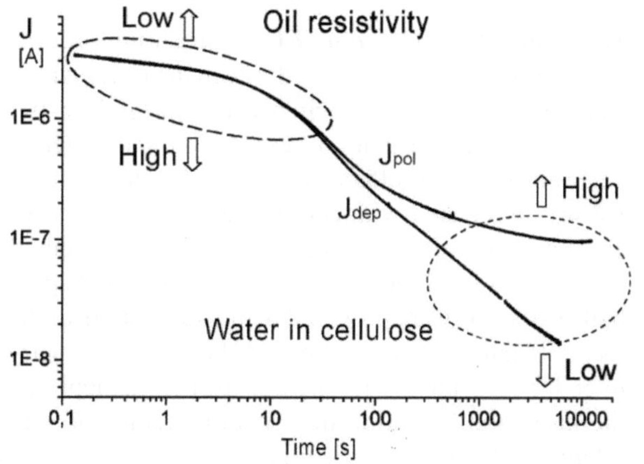

Fig. 9.16 Polarization and depolarization current plotted against time, with the water content in oil and in cellulose as parameter

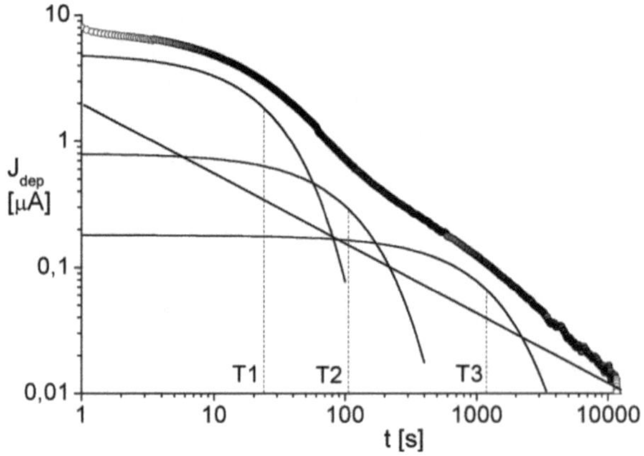

Fig. 9.17 Identification of the exponential components and their time constants

Specialized expert programs compare the i_{pol} and i_{dep} characteristics taken on an examined transformer to the respective curves obtained on pre-conditioned Transformerboard samples with a known water content and cellulose degradation. Transformer diagnostics in service often relays on a comparison of the characteristics taken on similar transformers, or at subsequent inspections of the same unit. Such procedure allows for identification of transformers that represent a higher operational risk, and scheduling their repairs or decommissioning.

An example of 270 MVA, 220/15 kV transformer is shown in Fig. 9.17, where the measured i_{dep} characteristic was analyzed and water content in cellulose determined by comparison to reference data from Transformer board samples.

9.4 Frequency Time Domain Spectroscopy

Frequency Time Domain (FDS) measuring procedure and the test circuit are the same as used for the time domain measurements, and shown in Fig. 9.15. An examined insulation capacitance *C* and $tg\delta$ low–frequency characteristics depend on temperature, insulation degradation and water content. This relation has been determined and an effect of increased water-content in cellulose and oil on *C* and $tg\delta$ was associated with certain frequency ranges (Fig. 9.18). Reference frequency characteristics of capacitance and $tg\delta$ have been determined in laboratory conditions for different insulation samples with controlled ageing and water content. They are built-in the software provided by manufacturers of the FDS measuring instrument to facilitate interpretation of recorded curves. Geometry of the examined winding has to be input together with the transformer nameplate data, and the program assesses the insulation condition in terms of ageing and water content. A database of the reference characteristics should preferably be developed for a given type and make of the paper–oil insulation, since the material characteristics do differ, and there are no universal reference curves.

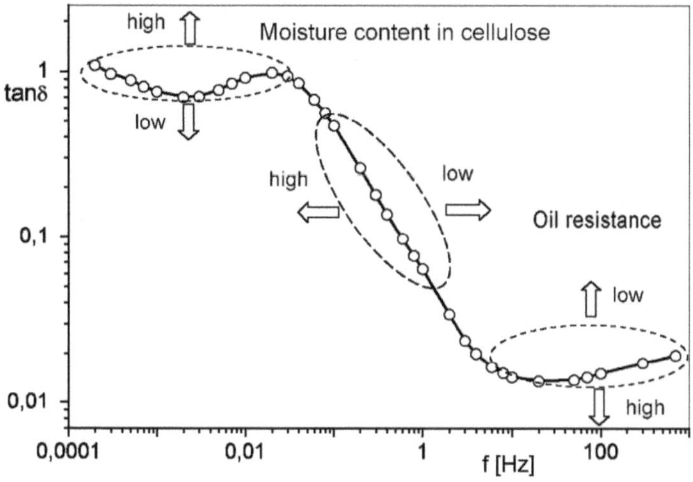

Fig. 9.18 Frequency characteristic of tgδ reveals water content and oil resistance in different frequency ranges

Another approach consists in using the Havriliiak–Negami equation and examining the relaxation characteristic of the examined insulation. Havriliiak–Negami Eq. (9.24) describes dielectric relaxation in polymers. This equation is used to analyse examined insulation.

$$\varepsilon(\omega) = \varepsilon' + j\varepsilon'' = \frac{\Delta\varepsilon}{(1 + (\omega\tau)^{\alpha})^{\beta}} + \varepsilon_{\infty} - j\left(\frac{\sigma}{\omega\varepsilon_0}\right)^N \tag{9.24}$$

where: $\Delta\varepsilon$—polarization, $\omega = 2\pi f$—pulsation, τ—relaxation time constant, σ—conductivity, α, β, N—coefficients, $\varepsilon(\omega)$, ε_{∞}, ε_o—complex, optical and static dielectric permittivity respectively. ***It is necessary to note that dielectric constant is used in engineering texts, but in reality ε is variable.***

Such analysis is presented on an example of 75 MVA, 110/10.5 kV transformer manufactured in 1988 by Siemens. Two relaxation time constants $\tau_1 = 12.1$ s and $\tau_2 = 294$ s were identified from *C* and *tgδ* characteristics (Figs. 9.19 and 9.20) taken on the main insulation at 20°C. The faster relaxation occurs at the oil–cellulose interface, and the slower one reflects properties of Transformerboard with a high water content.

Water content in cellulose can be assessed using nomograms used by the recovery voltage measurement (RVM) method, since it is based on the time constant of interfacial relaxation. The relaxation time–constant of solid insulating materials, such as Transformerboard, have to be obtained from the material samples examined in the laboratory.

Additional parameters derived from these characteristics are given in Table 9.1. Water content in the examined insulation was assessed at 3.2% from the time–constant of the interfacial relaxation, and at 3.3% from the relaxation time–constant

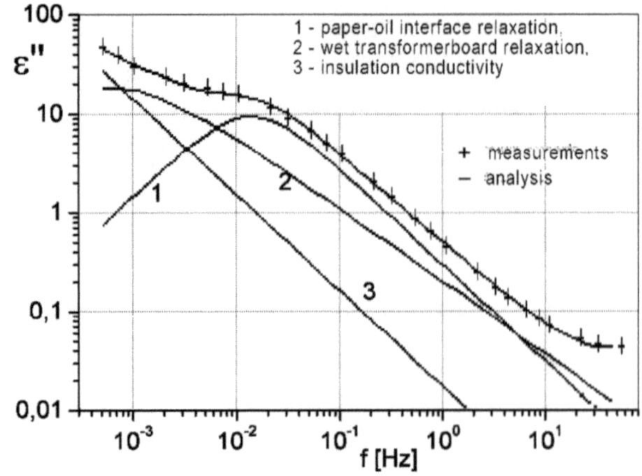

Fig. 9.19 Identification of time constants form tan δ frequency characteristics

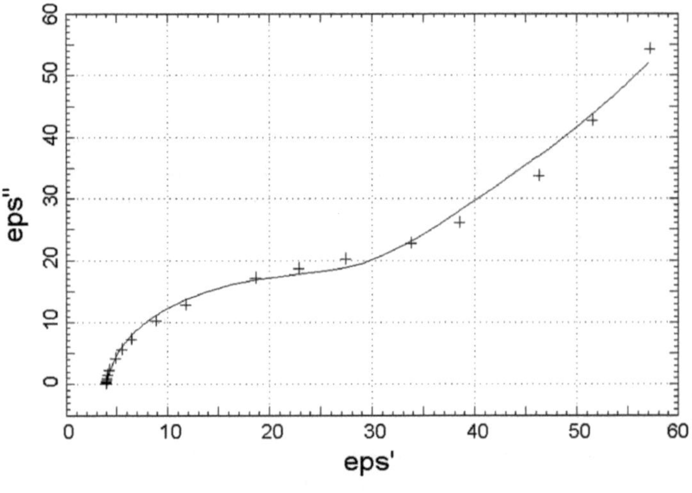

Fig. 9.20 Cole–Cole graph derived from measurements taken on 75 MVA, 110/10.5 kV transformer

measured on the Transformerboard samples, taking into account the temperature corrections.

An important advantage of the FSD methods lies in finding the insulation water–content even if the water–content in cellulose and in oil are not in equilibrium.

Table 9.1 Dielectric relaxation parameters from measurements taken on 5 MVA, 110/10.5 kV transformer

Parameter	$\sigma_0[s/cm]$	$\Delta\varepsilon_1$	$\tau_1[s]$	σ_1	$\Delta\varepsilon_2$	$\tau_2[s]$	σ_2
Value	8.4 e-15	20.9	12.1	1.00	61.7	294	0.76

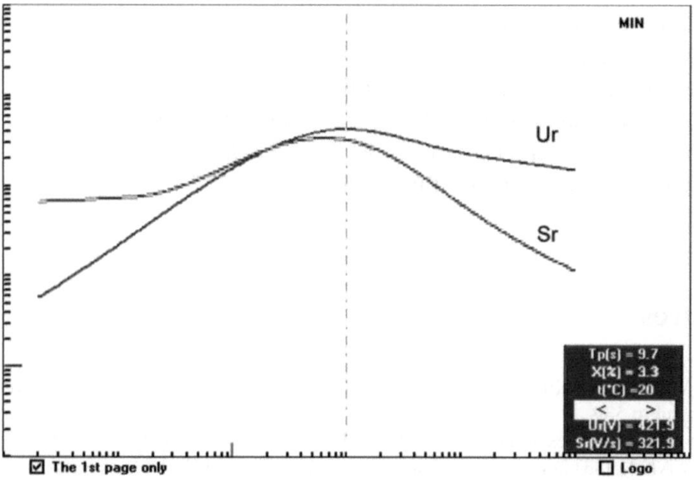

Fig. 9.21 Time-domain RVM characteristics measured on 75 MVA, 110/10.5 kV transformer insulation: Ur—the polarization recovery voltage, Sr—initial steepness of Ur characteristic

This method can separate the two polarization structures: at the paper-oil interface and in solid insulation, i.e. Transformerboard. It should be emphasized that the polarization behaviour of wet cellulose is relatively stable, but the properties of paper–oil interface depend on the water–content equilibrium and on products of oil decomposition.

The same transformer was also examined using *RVM method* and the main time-constant $T_p = 9.7\ s$ is comparable to $\tau_1 = 12.1\ s$ that corresponds to 3.3% water–content at 20 °C. In this case the moisture content in paper and oil are in equilibrium, as it is indicated by the well–defined, single peak of *Ur* and its slope *Sr* characteristics, as well as by the similar moisture-content obtained using *FDS method*. This situation is shown on Fig. 9.21. Apparently no oil–decomposition products are deposited on the Transformerboard surface.

So, new technologies of high voltage insulation diagnostic are very prospect. They will substitute all "typical" insulation diagnostic methods. RVM, PDS and FDS are based on dielectric spectroscopy phenomena. Basic fundamentals are detailed described in this paragraph.

One of the last tendencies in this direction is FDM using only.

Question and Tasks

1. Formulate the definition of dielectric spectroscopy.
2. What is dielectric response?
3. List main drawbacks of traditional insulation diagnostics technologies.
4. List main types of polarization in dielectrics.
5. Explain, what is the main principle of recovery voltage method?

6. Explain, what is the main principle of polarization and depolarization currents measurement method?
7. How is formed a depolarization current?
8. What is the main principle of frequency domain technology?
9. What is main difference between RVM and FDS?
10. Which way is most prospect in future?

References

1. Zaengl W, Kuffel E, Kuffel J (2000) High voltage engineering. Fundamentals—Butterworth—Heinemann, p 539
2. Electric Systems Issues at the turn of the 21st Century (2000) CIGRE ELECTRA, Special Issue
3. Jonscher AK (1983) Dielectric relaxation in solids—Chelsea Dielectrics Press
4. Zaengl WS (2001) Application of dielectric spectroscopy in time and frequency domain for high voltage power equipment (transformers, cables and etc.) Zaengl W.S. Application of dielectric spectroscopy in time and frequency domain for HV power equipment—Proceedings of 12th International Symposium on High Voltage Engineering (ISH 2001)—August 20–24,—Bangalore, India—Key Note Speech—Session 9—pp 76—85
5. Zaengl WS (2003) Application of dielectric spectroscopy in time and frequency domain for HV power equipment, part I—Theoretical Consideration—IEEE Electr Insul Mag 19(5): 5–19
6. CIGRE Working Group (2002) 15.01, Task Force "Dielectric response methods for diagnostics of power transformers"—Electra, pp 24–37

Chapter 10
Diagnostics of High Voltage Equipment

Abstract This chapter is devoted to the technology of diagnosing the main high-voltage equipment of power systems. As follows from the materials of the previous chapters, modern electric power industry uses high-voltage devices that differ in purpose, design, materials used, the spectrum and intensity of influencing factors, the degree of damage when they fail, etc. This requires a high level of qualification from a power engineer / electrician to select an adequate method and instruments for diagnostics, processing and interpretation of the results obtained. As a rule, the complexity of these tasks is proportional to the complexity of the apparatus itself. The authors considered it appropriate to arrange the materials of this chapter according to the principle "from more complex to less complex" as it was done in Chapter 2 (power transformers—high-voltage switching equipment—rotating machines, etc.).

10.1 Transformers and Other Oil-Filled Equipment

More than 38% of power transformer failures are due to defects in the insulation system. As follows from the materials of Chapter 6, a whole range of methods is used to diagnose the technical condition of high-voltage transformers and other oil-filled equipment, most of which are based on monitoring the dynamics of changes in the properties of transformer oil during operation or long-term storage.

Oil analysis allows not only to determine its quality indicators, which must comply with the requirements of regulatory and technical documentation, but also to reveal and identify most faults in high-voltage equipment, as follows from Table 10.1 on the example of power transformers.

The data given in Table 10.1 show that only by the concentration of various gases can one judge the type of incipient malfunction, as well as the severity of the damage. The oil contains about 70% of information about the condition of the equipment. On average, for oil-filled equipment, the frequency of inspection and testing of equipment is once every 2–4 years. Chemical methods for diagnosing the state of high-voltage equipment were considered in sufficient detail in Chapter 6. More information can be found in [1–3].

V. Y. Ushakov et al., *High-Voltage Equipment of Power Systems*, Power Systems,
https://doi.org/10.1007/978-3-031-38252-9_10

Table 10.1 Types of defects in relation to gas vapor concentrations

No.	The nature of the predicted defect	The ratio of the concentrations of characteristic gases			Technical specifications
		$\frac{C_2H_2}{C_2H_4}$	$\frac{CH_4}{H_2}$	$\frac{C_2H_4}{C_2H_6}$	
1	Fine	<0.1	0.1	≤1	Normal aging
2	PD with low energy density	<0.1	<0.1	≤1	Discharges in gas-filled cavities formed as a result of incomplete impregnation or humidity of the insulation
3	PD with high energy density		<0.1	≤1	Same as in point 2, leads to leaving a mark or breakdown of solid insulation
4	Low power discharges	<0.1	<0.1	1–3	Continuous sparking in oil between connectors of different potentials or floating potential. Breakdown of oil between solid materials
5	High power discharges		<0.1	≤3	Arc discharges; sparking; oil breakdown between windings or coils or between coils to ground
6	Low temperature thermal defect(<150 °C)	<0.1	<0.1	1–3	Overheating of an insulated conductor
7	Thermal defect in the low temperature range (150–300 °C)	<0.1	<0.1	<1	Local overheating of the core due to flux concentration. Increasing "hotspot" temperature
8	Thermal defect in the medium temperature range (300–700 °C)	<0.1	<0.1	1–3	The same as in paragraph 7, but with a further increase in the temperature of the "hot spot"
9	High temperature thermal defect (>700 °C)	<0.1		≥3	"Hot spot" in the core; overheating of copper due to eddy currents, poor contacts; circulating currents in the core or tank

Since power transformers are a rather complex and very important element of the power system, close attention is paid to the diagnostics of their condition. In this regard, most of the methods discussed in Chapter 5 are used to monitor the technical condition of power transformers.

10.2 Circuit Breakers and Other Switching Devices

High voltage testing requires "high voltage laboratory" capability with both power frequency "hi-pot" test capability, and also voltage surge ("lightning impulse" and "switching impulse") test capability. Testing is difficult, as the momentary test power requirements can exceed even that available on the transmission grid. Therefore, so-called "synthetic" techniques are used for this testing. Synthetic testing involves separate sources for the high current and high voltage, and only combining them during a very brief window of time. This greatly reduces the power requirements of simulating short circuit interruption.

The high current is typically supplied from dedicated "short-circuit generators," which are specialized machines in the 1000 to 3000 short-circuit MVA class. High voltage is supplied from a "high voltage synthetic circuit," which is a combination of capacitor banks, reactors, triggering, circuits, and computer controls. This synthetic circuit can produce a high voltage waveforms approximating real system recovery voltages.

In a real world, short-circuit interruption, a transient occurs following interruption which tries to reestablish the arc. This transient voltage must be synthesized by the test circuit and the required wave shapes differ widely between different breaker switching duties. A typical test involves initiating a short circuit with the high current circuit, and then at a "target" current zero simultaneously firing the high voltage synthetic circuit and switching out the high current circuit. Figures 10.1 and 10.2 are demonstrated common view and control panel of synthetic test facility.

The firing of the high voltage synthetic circuit provides correct short-circuit current *di*/*dt* at the target current zero, and also provides the correct recovery voltage

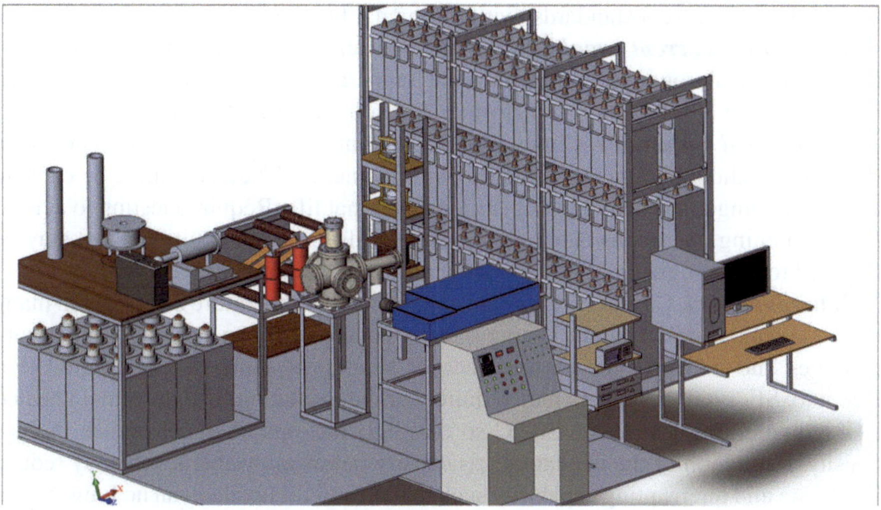

Fig. 10.1 Common view of synthetic test facility

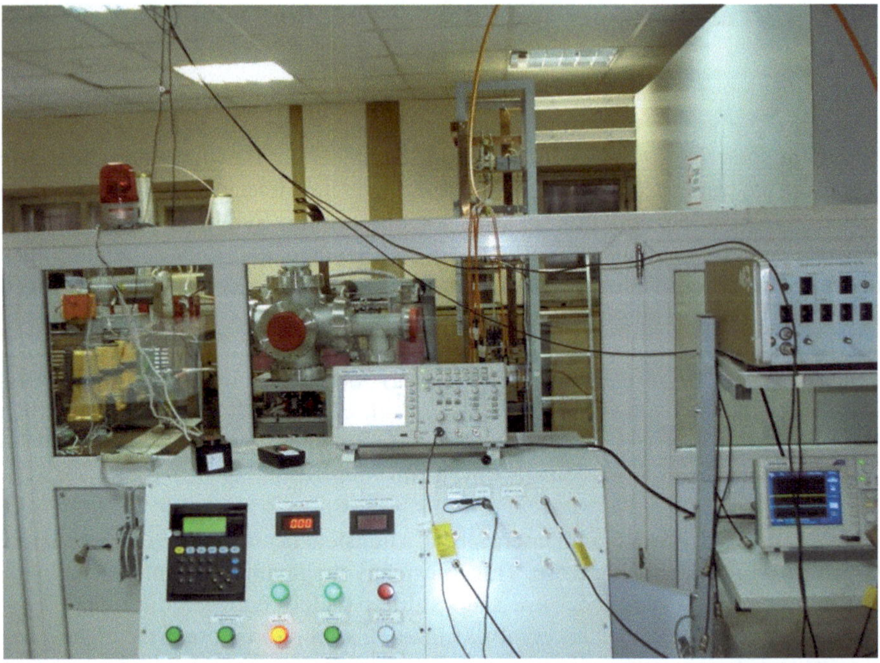

Fig. 10.2 Control panel of synthetic test facility

wave shape after successful interruption. Variations of the basic synthetic test circuit are used to simulate different switching and interrupting duties on the power system. Unit testing—testing only one interrupting break of a multi- break circuit breaker—is accepted by all current standards, with some qualification.

Continuous current capability. Circuit breakers must carry a rated continuous load current with- out exceeding allowable temperature limits. This is demonstrated by testing with a suitable test transformer and voltage/current regulator.

Mechanical endurance. Circuit breakers are intended to operate for many years in service without significant amounts of maintenance. They also must be capable of withstanding many switching operations over that life. Required testing to at least 2000 switching operations is required by standards, though circuit breakers may be required to meet higher values depending on the application and required ratings.

Environmental. Circuit breakers are applied in all temperature zones and in many severe seismic zones. Climatic testing is carried out to verify operation at temperature extremes and to verify the performance of heating systems, when required. For example, SF_6 circuit breakers may require that the insulating SF_6 gas remain heated to a temperature of at least –30 °C so that liquefactions does not reduce the gas density. Reduction of the insulating gas density below allowable levels may reduce insulating and interrupting capability such that the circuit breaker can no longer meet its rated performance.

Seismic withstand capability is demonstrated either by finite element analysis (FEA) or by shaker table testing, depending on the voltage class. Typically, testing is required at 245 kV and above, with FEA being acceptable below 245 kV. ***Production Tests***. Every circuit breaker is subjected to a series of routine production tests primarily intended to prove design conformance and quality. These tests typically include high voltage power frequency tests ("hi-pot"), mechanical operation and timing tests, fluid/gas leakage tests (when applicable), and control circuit operation verification.

Installation and field tests. These tests are carried out on-site according to specific users or man- ufacturer's instructions. Modern SF_6 breakers up to 245 kV are usually shipped fully assembled with a slight overpressure of SF_6, thus eliminating evacuation procedures on-site.

Service and Maintenance. With rising system voltages, currents, interrupting ratings, and the requirement for uninterrupted power supply, circuit breaker reliability becomes more and more important. Besides influencing factors of (1) design, (2) quality assurance, and (3) testing, which are mainly a responsibility of the circuit breaker manufacturer, maximum attention must be paid to the maintenance during service. Maintenance instructions for different makes and types of circuit breakers may differ considerably in details and volume, but all strive to obtain maximum breaker reliability despite longer maintenance intervals, smaller inventories of exchange parts, and shorter maintenance hours. Efforts are made to find the easiest way of handling service without influencing neighboring gear and consequently obtaining the lowest service costs. Utility maintenance shafts, standardizing groups, and circuit breaker developers have taken into account these requirements. The various steps from oil breaker to air-blast and finally the SF_6 and vacuum breakers indicate a considerable minimizing of maintenance combined with maximum reliability [4].

Variations of the basic synthetic test circuit are used to simulate different switching and interrupting duties on the power system. Unit testing—testing only one interrupting break of a multi-break circuit breaker—is accepted by all current standards, with some qualification.

10.3 High Voltage Turbogenerators and Other Rotating Machines

Once connected to a steam or gas turbine, each generator needs regular inspection to assess its technical conditions (TC). Such an inspection often entails the shutdown of an expensive unit, which can lead to undersupply of electricity to consumers and significant financial losses. In addition, the periodic removal of the rotor from the stator for technical inspection leads to a reduction in the life of the generator.

All this can be prevented by continuous monitoring of the condition of the turbogenerators. The technology belongs to the category of non-destructive and is based on a thorough theoretical study and practical basis. The essence of the technology is to analyze the PDs that occurs in the stator of the turbogenerator during operation. This technology allows specialists to perform the following actions:

- evaluate the generator TC without stopping the technological process;
- diagnose emerging problems and defects before they begin to affect the quality of electrical equipment;
- determine the terms of maintenance with high accuracy and include the equipment in the schedule of repair work.

Today, the PD monitoring system in high-voltage machines has a large number of devices (especially sensors).

10.4 Capacitors and Capacitor Banks

Diagnostics of high-voltage capacitors and capacitor banks includes the following procedures:

1. ***External Inspection (Visual Control)***

Cases and insulators of capacitors, the state of switching equipment of capacitor units and batteries are subjected to it. If a leak (drip or otherwise) of a liquid dielectric is detected, the capacitor is rejected regardless of the results of other tests.

2. ***Measurement of Insulation Resistance or Capacitor Bank***

The measurement is made between bushings closed between themselves and the grounded housing (case). Its value is not standardized. Measurement of insulation resistance is carried out with a megaohmmeter for a voltage of 2500 V.

3. ***Measuring the Resistance of the Discharge Resistor***

Its value should not exceed 100 MΩ.

4. ***Capacitance Measurement***

Capacitance is measured for each individual capacitor with its withdrawal from operation or under operating voltage (by measuring capacitive current or voltage distribution on series-connected capacitors). Capacitance measurement is mandatory after testing the capacitor with increased voltage.

5. ***Measurement of the Tangent of the Dielectric Loss Angle***

The measurement of tgδ is made on coupling capacitors, power take-off capacitors and voltage divider capacitors. The value of tg δ should not exceed 0.3% (at a temperature of 20 °C) at the first start-up and 0.8% in operation.

6. ***High Voltage Test***

The insulation against the case is tested with the capacitor terminals shorted. The magnitude and duration of voltage application are regulated by the factory instructions. The values of the test voltage of industrial frequency and rectified voltage are established by the relevant regulatory documents.

7. ***Capacitor Bank Test***

The test is carried out by turning on the battery three times at the rated voltage with monitoring of the current values by phases. The phase currents must not differ by more than 5%. Current control can be carried out using the instruments of the capacitor unit.

Tests and measurements of capacitors are carried out at an ambient temperature of at least +20 °C, in dry weather (or indoors). Humidity of the surrounding air is important when conducting high voltage tests, because condensate on capacitors and their insulators can lead to insulation breakdown and, accordingly, to equipment failure (both test and tested). Before carrying out high-voltage tests, capacitors should be wiped from dust, dirt and moisture. Atmospheric pressure does not have a special effect on the quality of the tests, but is recorded for entering data into the protocol.

Power engineers have a rich set of measuring instruments (bridges, multimeters, megaohmmeters) and devices for testing capacitors (test transformers, control devices, instrumentation and protection equipment).

10.5 High Voltage Bushings

High-voltage bushings are one of the elements of power devices, but due to their responsibility, design complexity and dimensions (see 2.5), they can be considered as one of the independent high-voltage devices that require monitoring and condition diagnostics.

To determine the state of high-voltage bushings, power engineers use complex control systems that allow them to monitor several key bushing insulation parameters. These include:

- Input capacity value;
- Presence of dependence of input parameters on temperature indicators;
- Dielectric loss tangent;
- Trends in parameter changes

Installation of systems for monitoring the state of bushings in itself is a rather difficult task. Nevertheless, it becomes much more complicated in the case of installation of control systems on equipment that has already been put into operation, when it is necessary to mount sensors from scratch, install diagnostic equipment, and lay various cable lines (signal, information, power) at the existing supply center.

These difficulties can be avoided by using the bushing status diagnostic system based on DB-2S wireless sensors. These devices are designed to control the conduction currents of the input insulation. They are easy to install on equipment (transformers, reactors, circuit breakers, etc.), whether it has recently arrived from a supplier or has been in operation for a long period of time.

DB-2S sensors are "smart" devices. All processes related to the fixation of signals, processing of the collected information and evaluation of the TC bushings are performed by the internal program of the sensor. At the output, a final expert opinion is formed on the real TC of the controlled high-voltage bushing.

According to experts, the presence of an intellectual component and ease of installation/dismantling of DB-2S wireless sensors make these measuring devices the best solution for organizing monitoring, operational diagnostics and evaluation of HV bushings of all types and operating voltage classes [5].

The design feature of the devices of this brand is completely wireless. The electronics built into the device is powered by the conduction current of the controlled input, and the output data about the TS of the controlled object is transmitted using wireless interfaces.

The main advantages of the DB-2S sensor include the following:

- Compact and protected version of the device. Thanks to these qualities, the device is quickly mounted on the input and switched on in the diagnostic mode for a short period of time when the transformer is taken out of service;
- No wires. The operability of the device is ensured by the conduction current of the bushing;
- The sensor controls two defective parameters—the current temperature of the inlet and the conduction current, which is used to calculate the value of the bushing capacitance during operation;
- Data on the actual state of the bushing is transmitted to the automated process control system using the LoRa ONE wireless interface with a coverage area of up to 1000 m;
- Encryption of the output data twice with two different keys, the so-called double encryption, ensures high information security of the monitoring system and the state of the bushings.

Experts consider insulation breakdown in one or more insulation gaps to be the most dangerous flaw in the insulation of the bushings. This can cause a short circuit between the conductive plates made of foil. The exclusion from the circuit of one series-connected capacitance of insulating gaps contributes to an increase in the total capacitance of the bushings.

The number of insulating gaps in the bushing insulation depends on its operating voltage: from 10 for bushings with an operating voltage of 110 kV and up to 50 for bushings with an operating voltage of 500 kV.

If the number of insulation gaps is large enough, closing the two plates will increase the current conductivity by 2–10% for bushings with an operating voltage of 500 and 110 kV, respectively. Therefore, when two insulating gaps are closed, the conduction current will double.

The second criterion for changing the bushing TC is a secondary parameter that shows the amount of voltage increase at the remaining insulating gaps in the event of a short circuit of the damaged one, since in this case the voltage is distributed between a smaller number of insulating gaps [6].

A feature of the operation of DB-2S measuring devices is that these sensors monitor only the value of the conduction current of the input and, based on this parameter, the input capacitance is determined. However, this does not take into account possible voltage changes on the windings of the transformer installation. The calculation scheme assumes that the voltage remains constant, corresponding to the nominal value.

However, in reality, this value is constantly changing. There may be several reasons:

- change in the load of consumers;
- making adjustments to the system configuration due to the connection / disconnection of lines, generators and individual consumers.

More complex systems for monitoring the state of bushings of high-voltage devices use voltages from measuring voltage transformers to take into account fluctuations in mains voltage in capacitance calculations. This technique improves the accuracy of expert opinions, but significantly complicates the monitoring system itself. For example, it is required to install additional devices that control the voltage of measuring transformers.

10.6 Gas-Insulated Switch Gears (GIS)

The **GIS-DM** (Gas Insulated System—Diagnostic Monitor) is designed for continuous monitoring of the TC of power equipment with SF_6 insulation—complete switchgears of various designs, circuit breakers, power and instrument transformers [7–9].

The assessment of the TC of electrical equipment is carried out using the analysis of the PD in isolation, which were recorded in the microwave frequency range. According to analysts, this PD registration method is the most sensitive and most effective for detecting defects in high-voltage insulation.

To register PD in SF_6 equipment, two types of measuring devices are used:

- *Advanced Computer System (ACS) high frequency transducer* Advanced Computer System. In fact, this is an electromagnetic antenna built into the equipment. As a rule, such sensors are installed on the inside of manhole covers. Typically, the sensor is mounted by specialists of the plant - the

manufacturer of SF_6 equipment at the stage of its manufacture. Installing ACS measuring devices on equipment that has already been put into operation is quite problematic;
- *active electromagnetic system(AES) brand PD sensor*. The measuring device is installed on the outer surface of the switchgear. Installation is carried out in the junction area of two tubular housings using an insulating gasket, which simultaneously acts as an internal supporting high-voltage insulator. The device registers electromagnetic pulses coming out from the inside of the machine through radio-transparent spacers separating individual switchgear housings. The AES brand sensor is easily installed on both new and existing power equipment.

Both types of sensors are connected to the measuring device of the GIS-DM monitoring system using coaxial cables. According to experts, a significant advantage of GIS-DM is the availability of software with a set of special functions, such as:

- carrying out additional digital processing of the recorded data for the most effective detuning from interference. To do this, various methods for estimating the digital parameters of pulses, amplitude comparison of pulses received from different measuring devices, as well as control of the time difference of their arrival are used. The functionality of the software product allows you to determine the location of the defect based on the difference in the time of arrival of pulses to each sensor;
- identification of the type (nature) of the problem, determination of the degree of its development. The automated expert diagnostic system PD-Expert copes well with this task. It uses an open data storage format. This format is available to various manufacturers of PDs measurement devices and software products. This must be kept in mind when choosing a diagnostic system, since it must be friendly in data storage formats. In addition, by comparing the information received from the sensors with the database of "defect samples" stored in the program memory, it is possible to determine the type of defect in the GIS insulation and assess the degree of its danger.
- implementation of the obtained diagnostic conclusions into the global monitoring and control systems, which allow to give an integral assessment of the technical equipment and evaluate the residual life of the electrical equipment complex. The role of the complex can be performed by a technological chain, which includes a switchgear control device, the entire supply center and even the path for supplying electrical energy to end-user facilities.

If in practice there is a need to create a monitoring system for large-scale, complex gas-insulated GIS, power engineers use several recording devices of the GIS-DM brand. However, it should be recognized that such a combination requires certain technical capabilities and specialized software solutions. The main requirements that ensure the integration of several devices into a single monitoring system include:

- synchronization of the processes of registration of PDs pulses coming from all sensors via an optical communication line. This allows you to more effectively tune out interference and determine the location of the defect in the insulation as accurately as possible;

- integration of all primary data collected by measuring instruments and transmitted over the communication line on one server or common computer of the control system. This technical solution provides for a comprehensive analysis of all PD pulses registered in a complete switchgear;
- realization of all diagnostic functions is possible because the devices are synchronized over an optical line with an accuracy of up to a nanosecond or up to tens of nanoseconds (in the case of using signals from GPS/GLONASS satellite navigation systems).

According to experts, the most important part of any monitoring system for high-voltage equipment is the expert system. It accomplishes several important tasks, including:

- determination of the current technical condition;
- detection of signs of a faulty state of electrical equipment;
- planning the timing and scope of future repairs;
- Evaluation of the effectiveness of the completed repair work.

There are two methods by which expert diagnostic systems for power equipment, including high-voltage equipment, are created:

1. *A specialized system* designed for a certain type of equipment. It is created on the basis of the knowledge of a practical expert or a group of such experts.
2. *A universal self-adjusting system* formed on the principle of neural networks or created on the basis of artificial intelligence. It is based on a universal software product. Practical experts are involved in the work within their competence only at the stage of adaptation or self-learning of the expert system.

The neural network is aimed at finding and diagnosing defects that have already been fixed earlier. Its functionality cannot detect and register defects that were not previously diagnosed, since it does not take into account the peculiarities of physical processes in controlled devices.

The fixed expert system is based on the most complete understanding by an experienced consultant (or a group of consultants) of the processes occurring in the equipment. The system also allows you to take into account the occurrence of repetitive and modified defects. In addition, it has great potential in terms of modernizing diagnostic algorithms.

As noted earlier, even a well-established system of planned preventive work today shows its imperfection. It does not fully meet the requirements of practice, it is characterized by the cost of such an approach and therefore needs to be transformed and adapted to modern trends to increase the percentage of power electrical equipment with an exhausted standard operating life. The best way out of this situation is the introduction of diagnostic and monitoring systems. The functionality of such systems allows real-time assessment of the current technical condition of machines, using non-destructive diagnostic methods to identify destructive processes in the early stages of their development and determine the remaining resource.

Question and Tasks

1. What defects can be found in oil-filled equipment based on the analysis of changes in the content and properties of transformer oil?
2. Resistance to what types of influences is assessed when diagnosing high-voltage circuit breakers?
3. List methods for diagnosing high-voltage circuit breakers (with brief explanations).
4. List the tests performed by manufacturers of rotating machines before delivery to their consumers (customers).
5. What is the monitoring of the technical condition of turbogenerators based on and what opportunities does it provide for specialists?
6. Name and briefly describe the 6 main methods for testing capacitors.
7. What are the key parameters for monitoring and diagnosing bushings?
8. Explain, why damage of one insulated layer of bushing leads to soon breakdown?
9. Name the principles of operation of sensors used in monitoring the technical condition of gas-insulated switchgear.
10. What is the purpose of the expert system in monitoring the technical condition of the switchgear?

References

1. Ushakov VYa, Mytnikov AV, Lavrinovich VA, Lavrinovich AV (2022) Transformer condition control—advanced and traditional technologies, Springer Verlag. p 150
2. Lyutikova MN, Korobeinikov SM, Busaev VV (2020) Determination of water content in liquid insulation of high-voltage equipment by different methods. Power Technol Eng 53(5):623–630
3. Lyutikova MN (2020) An improved electrochemical method for moisture determination in Mineral Oil. IEEE Trans Dielectr Electr Insul 27(6):2172–2178
4. Fink DG, Beaty HW (1987) Standard handbook for electrical engineers 16th Edition, McGraw-hill, p 2144
5. Warne DF (2005) Newnes electrical power engineers's handbook. Second Edition, Linarce House, Burlington, Amsterdam, p 477
6. Grigsby LL (2001) "Frontmatter" The electric power engineering handbook/ Ed. L.L. Grigsby Boca Raton: CRC PressLLL
7. Ignatov S Diagnostics and monitoring of high-voltage equipment. Modern technical and software solutions—https://marketelectro.ru/node/diagnostika-i-monitoring-vysokovoltnogo-oborudovaniya (in Russian)
8. Kurtenkov GE (1999) Fundamentals of designing insulation of high-voltage electrical equipment—Tomsk: NTL Publishing House, p 276 (In Russian)
9. Webster J (2003) Electrical measurement signal processing and displays. Publ. CRC, p 768

Chapter 11
Diagnostics of High-Voltage Cable Lines

Abstract This chapter describes various types of cable testing, such as insulation resistance test, rectified current surge test, power frequency surge test. The definition of the active resistance of conductors and the electrical performance of the conductors is given. Diagnostics and monitoring of power cable lines are described in detail. All stages of the creation and operation of high-voltage cable lines (CL)—cable production, route selection for laying, laying, diagnostics, repair and replacement are quite labour-consuming and expensive. The reasons for this should be mentioned as: (a) high quality requirements for electrical insulation, sealing and protection against mechanical damage, (b) large extent, (c) deep in the ground, (d) significant probability of mechanical damage in case of uncoordinated extraneous earthworks within the track. All of the above applies to the cable lines of marine laying. (Only their mechanical damage is of a different nature—mainly trawling).

11.1 Acceptance Testing of Power Cables

Existing cable can be tested and replaced if problems are found. One way to do this is to apply a DC voltage withstand test (approximately 3 times nominal RMS (Root Mean Squared) voltage). Since cables will either pass or not pass this test, information about the state of cable deterioration cannot be determined. Another popular method for cable testing is to inject a small signal into one end and check for reflections that will occur at partial discharge points. Other methods are measuring the power factor over a range of frequencies (dielectric spectroscopy), analyzing physical insulation samples in a lab for polymeric breakdown (degree of polymerization), and using cable indentors to test the hardness of the insulation [1].

In accordance with regulatory requirements (In Russia, this is the "Electrical Installation Rules (EIR)", the scope of acceptance tests of CLs includes the following work.

1. Checking the integrity and phasing of the cable cores.
2. Measurement of insulation resistance.
3. Test with increased voltage of the rectified current.
4. Power frequency high voltage test.

V. Y. Ushakov et al., *High-Voltage Equipment of Power Systems*, Power Systems,
https://doi.org/10.1007/978-3-031-38252-9_11

5. Determination of the active resistance of the conductors.
6. Determination of the electrical working capacity of the conductors.
7. Measurement of current distribution on single-core cables.
8. Checking protection against stray currents.
9. Test for the presence of undissolved air (impregnation test).
10. Testing of feeding units and automatic heating of end sleeves.
11. Control of the state of the anti-corrosion coating.
12. Checking the characteristics of the oil.
13. Measurement of ground resistance.

CL lines with voltage up to 1 kV are tested according to paragraphs 1, 2, 7, 13, above 1 kV and up to 35 kV—according to paragraphs 1–3, 6, 7, 11, 13, and with a voltage of 110 kV and above—to the fullest extent provided by this document.

11.1.1 Checking the Integrity and Phasing of the Cable Cores

Before the cable is put into operation, it is phasing. (The phases of the cable are ensured with the phases of the connected section of the electrical installation). The check is made by dialing using handsets or a megaohmmeter. Based on the check, the coloring of the cores is carried out in accordance with the coloring adopted at this installation.

After preliminary dialing, before the cable line is put into operation, it is phasing under voltage. To do this, a working voltage is applied from one end of the cable, and from the other end, the correspondence of the phases is checked by measuring the voltages between the same and different phases. Phasing is performed by voltmeters (in networks up to 1 kV) or voltmeters with voltage transformers, as well as using voltage indicators (in networks with voltages above 1 kV). The procedure for phasing in CLs of different voltages is approximately the same.

11.1.2 Measurement of Insulation Resistance

The operation is performed with a megohmmeter for a voltage of 2.5 kV. For power cables up to 1 kV, the insulation resistance must be at least 0.5 MΩ. For power cables above 1 kV, the insulation resistance is not standardized, but should be of the order of a dozen MΩ and higher. The measurement should be made before and after testing the cable with increased voltage. The method of measuring resistance and the instruments used in this case correspond to tests of insulation of electrical equipment with increased voltage [2–4].

Before starting the measurement of insulation resistance of CL, the following procedures must be performed:

1. Make sure that there is no voltage on the cable.
2. Ground the circuit under test while connecting the instrument.

After the end of the measurement, before disconnecting the ends from the device, it is necessary to remove the accumulated charge by applying grounding.

The cable must be discharged using a special discharge rod, first through the limiting resistance, and then short-circuited. Short cable runs up to 100 m can be discharged without limiting resistance.

When measuring the insulation resistance of long cable lines, it must be taken into account that they have a significant capacitance, so the megohmmeter readings should be read only after the cable is charged.

It is strictly forbidden to measure the insulation resistance on the CL if it, at least in a small area, passes near another live line.

11.1.3 Rectified Current High Voltage Test

Power cables with voltages above 1 kV are tested with increased rectified current voltage. The magnitude of the test voltages and the duration of application of the rated test voltage depend on the type of insulation and the level of the operating (rated) voltage. For cables with paper-oil and polymer insulation, the duration is 10 min, for cables with rubber insulation—5 min. The ratio of the test voltage to the nominal voltage decreases from 6 to 2 with an increase in the nominal voltage from 2 to 220 kV.

The procedure for testing with increased voltage of rectified current, as well as the installations and equipment for testing, are similar to those used in testing the insulation of high-voltage apparatus. The voltage shall rise smoothly to the test value and be kept constant throughout the entire test period. The rise of the test voltage for CL with voltage up to 10 kV is carried out within 1 min, and for CL 20–35 kV—at a rate of not more than 0.5 kV/s.

The measurement of cable leakage currents of 3–10 kV during tests with increased rectified voltage is carried out using microammeters connected either on the high voltage side of the test facility, or in the zero of the test transformer. When using the latter measurement scheme, readings may be distorted due to parasitic leakage currents. At the same time, their state is assessed not only by the absolute value of the leakage current, but also by taking into account the nature of the change in the leakage current over time, the asymmetry of the leakage currents in phases, the nature of the conservation and decline of the charge, etc. CL can be put into operation if the leakage currents have a stable value, but do not exceed 300 μA for lines with a rated voltage of up to 10 kV. For short cable lines (up to 100 m long) without couplings, the allowable leakage currents should not exceed 2–3 μA per 1 kV of test voltage. The asymmetry of the leakage currents in phases should not exceed 8–10, provided that the absolute values of the currents do not exceed the allowable ones.

Table 11.1 Power frequency test voltage values

Operating voltage, kV	Test voltage, kV	Test voltage in relation to the ground, kV	Test duration, min
110	220	130	5
220	500	288	5

With the normal quality and condition of the insulation of the power cable, the leakage current decreases depending on the duration of the application of the test voltage, and the more, the higher the quality of the insulation. For a power cable with defective insulation, the leakage current increases with time. With a noticeable increase in leakage current when testing a power cable, the duration of the test increases to 10–20 min. With a further increase in leakage, if it is not caused by termination defects, the test should be continued until the breakdown of the cable insulation [5–8].

During the test, the voltage from the rectified installation is applied to one of the cores of the cable under test. The remaining cores of the cable under test, as well as all cores of other parallel cables of this connection, must be securely connected to each other and grounded. For three-core cables, the insulation of each core is tested against the sheath and other earthed cores. For single-phase cables and cables with separated conductors, the insulation of the conductor relative to the metal sheath is tested. The cable is considered to have passed the test if no breakdown occurred, there were no sliding discharges and leakage current surges or its increase after reaching a steady value. After each test of the CL circuit, it must be discharged.

11.1.4 Power Frequency Overvoltage Test

Power frequency overvoltage test allowed to produce for CL 110–220 kV instead of testing with increased voltage of the rectified current. The values of the test voltage of industrial frequency are given in Table. 11.1.

The technique, installations and equipment for testing with increased voltage of industrial frequency are similar to those used in testing the insulation of high-voltage apparatus.

11.1.5 Determination of Active Resistance of Cable Cores

The active resistance of the cores is determined only for cable lines with a voltage of 35 kV and above. The active resistance of the cores of the cable line to direct current, reduced to 1 mm^2 of cross section, 1 m of length and a temperature of +20°

Table 11.2 Active resistance of cable cores to direct current at a temperature of +20 °C

Section, mm	Resistance, Ohm/km	Section, mm	Resistance, Ohm/km
16	1.15/1.95	95	0.194/0.33
25	0.74/1.26	120	0.153/0.26
35	0.52/0.88	150	0.122/0.207
50	0.37/0.63	185	0.099/0.168
70	0.26/0.44	240	0.077/0.131

Note the numerator indicates the values for the copper core, and the denominator for the aluminum core [2]

C, should not exceed 0.0179 Ω for a copper conductor and not more than 0.0294 Ω for an aluminum conductor.

The values of the active resistance of the cores of cables to direct current are presented in Table. 11.2.

11.1.6 Determination of the Electrical Performance of Cores

The procedure is performed for lines of 35 kV and above. The measured capacitance, reduced to specific values, should not differ from the factory test results by more than 5%. The capacitance of cable lines is measured by the ammeter-voltmeter method or by the bridge scheme. The first of them allows you to determine the capacitances with values $C \geq 0.1$ μF with high accuracy, which corresponds to the parameters of the cables. The scheme of measurements for this method is shown in Fig. 11.1.

Based on the results of measuring voltage and current, the capacitance, μF, is calculated by the formula:

$$C = \frac{I}{2 \times \pi \times f \times U} \times 10^{-6} \tag{11.1}$$

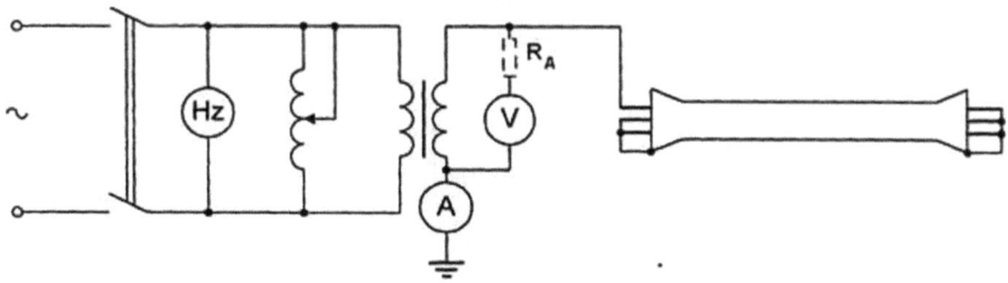

Fig. 11.1 Scheme for measuring the capacitance of cables using the ammeter-voltmeter method

where: I—capacitive current, A; U—voltage on the cable, V; f—voltage frequency in the network, Hz.

According to the measurement data, the specific capacitance C_{SC} of the cable is determined, μF/km, as:

$$C_{SC} = \frac{C}{L} \tag{11.2}$$

In the case when special equipment and instruments are required for measurement by the ampervoltmetric method, it is desirable to use the bridge method. Measurements are made according to the inverted scheme (the order of measurement must comply with the instructions). When choosing an AC bridge for measuring the specific linear capacitances of a cable of this voltage class, of course, it is necessary to take into account the limits of capacitance measurement by this bridge (For example, the linear capacitance of a cable of 35 kV and above is tenths of μF/km).

The results of measuring the uneven distribution of current in single-core cables should not exceed 10%. Measurements are made with portable instruments or clamp meters.

11.2 Diagnostics of Power Cables

Cables are characterized by defects caused by damage to the protective sheath due to mechanical stress or corrosion. In this case, the insulation is moistened, which leads to premature loss of electrical strength and cable failure. In cables impregnated with oil-paper insulation, branched (dendritic-like) discharges with deposition of "X-wax" can develop from the core to the sheath. Under working tensions, they develop very slowly—sometimes over many years. Their occurrence and development are associated with the appearance of gas voids due to the alternation of heating and cooling cycles—"cable breathing".

Defects in the form of cracks, cavities in the casting mass (mastic, epoxy compound) may occur in cable boxes. Terminations (cable outlets) can be contaminated by natural and industrial entrainment.

The main method for diagnosing the condition of cables is their test with increased rectified voltage.

For cables for a voltage level.

- of 3–10 kV, the test voltage (U_{test}) is (5–6) U_n,
- for a voltage of 10–35 kV U_{test} is (4–5) U_n,
- at U_n more than 35 kV U_{test} is approximately 2 U_n.

The use of rectified voltage for diagnosing the condition of cables is due to the following reasons.

1. Cable insulation has a higher dielectric strength with rectified voltage than with alternating voltage. Exposure to an alternating test voltage can lead to partial

damage to the insulation (drying or even charring) due to partial discharge and dielectric losses.
2. The rectified test voltage can be significantly higher than the AC voltage without damaging healthy insulation during the test. Testing with increased rectified voltage allows you to better detect defects, because. in this case, the undamaged insulation layer (with high resistance) bears most of the voltage, which facilitates the breakdown of the damaged layer and the detection of a defect. (As you know, the alternating voltage in multilayer insulation is distributed inversely proportional to the dielectric constant of the material).
3. The power of the installation when tested with rectified voltage is much less than when tested with alternating voltage.

Currently, in order to reduce the complexity and duration of testing, cable testing under load is used. The admissibility of such tests is because most CLs operate in systems with an isolated neutral, and the rectified voltage is applied to the neutral; otherwise, the neutral must be disconnected from earth. At the same time U_{test} is the sum of the constant and variable components with the amplitude U_{ph}. The presence of a variable component can cause ionization processes, so the amplitude U_{test} in such tests, they are reduced by 20–30%. During these tests, only phase insulation is controlled, i.e. isolation from ground. After the defect is detected and a breakdown is carried out, "burning" is carried out at the defect site using a powerful rectifier to create a continuous conductive bridge to quickly find the fault location with a resistance of units and fractions of an ohm.

There are various methods for locating cable faults, but the most common are:

1. Pulse sounding
2. Induction method
3. Acoustic method.

The first method is based on measuring the time interval between the moment of sending a probing pulse to the CL and the moment of arrival of the signal reflected from the fault. The accuracy of determining the location of signal damage is a few meters, the distance of the damage site is approximately 10 km. (See Chapter 8 for details).

The second method is based on the fact that the electromagnetic radiation generated at the defect site when a high-frequency current is passed through the cable is received as noise in the headphones of a telephone equipped with a special receiving device. When bypassing the route, the operator can determine with great accuracy the location of cable damage by the maximum noise in the headphones.

The third method is to listen to sound vibrations above the cable fault, which are created by a spark discharge from a pulse generator, perceived as a powerful blow. Sound can be heard on the surface of the earth with the help of a simple device—a stethoscope or more complex devices. (The second and third methods are discussed in detail in paragraph 6.5). Apparatus and devices for testing, burning and determining the location of damage in power cables are mounted in mobile testing laboratories or used separately, taking into account the nature of the detected defects in the cables, the length of the latter, the organization of maintenance, repair and other factors [8–12].

11.3 Monitoring Cable Lines

On the example of one of the modern multi-channel systems for continuous monitoring of cable insulation testing system (Cables Diagnostics Monitor—CDM), one can evaluate the capabilities of such systems that allow solving the following tasks:

- Carry out continuous monitoring of the state of insulation of a CL, identify destructive processes, determine the type of defect and establish the degree of its development, as well as the level of danger during further operation.
- Automatically localize places with detected defects. This function is implemented in real time, under operating voltage. The instrument's internal memory is sufficient to store data for an extended period. All collected information and diagnostic conclusions are promptly transferred to an automated process control system of a higher level.
- Minimize the total cost of the diagnostic system, which is calculated for one controlled cable;
- Maximize the efficiency of picking up high-frequency noise pickups on the controlled equipment;
- On the basis of the method of registration and analysis of PD, control the state of the CL not at once, but constantly, in all modes of operation;
- Analyze the collected information, promptly develop diagnostic solutions and make recommendations regarding the further work of the CL.

Such systems allow you to monitor the state of a few to dozens of cable lines with an operating voltage of tens of kilovolts [12–14].

Monitoring of controlled cable lines is carried out online using built-in algorithms. In the case of fixing the achievement of the specified limit values, data on the critical state of the cable are displayed on the front panel of the protective cabinet. Operational personnel are informed about this by lighting up light-emitting diodes. Based on the results of the study, the system automatically generates protocols for the status of all controlled CLs. To control the CL, it is necessary to perform the following procedures.

- Preliminarily install PD recorders (primary sensors) on all cable lines on the earthing leads of the terminations (for example, RFCT-7 series). In fact, these are measuring current transformers that operate effectively in the high-frequency range. Connect the primary sensors using coaxial cables of the same length to the CDM.
- Create conditions for the system to work in automatic mode, for which a controlled cable assembly scheme is generated on the computer, which is loaded into the CDM device before starting work. A circuit configuration is created using a special software product—a constructor, which allows you to draw cable diagrams and indicate the location of the sensors. In the process of creating a circuit, detailed technical information is entered that is necessary to identify a damaged cable, assess its current technical condition, determine the type of insulation damage and the degree of danger during further operation.

One of the most difficult tasks in the registration of PD in high-voltage lines, experts call detuning from external interference, under the influence of which the sensitivity of monitoring systems is greatly reduced. To detect defects at early stages, the device provides a number of detuning tools, which contains a set of the following algorithms:

- Joint analysis of the time of occurrence of impulses and the phase of the supply voltage;
- Time of arrival. The algorithm is a detuning taking into account the time of arrival of pulses from different CLs with a resolution of a few nanoseconds;
- Amplitude disassembly of PD pulses, which allows to remove from the analysis signals received from neighboring CLs;
- Analysis of the frequency properties of each impulse. This algorithm makes it possible to separate the interference pulses from the repetitive PD pulses;
- Measurement of air temperature and humidity in a controlled complete switchgear, which provides more informative diagnostic conclusions of the device.

Another no less important task of the built-in algorithms is to determine the type of damage to the cable insulation and to diagnose the degree of danger of this defect. This task can be solved using the PD-Expert system integrated into the device. Its functionality allows you to automatically evaluate the level of PD and determine the type of a particular defect in the CL insulation. In addition, the system allows you to separate several faults of the same type located in different places of one cable. The CDM system provides a function to detect the location of a defect in the line. The role of the probing is performed by an impulse from the PD, which occurs at the site of insulation damage. In the absence of pulses of sufficient amplitude, pulses from induced external interference can be used as probing pulses. The diagnostic capabilities of the program can be significantly expanded with an online reflectometer, which is usually included in the device package [14–16].

To control the technical condition of CLs with an operating voltage of 110 kV and higher, more complex information monitoring systems are used.

Since, during the operation of cable lines, mainly concentrated defects occur in their insulation, they are eliminated by cutting a piece of cable of a certain length and replacing it with a new one.

Question and Tasks

1. Name the main procedures provided for by the program of acceptance tests of power cables.
2. How are high-voltage cable tests performed? What are the advantages of rectified current test voltage?
3. Name the main methods for finding the place of cable damage.
4. What is the procedure for measuring cable insulation resistance?
5. What opportunities does a multi-channel system for monitoring the technical condition of the insulation of cable lines provide?

References

1. Tobias N (2015) Testing and diagnostics on medium voltage underground cables based on VLF, p 152
2. Kurtenkov GE (1999) Fundamentals of insulation design of high-voltage electrical equipment: textbook.—Tomsk: NTL Publishing House, p 276. (in Russian)
3. Ignatov S Diagnostics and monitoring of high-voltage equipment. Modern technical and software solutions https://marketelectro.ru/node/diagnostika-i-monitoring-vysokovoltnogo-oborudovaniya. (in Russian)
4. IEC60502 (2014) International standard, power cables with extruded insulation and their accessories for rated voltages from 1kV up to 30kV, IEC, Geneva, Switzerland
5. Kalkner RP (2003) PD-Testing of service aged joints in XLPE-insulated medium voltage cables at test voltages with variable shape and frequency, International Symposium of High Voltage Engineering, Netherlands
6. Fletcher CL, Hampton N, Hernandez JC, Hesse J, Pearman MG, Perker J, Wall CT, Zenger W (2011) First practical utility implementations of monitored withstand diagnostics in the USA, 8th International Conference on Insulated Power Cables, JiCable 2011, France
7. Oyegoke B, Hyvönen P, Aro M (2001) Dielectric response measurement as diagnostic tool for power cable systems, Espoo, Finland
8. Wadhwa CL (2007) High voltage engineering, Second edition, New Age International Publishers, p 312
9. Zaengl W, Kuffel E, Kuffel J (2000) High voltage engineering. Fundamentals.—Butterworth—Heinemann, p 539
10. Fink DG, Beaty HW (2006) Standard handbook for electrical engineers, 15th edn. McGraw-Hill
11. Fraden J (2004) Handbook of mjdern sensors: physics, design, and applications. Springer, 3rd edn. p 590
12. Moore G (1998) Electric cables handbook, 3rd edn. BICC Cables, Blackwell Science (ISBN 0632040750)
13. Faremo H et al (2006) Improved productivity for HV cable manufacture. CIGRE. Paris Paper B-109
14. Campbell A (2003) Undergrounding HV and EHV-Window of opportunity opens. ICF Congress, pp 169–193
15. Henningsen C, Muller K, Polster K, Schroth R (1998) New 400 кВ XLPE long distance cable systems, their first application for the power supply of Berlin. CIGRE, 21–109
16. Brincourt T, Regaudie V (1999) Evaluation of different diagnostic methods for the French Underground MV Network. Jicable, pp 451–456

Chapter 12
Diagnostics of Insulating Structures of Overhead Power Lines and Outdoor Substations

Abstract This chapter is devoted to the insulation structures of overhead power lines and outdoor substations. Various designs of insulators for overhead power lines are considered. Analyzed outdoor substations and features of their working conditions, in particular, lightning surges, other natural adverse factors (rain, fog, snow, dust storms, ice, etc.).

12.1 Brief Information About Insulators for Overhead Transmission Lines and Outdoor Substations and the Features of their Working Conditions

For many decades, the main (and practically the only) insulating structure for high-voltage transmission lines was porcelain or glass disc insulators (Fig. 12.1), from which strings were assembled for hanging wires, and for outdoor substations—rod support and bushing insulators [1–3].

By design and purpose, the following types of insulators for transmission lines and outdoor substations are distinguished:

- Suspended (Figs. 12.1, 12.2, 12.4);
- Support;
- Bushings (Fig. 12.3);
- Rod;
- Pin.

Over the years, the level of operating voltage increased, and the requirements for the reliability and safety of operation of overhead transmission lines (OTL) and substations became more stringent and, as a response, the range of dielectric materials expanded (in addition to porcelain and glass, polymer and composite materials were widely used, Fig. 12.2), the design of suspension and layout (Fig. 12.3) and support insulators was modified, the requirements for testing and diagnosing their condition increased [3].

V. Y. Ushakov et al., *High-Voltage Equipment of Power Systems*, Power Systems,
https://doi.org/10.1007/978-3-031-38252-9_12

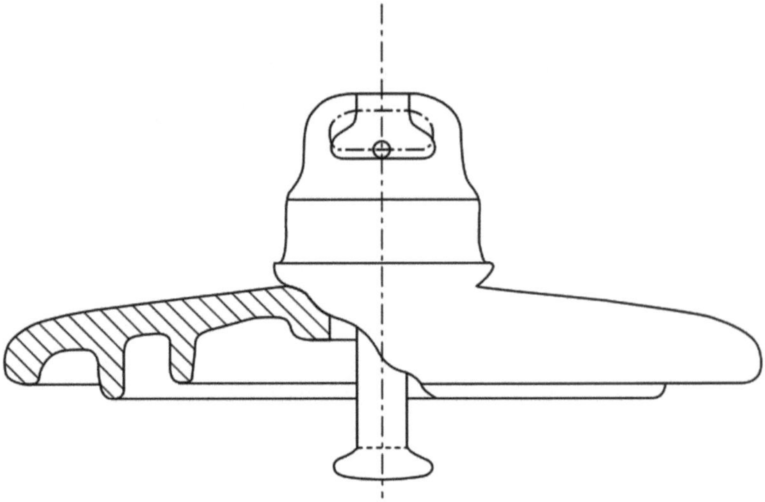

Fig. 12.1 Typical constructions of ceramic type standard suspension insulator—single element string of insulators

The insulation of OTLs and outdoor substations operates under the influence of a number of specific unfavorable factors such as:

- lightning and switching surges,
- operating voltage surges,
- air pollution and surface wetting due to precipitation,
- temperature fluctuations (daily and seasonal),
- exposure to solar radiation.

All this must be taken into account when choosing methods and instruments of testing, diagnostics and monitoring.

The value of the discharge voltage when exposed to a pulsed voltage depends on the shape of the voltage acting on the insulation. In order to be able to compare the impulse responses obtained in different laboratories with each other, the waveform for impulse tests is normalized and is in accordance with the waveform of atmospheric surges acting on the insulation as a result of lightning discharges [2].

The standard wave is characterized by a rapid rise in voltage to a maximum (wave front) and a slower decay (wave tail). Since the wave front smoothly turns into a recession, it is difficult to determine the beginning of the wave, the duration of the front and the duration of the wave from oscillograms without their special processing. The procedure for determining the parameters of a standard wave is regulated and is as follows.

The overvoltages that have their origin in atmospheric discharges in which case they are called *external* or *lightning overvoltages* or they are generated internally by connecting or disconnecting the system, or due to the system fault initiation or extinction. The latter types are called *internal overvoltages.*

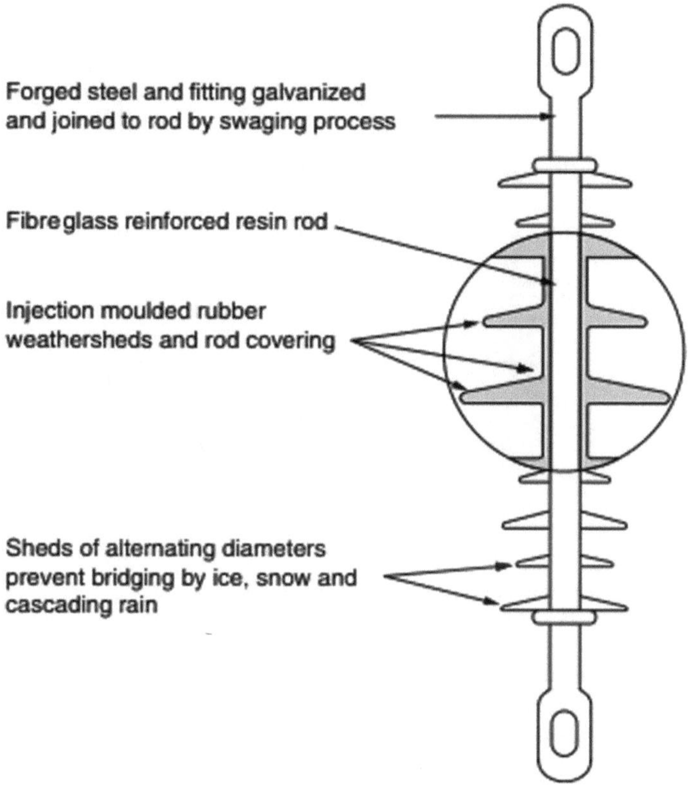

Fig. 12.2 Example of non-ceramic insulator

Fig. 12.3 Photo of lay-out

This class my be futher subdivided into *temporary overvoltages*, if they are oscillatory of power frequency or harmonic, and *switching overvoltages*, if they are heavily damped and short duration. Temporary overvoltages occur almost without exception under no load or very light load conditions. Because of their common origin the temporary and switching overvoltages occur together and their combined effect has to be taken into account in the design of h.v. systems insulation. "Basic Insulation Levels (also commonly known as "Impulse Withstand Level" defined as the impulse

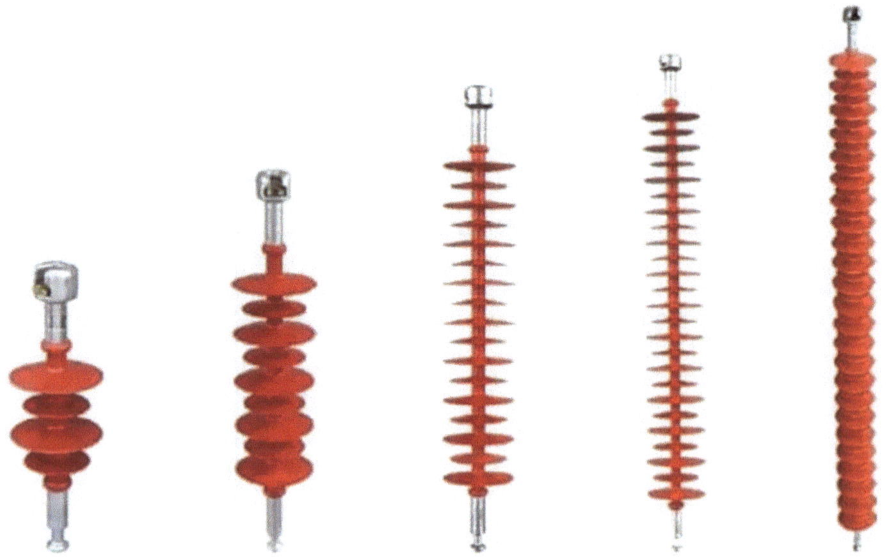

Fig. 12.4 Polymer suspension insulators of various designs

voltage which the insulation of any electrical equipment for a given rated voltage must be able to withstand.

The magnitude of the lightning overvoltages remains essentially independent of the system's design, whereas that of switching overvoltages increases with increasing the operating voltage of the system. Hence, with increasing the system's operating voltage a point is reached when the switching overvoltages become the dominant factor in designing the system's insulation. Up to approximately 220–300 kV, the system's insulation has to be designed to withstand primarily lightning surges. Above that voltage, both lightning and switching surges have to be considered. For u.h.v. systems, 750–765 kV and above switching overvoltages in combination with insulator contamination become the predominating factor in the insulating design. For the study of the overvoltages occurring in power systems, a thorough knowledge of surge propagation laws is needed which can be found in a number of textbooks and will not be discussed here.

12.2 Simulating Lightning and Switching Surges for Testing

The data on lightning strokes and voltages has formed the basis for establishing the standard impulse or lightning surge for testing equipment in laboratories. Exhaustive measurement of lightning currents and voltages and long experience have formed the basis for establishing and accepting what is known as the standard surge or "impulse" voltage to simulate external or lightning overvoltages.

In power transmission systems with voltages of 220–300 kV and above, the electrical strength of the insulation to switching overvoltages becomes important for insulation design. A variety of overvoltages surge wave shapes and the correspondingly large range of flashover values make it difficult to choose a standard shape of switching impulses. Many tests have shown that the flashover voltage for various geometrical arrangements under unidirectional switching surge voltages decreases with increasing the front of duration of the surge, reaching the lowest value somewhere in the range between 100 and 500 ms. The time to half-value has less effect upon the breakdown strength because flashover usually takes place before or at the crest of the wave.

The value of the discharge voltage when exposed to a pulsed voltage depends on the shape of the voltage acting on the insulation. In order to be able to compare the impulse responses obtained in different laboratories with each other, the waveform for impulse tests is normalized and is in accordance with the waveform of atmospheric surges acting on the insulation as a result of lightning discharges [2–5].

The standard wave is characterized by a rapid rise in voltage to a maximum (wave front) and a slower decay (wave tail). Since the wave front smoothly turns into a recession, it is difficult to determine the beginning of the wave, the duration of the front and the duration of the wave from waveforms without their special processing. The procedure for determining the parameters of a standard wave is regulated and is as follows.

To determine the wave front, it is necessary to draw a straight line on the oscillogram, parallel to the time axis and passing through the wave maximum (U_{max} in Fig. 12.5 point 1). Then draw two straight parallel U_{max} at the level of 0.3 U_{max} and 0.9 U_{max}. Through the points of intersection of the lines 0.3 U_{max} and 0.9 U_{max} with the wave front (points 2 and 3 in Fig. 12.6), draw line A until it intersects with the time axis (point t_0 in Fig. 12.6), which is called the conditional beginning of the wave) and the straight line U_{max} (point t_F in Fig. 12.5). The distance from t_0 to t_F expressed on a time scale is called the wave front (see Fig. 12.5).

As a standard wave, a wave with parameters is taken:

- wave front duration (t_F)—1.2 ± 0.4 μs
- wave duration (t_W)—50 ± 10 μs.

During the operation of protective equipment against atmospheric surges, as well as during the breakdown of insulating gaps near the busbars of the substation, the so-called shear waves may appear on the insulation of electrical equipment, which pose a danger to the operation of the insulation. This danger is associated with a high rate of voltage change, leading to high gradient overvoltages in the windings of transformers, reactors and electrical machines.

In this regard, such equipment is tested with a standard shear wave having a duration of 2.0 ± 0.5 μs and representing the initial stage of a full standard wave 12.6. The duration of the cut wave is defined as the time interval from the conditional

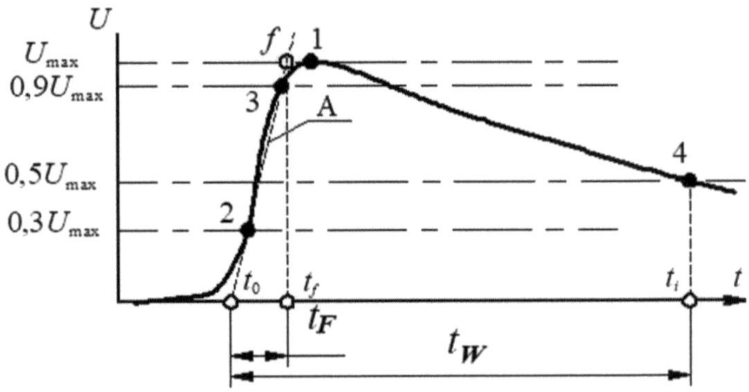

Fig. 12.5 The shape of the full test wave

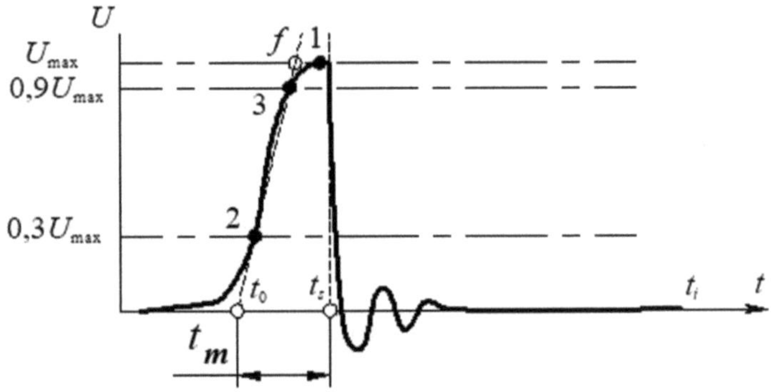

Fig. 12.6 The shape of the cut test wave

beginning of the wave (point t_0 in Fig. 12.6) to the projection onto the time axis of the point of the beginning of a sharp decline (cutoff) of the wave (point t_s in Fig. 12.6).

The switching surge values of crest between 100 and 500 ms falling well below the corresponding power frequency flashover values. There is a rapid fall in switching surge breakdown strength with increasing the gap length. This effect leads to costly design clearances, especially in the ultra h.v. regions. Nearly all gap configurations are of practical interest, positive switching impulses result in lower flashover voltage than negative ones. The flashover behavior of external insulations with different configurations under positive switching impulse stress is therefore most important. The switching surge voltage breakdown increases with increasing the air humidity.

The general designation for a standard switching impulse is given by IEEE as 250/2500 microseconds.

12.3 Voltage–Time Characteristic

The dependence of the maximum discharge voltage on the duration of the pulse is called the voltage–time characteristic (VTC) of the insulation. Since the onset and rate of development of ionization processes depend on the voltage value, the VTC depend on the pulse shape. In order to unify the tests and the possibility of comparing insulating structures, a standard lightning impulse with a front duration (voltage increase) and an impulse duration is set and is designated 1.2/50 μs (Fig. 12.7).

To experimentally determine the VTC, standard-shaped pulses are applied to the gap under study. At each value of the maximum pulse voltage, a series of experiments is performed. Due to the statistical spread of the discharge time, the VTC is obtained in the form of a region of points (Fig. 12.8), for which the average curve and the boundaries of the discharge time spread are indicated.

The form of the VTC depends on the degree of inhomogeneity of the electric field in the gap. For gaps with a uniform or slightly inhomogeneous field, the VTC weakly depends on (Fig. 12.9, curve 1), and only at discharge times of the order of 1 μs or less does the discharge voltage increase. This is due to the fact that the discharge in such intervals is formed in a very short time at a voltage equal to the initial value and

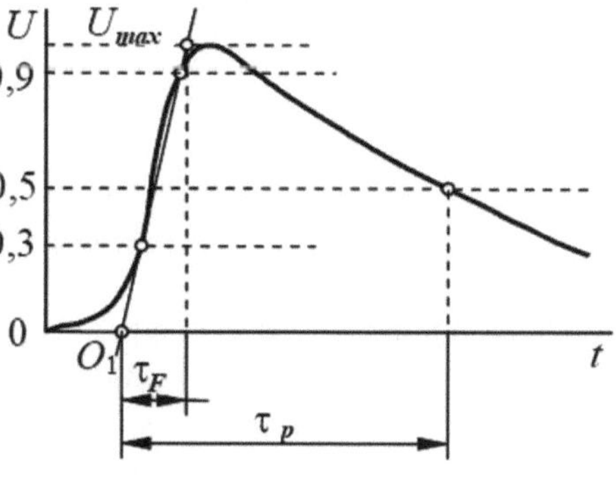

Fig. 12.7 Determining voltage pulse parameters

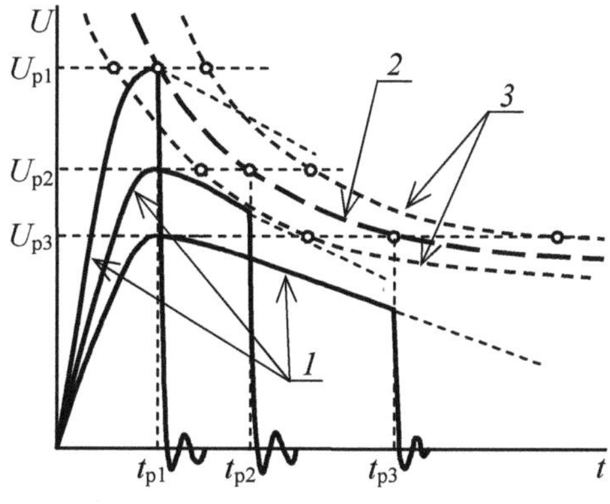

Fig. 12.8 Construction of the voltage–time characteristic of insulation according to experimental data (lightning impulses): 1—voltage impulses; 2—the curve of the average values of the breakdown voltage; 3—breakdown voltage spread boundaries

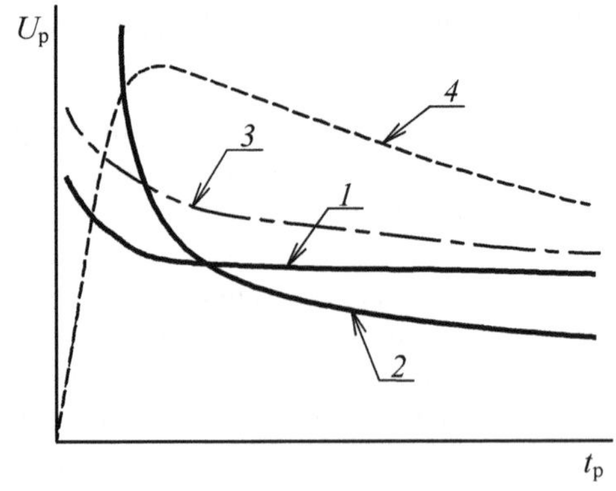

Fig. 12.9 VTC of protective arresters and insulation: 1—VTC of valve arrester (homogeneous field); 2—VTC of a tubular spark gap (sharply inhomogeneous field); 3—VTC of the protected object; 4—voltage pulse

there is no corona. The noted properties of the VTC make it possible to use the gap between spherical electrodes, which create an almost uniform field, if the distance between the electrodes is less than their radius, as a universal device for measuring the maximum voltage values.

The VTC of gaps with a sharply inhomogeneous field (Fig. 12.9, curve 2) have a rather large steepness, since in such gaps the discharge formation time strongly depends on the value of the applied voltage. For such gaps with lightning impulses, higher discharge voltages ($U_{discharge}$) are characteristic than with an alternating voltage of industrial frequency of 50 Hz (U_{AC}). Attitude is called the momentum factor:

$$K_{pulse} = \frac{U_{disch\,\mathrm{arg}\,e}}{U_{AC}} \tag{12.1}$$

Gaps with uniform and weakly inhomogeneous fields have a momentum coefficient in almost the entire range of discharge times.

VTCs are widely used to coordinate the insulation of high-voltage equipment, i.e. for protection against lightning and switching surges. For this purpose, an air arrester (for example, a valve arrester) with a flat VTC is switched on in parallel with the protected object. Reliable protection will be ensured if the VTC of the arrester (Fig. 12.9, curve 1) lies below the VTC of the protected equipment (curve 3) over the entire range of times of the applied voltage [6].

12.4 Insulation Coordination

12.4.1 Statistical Approach to Insulation Coordination

Insulation coordination is the correct economically justified choice of insulation levels and levels of test voltages of individual high-voltage objects and the coordination of these levels with the characteristics of protective arresters (devices). Large insulation margins are not economically viable. Insufficient insulation stocks are unacceptable due to loss of supply. The strategy here is to select the insulation levels of the elements according to the conditions of reliable long-term operation in normal operating mode (according to allowable operating phase voltage—U_{aop} and select the levels of protective devices taking into account possible overvoltages (their magnitude, multiplicity, duration, probability of occurrence), that is, limit them to normally permissible or, as they say, "bring isolation back to normal." Such a strategy gives optimal results from an economic point of view.

To coordinate the insulation, it is necessary to know the possible multiplicity of switching and lightning overvoltages, as well as the probability of their occurrence in the system [7].

12.4.2 Correlation Between Insulation and Protection Levels

The main types of insulation in the electrical system:

- air surrounding the external elements of power transmission lines and substations;
- solid insulation of power lines,
- isolation of high-voltage equipment of substations (transformers, cirquit breakers and other switching devices, rotative machines, bus supports, etc.

Each of them has its own isolation level.

The insulation level is the ratio of the 50% discharge voltage ($U_{d50\%}$) of one insulating element (air gap, strings, support insulator, internal insulation of a transformer, rotating machine) to the amplitude of the highest continuously allowable operating phase voltage (U_{aop}):

$$K_{il} = \frac{U_{d50\%}}{U_{aop}} \tag{12.2}$$

where

- k_{il}—is the isolation level;
- $U_{d50\%}$—50% discharge voltage of the insulating element;
- U_{aop}—the highest long-term permissible operating phase voltage of the network.

Insulation levels decrease from air to transformer and machine for economic reasons, since it is much cheaper to make a higher discharge voltage for an air gap than for transformer or machine insulation. With an increase in the rated voltage, for economic reasons, the insulation levels also decrease [8].

Thus, there are conflicting requirements. On the one hand, insulation levels decrease from air insulation of OTLs and substations to insulation of machines. On the other hand, the importance (role) of insulation for the apparatus and the inadmissibility of shutting down the system increases from insulation of machines to air insulation of OTLs and substations, i.e., the reliability of the system should increase with a decrease in the level of insulation. A similar situation is observed when considering insulation of various voltage classes. The level of insulation decreases from high voltage (HV) classes to ultra-high and ultra-high voltage (EHV and UHV), and the reliability of systems must increase from HV to EHV and from EHV to UHV. This is the essence of coordinating isolation levels and isolation protection levels.

Insulation protection level:

$$K_{pd} = \frac{U_{pd}}{U_{aop}} \tag{12.3}$$

where U_{pd} is the voltage of operation of the protective device. With proper insulation coordination $k_{pd} < k_i$.

Insulation coordination is the coordination of the characteristics of electrical equipment with those of a protective device in such a way that the insulation of the equipment is protected from excessive overvoltages. In a substation, for example, the insulation of transformers, circuit breakers, bus supports, etc., should have insulation strength in excess of voltage provided by protective device.

Electric systems insulation designers have two options available to them: choose insulation levels for components that would withstand high voltages, work out protective devises that could be installed at the sensitive points in the system that would limit overvoltages there. The first alternative is unacceptable spatially for e.h.v. and u.h.v. operating levels because of the excessive insulation required. Hence, there has been great incentive to develop and use protective devices. The actual relationship between the insulation levels and protective levels is a question of economics. Conventional methods of insulation coordination provide a margin of protection between electrical stress and electrical strength based on predicated maximum overvoltage and minimum strength, the maximum strength being allowed by the protective devices. ***Insulation level*** is defined by the values of test voltages, which the insulation of equipment under test must be able to withstand.

In today's systems for voltages up to 220–245 kV the tests are limited to lightning impulses (1.2/50 microseconds) and 1-min power frequency tests. Above 300–330 kV tests include an addition of switching impulse voltage (250/2500 ms).

12.5 Control of Thermal and Mechanical Resistance

As noted above, the suspension, support and bushing insulators of OTLs and outdoor substations of very high voltage classes are made mainly of porcelain and tempered glass, for medium and low voltage classes—of polymers (mainly silicone rubber).

For *porcelain insulators*, characteristic damages are through and surface cracks from mechanical loads arising from the difference in the values of the temperature coefficient of expansion of porcelain and reinforcement metal, as well as damage to the glaze from sunlight or from surface discharges [3, 9].

For *glass insulators*, in case of mechanical damage or electrical breakdown, the glass part crumbles into pieces-cells. However, the rest of the insulator in the cap retains sufficient mechanical strength, and the string does not disengage.

For OTLs and outdoor substations with porcelain and glass insulators operating in areas with large daily and seasonal temperature fluctuations, serious problems are caused by significant differences in the temperature coefficients of volume and linear expansion of solid insulation and metal fittings.

To check for heat resistance, the insulator is immersed in water with a temperature of 80–90° C for a time $T = 15 + 0.7\ G$ (G is the mass of the insulator, kg), but not more than 1 h, then—in water with a temperature of 20–30° C. This test is repeated twice.

To identify defective insulators at the factory and (or) immediately before installation, their resistance is measured with a megaohmmeter and subjected to high voltage tests.

Question and Tasks

1. Explain the procedure for testing string insulator.
2. Describe the arrangement of insulators for performing various tests.
3. List out various tests to be carried out on insulator and give a brief account of each test.
4. Write a short note on the cable sample preparation before it is subjected to various tests.
5. List out various tests to be carried out on a cable and give a brief account of each test.
6. Explain briefly various tests to be carried out on a bushing.
7. Explain the function of discharge device used in a power capacitor and explain the test for efficiency of this device.
8. Describe surge waveforms for overvoltage tests.
9. Explain briefly impulse testing of power transformer.
10. Describe various mechanical and thermal tests of string insulator.
11. What does it mean insulation coordination?
12. How do you understand the term “bring isolation to normal”?

References

1. Ushakov VYa (2001) High-voltage engineering. Textbook. Tomsk, TPU Press, p 253
2. Bazelyan EM, Raizer YuP (1997) Spark Discharge, CRC Press, No 4, Boca, Ration, p 294
3. Kuffel E, Zaengel WS (1984) High-Voltage Engineering (Fundamentals), Pergamon Press, p 488
4. Diagnostics of electrical equipment of power stations and substations: textbook / A. I. Khalyasmaa et al. (2015) Ekaterinburg: publishing house of the Ural Federal University, p 64 (in Russian)
5. Kurtenkov GE (1999) Fundamentals of insulation design of high-voltage electrical equipment: textbook.—Tomsk: NTL Publishing House, p 276 (in Russian).
6. Warne DF (2005) Newnes electrical power engineer's handbook, Second edition, Linarce House, Burlington, p 477
7. Wadhwa CL (2007) High voltage engineering, Second Edition, New Age International Publishers, p 312
8. Fink DG, Beaty HW (2008) Standard handbook for electrical engineers, 15th edn. McGraw-Hill, p 2000
9. van Der Cluis L (2001) Transients in power systems. Willeyns Ltd., Publ, p 222

Conclusion

A big variety of possible defects developing in operating high voltage equipment requires the fast identification of dangerous fails during diagnostics. Currently, there is a common tendency of shift from the strategy of periodically conducted preventive work to preventive system, which only depends on the current condition state of the transformer. An effective preventive system ensures an increase in the reliability of equipment operation by reducing the number and duration of emergency and repair failures, timely withdrawal for repair and thereby extending the service life of the high voltage equipment. Prevention of emergency failure of high voltage equipment allows obtaining such a significant economic effect that it becomes advisable to use even very complex means of measuring, processing data and issuing control results.

The organization of the optimal operation of transformers with a long service life requires the implementation of programs for a mass survey of the high voltage equipment fleet. Many countries have adopted multistep survey programs containing the minimum required set of methods for assessing the condition of high voltage equipment.

The effectiveness of preventive maintenance according to the state is confirmed by the practice of mass surveys carried out recently. A full range of measures to assess the state of the transformer and determine its operability includes continuous and periodic monitoring of parameters in operation and detailed examinations in the disconnected state.

It is especially difficult to decide when it is necessary to leave the high voltage equipment in operation, despite the presence of a defect. Also, high qualification of technician staff is necessary to make a decision on where to control the high voltage equipment—on working destination or at the high voltage equipment. The materials presented in this textbook will help to solve the abovementioned problems and receive appropriate qualification in high voltage equipment diagnostics.

V. Y. Ushakov et al., *High-Voltage Equipment of Power Systems*, Power Systems,
https://doi.org/10.1007/978-3-031-38252-9

Appendix A

ABCB	Air Blast Circuit Breaker
AC	Alternative Current
ACB	Air Circuit Breaker
ACM	Acoustic Control Methods
ACP	Amount of Caustic Potassium
ADC	Analog-to-Digital Converters
ASTM	American Society for Testing of Materials
ATC	Assessing the Technical Condition
BOCB	Bulk Oil Circuit Breaker
BSEC	Black Sea Economic Community
CADG	Chromatographic Analysis of Dissolved Gases
CB	Circuit breaker
CBM	Condition Based Maintenance
CDM	Cables Diagnostics Monitor
CIGRE	International Council on Large Electric Systems
CLPE	Cross-Linked Polyethylene
CO	Controlled Object
CMEA	Council for Mutual Economic Assistance
CT	Current transformer
DB	Database
DBPC	Ditertiary Butyl Paracresol
DC	Direct Current
DCB	Disconnecting Circuit Breaker
DETC	De-Energized Tap Changer
DGC	Distribution Grid Companies
EHV	Extra—high Voltage
EMF	Electromotive Force
EPG	Electric Power Grid
ENTSO-E	European Network of Transmission System Operators for Electricity
EPR	Ethylene-Propylene Rubber

V. Y. Ushakov et al., *High-Voltage Equipment of Power Systems*, Power Systems,
https://doi.org/10.1007/978-3-031-38252-9

ERCOT	Electric Reliability Council of Texas
FEA	Finite Element Analysis
FDS	Frequency Domain Spectroscopy
FFTA	Fast Fourier Transform Algorithm
FOCS	Fiber Optic Current Sensor
FRA	Frequency Response Analysis
FTDS	Frequency Time Domain Spectroscopy
GIS	Gas Insulated Switchgear
GOST	Russian Abbreviation for name of the document: state standard
HV	High-Voltage
HPP	Hydroelectric Power Plant
HTS	High-Temperature Superconductivity
HVDC	High-Voltage Direct Current
HPP	Hydroelectric Power Plant
IEC	International Electrotechnical Commission
IEEE	Institute of Electrical and Electronics Engineers
IPS	Interconnected Power System
IRD	Infrared Diagnostics
IT	Instrument Transformer
KB	Knowledge Base
LTC	Load Tap Changer
LVPD	Low-Voltage Pulsed Diagnostics
MOCB	Minimum Oil Circuit Breaker
NCIT	Non Conventional Instrument Transformers
NEMA	National Electrical Manufacturers Association
NERC	North American Electric Reliability Council
NDT	Non-Destructive Testing
NPP	Nuclear Power Plant
OBB	Oil Barrier Bushing
OCB	Oil Circuit Breaker
OLVR	On-load Voltage Regulators
OTL	Overhead Transmission Line
PC	Poly Carbonate
PCBs	Polychlorinated Biphenyls
PD	Partial Discharge
PDC	Polarization and Depolarization Current
PDG	Pulsed Defectography
PEI	Polyetherimine (the trade name Ultem)
PP	Power Plant
PVC	Polyvinyl Chloride
REI	Rules for Electrical Installations
RVM	Ruby Version Manager (software platform)
SC	Short Circuit
SDPS	State District Power Station
SHV	Super-High Voltage

SWE	Switching (control) Without Excitation
TBM	Time—Based Maintenance
TESIS	Trance-European Synchronously Interconnected System
TEV	Transient Earth Voltage (sensor)
TC	Technical Conditions
TCM	Thermal Control Methods
TPP	Thermal Power Plant
TR	Transformation Ratio
UCTE	Union for the Coordination of Transmission of Electricity
UES	Unified Energy System
UHV	Ultra—High Voltage
URD	Underground Residential Distribution
USSR	Union of Soviet Socialist Republics
VAC	Volt-Ampere Characteristic
VPD	Vibration Preventive Diagnostics
VTC	Voltage-Time Characteristic
WSCC	Western States Coordination Council
XLPE	Cross-linked Polyethylene

Appendix B

See Figs. B.1, B.2, B.3, B.4, B.5, B.6 and B.7.

V. Y. Ushakov et al., *High-Voltage Equipment of Power Systems*, Power Systems,
https://doi.org/10.1007/978-3-031-38252-9

Fig. B.1 Fragment of the laboratory for testing high-voltage equipment

Fig. B.2 Installation for testing with direct voltage 2250 kV at a current of 100 mA; long working hours

Fig. B.3 AC test setup 1100 kV, 1.25 A in the high-voltage hall of the switching station of Siemens AG, Berlin

Fig. B.4 Test transformer 600 kV, 3.33 A; long working hours

Fig. B.5 An installation that allows superimposing an alternating voltage of 500 kV, 50 Hz on a direct voltage of 1200 kV

Fig. B.6 Alternating voltage test facility 1200 kV, 1.25 A. Transformers with aluminum tanks, operating period 4 h at 35 °C

Fig. B.7 3000 kV ladder impulse voltage generator in the high-voltage laboratory of Tomsk Polytechnic University